THE COMPUTER VISION WORKSHOP

Develop the skills you need to use computer vision algorithms in your own artificial intelligence projects

Hafsa Asad, Vishwesh Ravi Shrimali, and Nikhil Singh

THE COMPUTER VISION WORKSHOP

Copyright © 2020 Packt Publishing

All rights reserved. No part of this course may be reproduced, stored in a retrieval system, or transmitted in any form or by any means, without the prior written permission of the publisher, except in the case of brief quotations embedded in critical articles or reviews.

Every effort has been made in the preparation of this course to ensure the accuracy of the information presented. However, the information contained in this course is sold without warranty, either express or implied. Neither the authors, nor Packt Publishing, and its dealers and distributors will be held liable for any damages caused or alleged to be caused directly or indirectly by this course.

Packt Publishing has endeavored to provide trademark information about all of the companies and products mentioned in this course by the appropriate use of capitals. However, Packt Publishing cannot guarantee the accuracy of this information.

Authors: Hafsa Asad, Vishwesh Ravi Shrimali, and Nikhil Singh

Reviewers: Richmond Alake, Sayantani Basu, John Wesley Doyle, Tim Hoolihan, Harshil Jain, Amar Kumar, Shiva Prasad, Geetank Raipura, Shovon Sengupta, and Sanchit Singh

Managing Editors: Anush Kumar Mehalavarunan, Bhavesh Bangera, and Rutuja Yerunkar

Acquisitions Editors: Manuraj Nair, Royluis Rodrigues, Kunal Sawant, Anindya Sil, and Archie Vankar

Production Editor: Shantanu Zagade

Editorial Board: Megan Carlisle, Samuel Christa, Mahesh Dhyani, Heather Gopsill, Manasa Kumar, Alex Mazonowicz, Monesh Mirpuri, Bridget Neale, Dominic Pereira, Shiny Poojary, Abhishek Rane, Brendan Rodrigues, Erol Staveley, Ankita Thakur, Nitesh Thakur, and Jonathan Wray

First published: July 2020
Production reference: 2230221
ISBN: 978-1-80020-177-4
Published by Packt Publishing Ltd.
Livery Place, 35 Livery Street
Birmingham B3 2PB, UK

WHY LEARN WITH A PACKT WORKSHOP?

LEARN BY DOING

Packt Workshops are built around the idea that the best way to learn something new is by getting hands-on experience. We know that learning a language or technology isn't just an academic pursuit. It's a journey towards the effective use of a new tool—whether that's to kickstart your career, automate repetitive tasks, or just build some cool stuff.

That's why Workshops are designed to get you writing code from the very beginning. You'll start fairly small—learning how to implement some basic functionality—but once you've completed that, you'll have the confidence and understanding to move onto something slightly more advanced.

As you work through each chapter, you'll build your understanding in a coherent, logical way, adding new skills to your toolkit and working on increasingly complex and challenging problems.

CONTEXT IS KEY

All new concepts are introduced in the context of realistic use-cases, and then demonstrated practically with guided exercises. At the end of each chapter, you'll find an activity that challenges you to draw together what you've learned and apply your new skills to solve a problem or build something new.

We believe this is the most effective way of building your understanding and confidence. Experiencing real applications of the code will help you get used to the syntax and see how the tools and techniques are applied in real projects.

BUILD REAL-WORLD UNDERSTANDING

Of course, you do need some theory. But unlike many tutorials, which force you to wade through pages and pages of dry technical explanations and assume too much prior knowledge, Workshops only tell you what you actually need to know to be able to get started making things. Explanations are clear, simple, and to-the-point. So you don't need to worry about how everything works under the hood; you can just get on and use it.

Written by industry professionals, you'll see how concepts are relevant to real-world work, helping to get you beyond "Hello, world!" and build relevant, productive skills. Whether you're studying web development, data science, or a core programming language, you'll start to think like a problem solver and build your understanding and confidence through contextual, targeted practice.

ENJOY THE JOURNEY

Learning something new is a journey from where you are now to where you want to be, and this Workshop is just a vehicle to get you there. We hope that you find it to be a productive and enjoyable learning experience.

Packt has a wide range of different Workshops available, covering the following topic areas:

- Programming languages
- Web development
- Data science, machine learning, and artificial intelligence
- Containers

Once you've worked your way through this Workshop, why not continue your journey with another? You can find the full range online at http://packt.live/2MNkuyl.

If you could leave us a review while you're there, that would be great. We value all feedback. It helps us to continually improve and make better books for our readers, and also helps prospective customers make an informed decision about their purchase.

Thank you,
The Packt Workshop Team

Table of Contents

Preface   i

Chapter 1: Basics of Image Processing   1

Introduction 2

NumPy Arrays 4

    Exercise 1.01: Creating NumPy Arrays 5

Pixels in Images 11

    Pixel Location – Image Coordinate System 12

    Image Properties 14

        Size of the Image 14

        Color Spaces and Channels 16

    Pixel Values 19

Introduction to OpenCV 25

    Images in OpenCV 25

    Important OpenCV Functions 27

    Exercise 1.02: Reading, Processing, and Writing an Image 30

    Using Matplotlib to Display Images 35

    Accessing and Manipulating Pixels 38

    Exercise 1.03: Creating a Water Effect 39

    Activity 1.01: Mirror Effect with a Twist 46

Summary 50

Chapter 2: Common Operations When Working with Images — 53

- Introduction — 54
- Geometric Transformations — 55
  - Image Translation — 56
  - Exercise 2.01: Translation Using NumPy — 58
  - Image Rotation — 61
    - Finding the Rotation Matrix — 62
    - Finding the Size of the Output Image — 64
  - Image Resizing — 67
  - Affine Transformation — 73
  - Exercise 2.02: Working with Affine Transformation — 75
  - Perspective Transformation — 79
  - Exercise 2.03: Perspective Transformation — 80
- Image Arithmetic — 84
  - Image Addition — 84
  - Exercise 2.04: Performing Image Addition — 85
  - Image Multiplication — 92
  - Exercise 2.05: Image Multiplication — 92
- Binary Images — 96
  - Exercise 2.06: Converting an Image into a Binary Image — 98
  - Bitwise Operations on Images — 102
  - Exercise 2.07: Chess Pieces — 103
  - Masking — 109
  - Activity 2.01: Masking Using Binary Images — 111
- Summary — 114

Chapter 3: Working with Histograms — 117

Introduction 118

Introduction to Matplotlib 119

    Displaying Images with Matplotlib 120

    Plotting Histograms with Matplotlib 121

    Exercise 3.01: Plotting a Sample Image and Its Histogram
    with 256 bins 122

    Exercise 3.02: Plotting a Sample Image and Its Histogram
    with 10 bins 125

Histograms with OpenCV 131

    User-Selected ROI 136

    Exercise 3.03: Creating a Mask Image Using a User-Selected ROI 137

    A Comparison of Some Sample Histograms 141

    What Is Histogram Equalization? 142

    Exercise 3.04: Histogram Equalization of a Grayscale Image 142

    Contrast Limited Adaptive Histogram Equalization (CLAHE) 148

    Exercise 3.05: Application of CLAHE on a Grayscale Image 152

    Activity 3.01: Enhancing Images Using Histogram
    Equalization and CLAHE 159

    Exercise 3.06: Histogram Equalization in the BGR Color Space 161

    The Histogram Equalization of Color Images
    Using the HSV and LAB Color Spaces 165

    Exercise 3.07: Histogram Equalization in the HSV Color Space 166

    Exercise 3.08: Histogram Equalization in the LAB Color Space 170

    Activity 3.02: Image Enhancement in a User-Defined ROI 176

Summary 178

Chapter 4: Working with contours — 181

Introduction 182

Contours – Basic Detection and Plotting 182

> Exercise 4.01: Detecting Shapes and Displaying Them
> on BGR Images 188
>
> Exercise 4.02: Detecting Shapes and Displaying Them
> on Black and White Images 195
>
> Exercise 4.03: Displaying Different Contours
> with Different Colors and Thicknesses 197
>
> Drawing a Bounding Box around a Contour 200
>
> Area of a Contour 201
>
> Difference between Contour Detection and Edge Detection 202

Hierarchy 204

> Exercise 4.04: Detecting a Bolt and a Nut 210
>
> Exercise 4.05: Detecting a Basketball Net in an Image 216

Contour Matching 224

> Exercise 4.06: Detecting Fruits in an Image 228
>
> Exercise 4.07: Identifying Bananas from the Image of Fruits 232
>
> Exercise 4.08: Detecting an Upright Banana
> from the Image of Fruits 239
>
> Activity 4.01: Identifying a Character on a Mirrored Document 240

Summary 243

Chapter 5: Face Processing in Image and Video — 245

Introduction 246

Introduction to Haar Cascades 247

> Using Haar Cascades for Face Detection 250
>
> Exercise 5.01: Face Detection Using Haar Cascades 256

Detecting Parts of the Face 262

Exercise 5.02: Eye Detection Using Cascades 264

Clubbing Cascades for Multiple Object Detection 270

Activity 5.01: Eye Detection Using Multiple Cascades 271

Activity 5.02: Smile Detection Using Haar Cascades 274

GrabCut Technique 276

Exercise 5.03: Human Body Segmentation
Using GrabCut with Rectangular Mask 281

Exercise 5.04: Human Body Segmentation Using Mask and ROI 287

Activity 5.03: Skin Segmentation Using GrabCut 298

Activity 5.04: Emoji Filter 301

Summary 305

Chapter 6: Object Tracking   307

Introduction 308

Naïve Tracker 309

Exercise 6.01: Object Tracking Using Basic Image Processing 311

Non-Deep Learning-Based Object Trackers 317

Kalman Filter – Predict and Update 317

Meanshift – Density Seeking Filter 319

CAMshift – Continuously Adaptive Meanshift 322

The OpenCV Object Tracking API 325

Object Tracker Summary 328

Exercise 6.02: Object Tracking Using the Median Flow
and MIL Trackers 330

Installing Dlib 338

Object Tracking Using Dlib 339

Exercise 6.03: Object Tracking Using Dlib 342

    Activity 6.01: Implementing Autofocus Using Object Tracking 346

Summary 350

Chapter 7: Object Detection and Face Recognition   353

Introduction 354

Face Recognition 355

    Face Recognition Using Eigenfaces 357

    Principal Component Analysis 357

    Eigenfaces 358

    Exercise 7.01: Facial Recognition Using Eigenfaces 362

    Limitations of the Eigenface Method 370

    Fisherface 371

    Exercise 7.02: Facial Recognition Using the Fisherface Method 376

    Local Binary Patterns Histograms 383

    Exercise 7.03: Facial Recognition Using the LBPH Method 388

Object Detection 395

    Single Shot Detector 395

    MobileNet 399

    Exercise 7.04: Object Detection Using MobileNet SSD 400

    Object Detection Using the LBPH Method 405

    Exercise 7.05: Object Detection Using the LBPH Method 406

    Haar Cascades 413

    Exercise 7.06: Object Detection Using Haar-Based Features 414

    Activity 7.01: Object Detection in a Video Using MobileNet SSD 418

    Activity 7.02: Real-Time Facial Recognition Using LBPH 421

Summary 423

Chapter 8: OpenVINO with OpenCV — 425

Introduction — 426

Exploring the OpenVINO Toolkit — 427

    Components of the OpenVINO Toolkit — 428

    Installing OpenVINO for Ubuntu — 430

    OpenVINO as a Backend in OpenCV — 434

    The Need for Pre-Trained Models — 436

    OpenVINO Model Zoo — 437

    Exercise 8.01: Downloading the Pedestrian and Vehicle Detection Model — 438

    Model Specifications — 442

    Image Transforms Using OpenCV — 445

    Exercise 8.02: Image Preprocessing Using OpenCV — 446

Model Conversion Using Model Optimizer — 449

    Introduction to OpenVINO's Inference Engine — 452

    Exercise 8.03: Vehicle and Pedestrian Detection — 453

    Activity 8.01: Face Detection Using OpenVINO and OpenCV — 459

Summary — 463

Appendix — 465

Index — 539

PREFACE

ABOUT THE BOOK

Computer Vision (CV) has become an important aspect of AI technology. From driverless cars to medical diagnostics and monitoring the health of crops to fraud detection in banking, computer vision is used across all domains to automate tasks. *The Computer Vision Workshop* will help you understand how computers master the art of processing digital images and videos to mimic human activities.

Starting with an introduction to the OpenCV library, you'll learn how to write your first script using basic image processing operations. You'll then get to grips with essential image and video processing techniques such as histograms, contours, and face processing. As you progress, you'll become familiar with advanced computer vision and deep learning concepts, such as object detection, tracking, and recognition, and finally shift your focus from 2D to 3D visualization. This CV workshop will enable you to experiment with camera calibration and explore both passive and active canonical 3D reconstruction methods.

By the end of this book, you'll have developed the practical skills necessary for building powerful applications to solve computer vision problems.

AUDIENCE

If you are a researcher, developer, or data scientist looking to automate everyday tasks using computer vision, this workshop is for you. A basic understanding of Python and deep learning will help you to get the most out of this workshop.

ABOUT THE CHAPTERS

Chapter 1, *Basics of Image Processing*, introduces you to the basic building blocks of images – pixels. You will get hands-on experience in how to access and manipulate images using NumPy and OpenCV.

Chapter 2, *Common Operations When Working with Images*, introduces the idea of geometric transformations as matrix multiplication and shows how to carry them out using the OpenCV library. This chapter also details the arithmetic operations involved when working with images.

Chapter 3, *Working with Histograms*, introduces the concept of histogram equalization, a powerful tool to enhance the contrast and improve the visibility of dark objects in an image.

Chapter 4, *Working with Contours*, will familiarize you with the concept of contours. You will learn how to detect contours and nested contours, which can greatly assist you in object detection tasks.

Chapter 5, *Face Processing in Image and Video*, explains the concept of Haar Cascade classifiers for carrying out face detection, face tracking, smile detection, and skin detection to build Snapchat-type filters.

Chapter 6, *Object Tracking*, introduces various object trackers like GOTURN, MIL, Kalman Trackers, and more, that help track the detected object across various frames in videos.

Chapter 7, *Object Detection and Face Recognition*, teaches you how to implement various face recognition techniques that help in recognizing the identity of persons in a given image. This chapter will also help you to implement object-detection techniques that are used to detect and recognize objects in a given image or video. This chapter covers traditional machine learning and deep learning algorithms using the OpenCV module.

Chapter 8, *OpenVINO with OpenCV*, introduces a new toolkit by Intel – OpenVINO – which can be used for optimizing and enhancing the program performance built using the OpenCV library.

CONVENTIONS

Code words in text, database table names, folder names, filenames, file extensions, pathnames, dummy URLs, user input, and Twitter handles are shown as follows:

"The **np.random.rand** function, on the other hand, only needs the shape of the array. For a 2D array, it will be provided as **np.random.rand(number_of_rows, number_of_columns)**."

Words that you see on the screen (for example, in menus or dialog boxes) appear in the same format.

A block of code is set as follows:

```
# Display the image
cv2.imshow("Lion",img)
cv2.waitKey(0)
cv2.destroyAllWindows()
```

New terms and important words are shown like this:

"The world of **artificial intelligence** (**AI**) is impacting how we, as humans, can use the power of smart computers to perform tasks much faster, more efficiently, and with minimal effort."

Long code snippets are truncated and the corresponding names of the code files on GitHub are placed at the top of the truncated code. The permalinks to the entire code are placed below the code snippet. These should appear as follows:

Activity5.03.ipynb

```
# OpenCV Utility Class for Mouse Handling
class Sketcher:
    def __init__(self, windowname, dests, colors_func):
        self.prev_pt = None
        self.windowname = windowname
        self.dests = dests
        self.colors_func = colors_func
        self.dirty = False
        self.show()
        cv2.setMouseCallback(self.windowname, self.on_mouse)
```

The complete code for this step can be found at https://packt.live/2Bv4wU3.

Key parts of code snippets are highlighted as follows:

```
foreground = cv2.imread("../data/zebra.jpg")
foreground = cv2.cvtColor(foreground, cv2.COLOR_BGR2GRAY)
```

CODE PRESENTATION

Lines of code that span multiple lines are split using a backslash (\ ). When the code is executed, Python will ignore the backslash, and treat the code on the next line as a direct continuation of the current line.

For example:

```
history = model.fit(X, y, epochs=100, batch_size=5, verbose=1, \
                    validation_split=0.2, shuffle=False)
```

Comments are added into code to help explain specific bits of logic. Single-line comments are denoted using the # symbol, as follows:

```
# Print the sizes of the dataset
print("Number of Examples in the Dataset = ", X.shape[0])
print("Number of Features for each example = ", X.shape[1])
```

Multi-line comments are enclosed by triple quotes, as shown below:

```
"""
Define a seed for the random number generator to ensure the
result will be reproducible
"""
seed = 1
np.random.seed(seed)
random.set_seed(seed)
```

SETTING UP YOUR ENVIRONMENT

Before we explore the book in detail, we need to set up specific software and tools. In the following section, we shall see how to do that.

DOWNLOADING ANACONDA INSTALLER

Anaconda is a Python package manager that easily allows you to install and use the libraries needed for this book. To download Anaconda, head over to this link: https://www.anaconda.com/products/individual. It will open up to the following page:

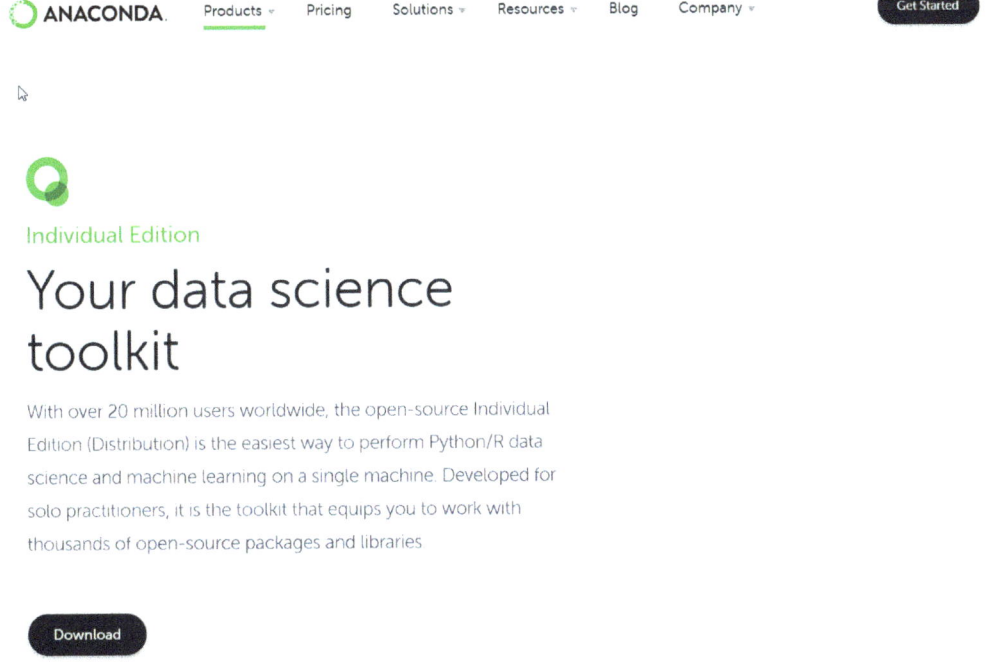

Figure 0.1: Anaconda Website

Click the **Download** button. It will take you to the bottom of the page where you can select the Anaconda installer based on your operating system. Ensure that you click the correct "`-bit`" version depending on your computer system, either 32-bit or 64-bit. You can find this information in the `System Properties` window of your operating system.

> **IMPORTANT**
>
> Make sure that you download the installer for **Python 3.7** only, as that's the Python version we are going to use in this book.

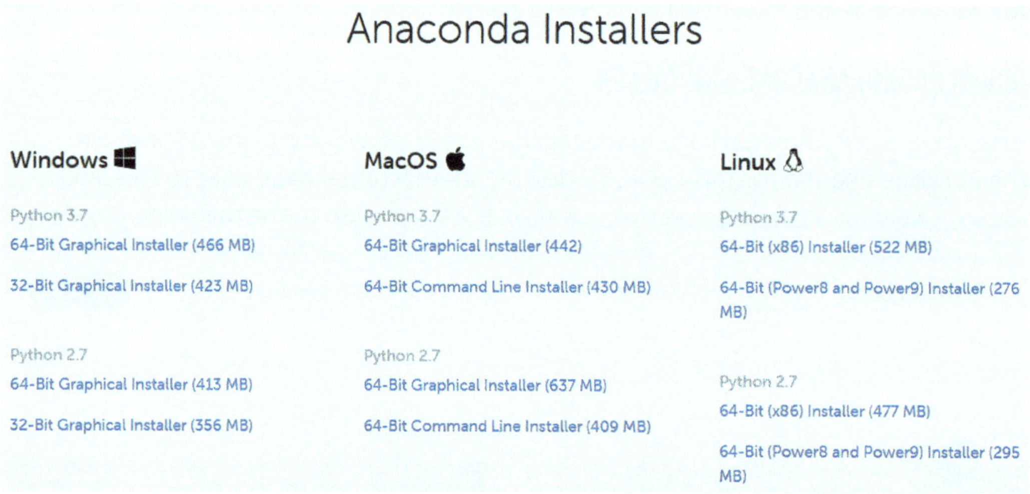

Anaconda Installers

Windows	MacOS	Linux
Python 3.7	Python 3.7	Python 3.7
64-Bit Graphical Installer (466 MB)	64-Bit Graphical Installer (442)	64-Bit (x86) Installer (522 MB)
32-Bit Graphical Installer (423 MB)	64-Bit Command Line Installer (430 MB)	64-Bit (Power8 and Power9) Installer (276 MB)
Python 2.7	Python 2.7	
64-Bit Graphical Installer (413 MB)	64-Bit Graphical Installer (637 MB)	Python 2.7
32-Bit Graphical Installer (356 MB)	64-Bit Command Line Installer (409 MB)	64-Bit (x86) Installer (477 MB)
		64-Bit (Power8 and Power9) Installer (295 MB)

Figure 0.2: Available Anaconda installers

Based on your operating system, detailed instructions on how to install Anaconda are provided in the sections that follow.

INSTALLING ANACONDA ON WINDOWS

1. Once you have downloaded the file, click it to launch the installer.
2. Agree to the license terms and click **Next**.
3. Select installation for "**Just Me**" and click **Next**.
4. Select "`Add Anaconda3 to my PATH environment variable`" and "`Register Anaconda3 as my default Python 3.7`" and click **Install**.

5. Click **Next** and wait for the installation to complete.
6. Click **Finish**.

INSTALLING ANACONDA ON LINUX

1. Locate the Python 3.7 Anaconda Installer `.sh` file that you downloaded.
2. Use `bash ~/Downloads/anaconda3-2020.02-linux-x86_64.sh/` to install Anaconda.
3. Enter **Yes** to agree to the license terms and conditions.
4. You can use the default answers for the rest of the questions by pressing *Enter*.
5. Once the installation is complete, close the current terminal.

INSTALLING ANACONDA ON MACOS X

1. Double-click the downloaded file to launch the installer.
2. Click **Yes** to agree to the license terms and conditions.
3. Click the `Install` button to start the installation.
4. Select "`Install for me only`" to install Anaconda for the current user only and click `Continue`.
5. The preceding steps will install Anaconda on your system. Finally, click `Close` to close the installer.

INSTALLING OTHER LIBRARIES

`pip` comes pre-installed with Anaconda. Once Anaconda is installed on your machine, all the required libraries can be installed using `pip`, for example, `pip install numpy`. Alternatively, you can install all the required libraries using `pip install -r requirements.txt`. You can find the `requirements.txt` file at https://packt.live/3e5kj9n.

The exercises and activities will be executed in Jupyter Notebooks. Jupyter is a Python library and can be installed in the same way as the other Python libraries – that is, with `pip install jupyter`, but fortunately, it comes pre-installed with Anaconda. To open a notebook, simply run the command `jupyter notebook` in the Terminal or Command Prompt.

ACCESSING THE CODE FILES

You can find the complete code files of this book at https://packt.live/31VCegb.

The high-quality color images used in this book can be found at https://packt.live/32tsCJK.

If you have any issues or questions about installation, please email us at workshops@packt.com.

1

BASICS OF IMAGE PROCESSING

OVERVIEW

This chapter serves as an introduction to the amazing and colorful world of image processing. We will start by understanding images and their various components – pixels, pixel values, and channels. We will then have a look at the commonly used color spaces – **red**, **green**, and **blue** (**RGB**) and **Hue**, **Saturation**, and **Value** (**HSV**). Next, we will look at how images are loaded and represented in OpenCV and how can they be displayed using another library called Matplotlib. Finally, we will have a look at how we can access and manipulate pixels. By the end of this chapter, you will be able to implement the concepts of color space conversion.

INTRODUCTION

Welcome to the world of images. It's interesting, it's wide, and most importantly, it's colorful. The world of **artificial intelligence** (**AI**) is impacting how we, as humans, can use the power of smart computers to perform tasks much faster, more efficiently, and with minimal effort. The idea of imparting human-like intelligence to computers (known as AI) is a really interesting concept. When the intelligence is focused on images and videos, the domain is referred to as **computer vision**. Similarly, **natural language processing** (**NLP**) is the AI stream where we try to understand the meaning behind the text. This technology is used by major companies for building AI-based chatbots designed to interact with customers. Both computer vision and NLP share the concepts of deep learning, where we use deep neural networks to complete tasks such as object detection, image classification, word embedding, and more. Coming back to the topic of computer vision, companies have come up with interesting use cases where AI systems have managed to change entire scenarios. Google, for example, came up with the idea of Google Goggles, which can perform several operations, such as image classification, object recognition, and more. Similarly, Tesla's self-driving cars use computer vision extensively to detect pedestrians and vehicles on the road and to detect the lane on which the car is moving.

This book will serve as a journey through the interesting components of computer vision. We will start by understanding images and then go over how they can be processed. After a couple of chapters, we will jump into detailed topics such as histograms and contours and finally go over some real-life applications of computer vision – face processing, object detection, object tracking, 3D reconstruction, and so on. This is going to be a long journey, but we will get through it together.

We love looking at high-resolution color photographs, thanks to the gamut of colors they offer. Not so long ago, however, we had photos printed only in black and white. However, those "black-and-white" photos also had some color in them, the only difference being that the colors were all shades of gray. The common thing that's there in all these components is the vision part. That's where computer vision gets its name. *Computer* refers to the fact that it's the computer that is processing the visual data, while *vision* refers to the fact that we are dealing with visual data – images and videos.

An image is made up of smaller components called *pixels*. A video is made up of multiple frames, each of which is nothing but an image. The following diagram gives us an idea of the various components of videos, images, and pixels:

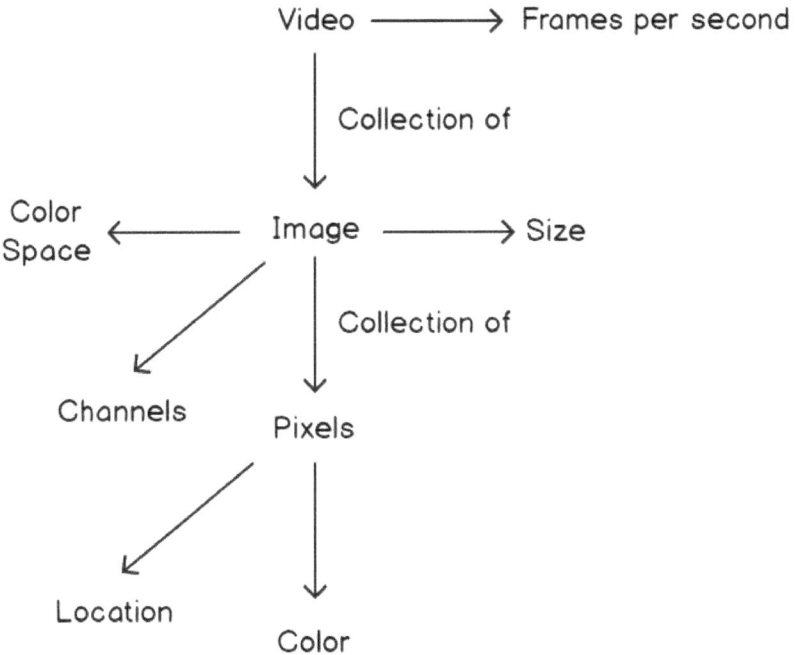

Figure 1.1: Relationships between videos, images, and pixels

In this chapter, we will focus only on images and pixels. We will also go through an introduction to the OpenCV library, along with the functions present in the library that are commonly used for basic image processing. Before we jump into the details of images and pixels, let's get through the prerequisites, starting with NumPy arrays. The reason behind this is that images in OpenCV are nothing but NumPy arrays. Just as a quick recap, NumPy is a Python module that's used for numerical computations and is well known for its high-speed computations.

NUMPY ARRAYS

Let's learn how to create a NumPy array in Python.

First, we need to import the NumPy module using the `import numpy as np` command. Here, `np` is used as an alias. This means that instead of writing `numpy.function_name`, we can use `np.function_name`.

We will have a look at four ways of creating a NumPy array:

- Array filled with zeros – the `np.zeros` command
- Array filled with ones – the `np.ones` command
- Array filled with random numbers – the `np.random.rand` command
- Array filled with values specified – the `np.array` command

Let's start with the `np.zeros` and `np.ones` commands. There are two important arguments for these functions:

- The shape of the array. For a 2D array, this is `(number of rows, number of columns)`.
- The data type of the elements. By default, NumPy uses floating-point numbers for its data types. For images, we will use unsigned 8-bit integers – `np.uint8`. The reason behind this is that 8-bit unsigned integers have a range of **0** to **255**, which is the same range that's followed for pixel values.

Let's have a look at a simple example of creating an array full of zeros. The array size should be **4x3**. We can do this by using `np.zeros(4,3)`. Similarly, if we want to create a **4x3** array full of ones, we can use `np.ones(4,3)`.

The `np.random.rand` function, on the other hand, only needs the shape of the array. For a 2D array, it will be provided as `np.random.rand(number_of_rows, number_of_columns)`.

Finally, for the `np.array` function, we provide the data as the first argument and the data type as the second argument.

Once you have a NumPy array, you can use `npArray.shape` to find out the shape of the array, where `npArray` is the name of the NumPy array. We can also use `npArray.dtype` to display the data type of the elements in the array.

Let's learn how to use these functions by completing the first exercise of this chapter.

EXERCISE 1.01: CREATING NUMPY ARRAYS

In this exercise, we will get some hands-on experience with the various NumPy functions that are used to create NumPy arrays and to obtain their shape. We will be using NumPy's **zeros**, **ones**, and **rand** functions to create the arrays. We will also have a look at their data types and shapes. Follow these steps to complete this exercise:

1. Create a new notebook and name it **Exercise1.01.ipynb**. This is where we will write our code.

2. First, import the NumPy module:

   ```
   import numpy as np
   ```

3. Next, let's create a 2D NumPy array with 5 rows and 6 columns, filled with zeros:

   ```
   npArray = np.zeros((5,6))
   ```

4. Let's print the array we just created:

   ```
   print(npArray)
   ```

   The output is as follows:

   ```
   [[0. 0. 0. 0. 0. 0.]
    [0. 0. 0. 0. 0. 0.]
    [0. 0. 0. 0. 0. 0.]
    [0. 0. 0. 0. 0. 0.]
    [0. 0. 0. 0. 0. 0.]]
   ```

5. Next, let's print the data type of the elements of the array:

   ```
   print(npArray.dtype)
   ```

   The output is **float64**.

6. Finally, let's print the shape of the array:

   ```
   print(npArray.shape)
   ```

   The output is **(5, 6)**.

7. Print the number of rows and columns in the array:

> **NOTE**
> The code snippet shown here uses a backslash (\ ) to split the logic across multiple lines. When the code is executed, Python will ignore the backslash, and treat the code on the next line as a direct continuation of the current line.

```
print("Number of rows in array = {}"\
      .format(npArray.shape[0]))

print("Number of columns in array = {}"\
      .format(npArray.shape[1]))
```

The output is as follows:

```
Number of rows in array = 5
Number of columns in array = 6
```

8. Notice that the array we just created used floating-point numbers as the data type. Let's create a new array with another data type – an unsigned 8-bit integer – and find out its data type and shape:

```
npArray = np.zeros((5,6), dtype=np.uint8)
```

9. Use the **print()** function to display the contents of the array:

```
print(npArray)
```

The output is as follows:

```
[[0 0 0 0 0 0]
 [0 0 0 0 0 0]
 [0 0 0 0 0 0]
 [0 0 0 0 0 0]
 [0 0 0 0 0 0]]
```

10. Print the data type of **npArray**, as follows:

```
print(npArray.dtype)
```

The output is **uint8**.

11. Print the shape of the array, as follows:

    ```
    print(npArray.shape)
    ```

    The output is **(5, 6)**.

    > **NOTE**
    >
    > The code snippet shown here uses a backslash (\ ) to split the logic across multiple lines. When the code is executed, Python will ignore the backslash, and treat the code on the next line as a direct continuation of the current line.

12. Print the number of rows and columns in the array, as follows:

    ```
    print("Number of rows in array = {}"\
          .format(npArray.shape[0]))

    print("Number of columns in array = {}"\
          .format(npArray.shape[1]))
    ```

    The output is as follows:

    ```
    Number of rows in array = 5
    Number of columns in array = 6
    ```

13. Now, we will create arrays of the same size, that is, **(5,6)**, and the same data type (an unsigned 8-bit integer) using the other commands we saw previously. Let's create an array filled with ones:

    ```
    npArray = np.ones((5,6), dtype=np.uint8)
    ```

14. Let's print the array we have created:

    ```
    print(npArray)
    ```

    The output is as follows:

    ```
    [[1 1 1 1 1 1]
     [1 1 1 1 1 1]
     [1 1 1 1 1 1]
     [1 1 1 1 1 1]
     [1 1 1 1 1 1]]
    ```

15. Let's print the data type of the array and its shape. We can verify that the data type of the array is actually a **uint8**:

    ```
    print(npArray.dtype)
    print(npArray.shape)
    ```

    The output is as follows:

    ```
    uint8
    (5, 6)
    ```

16. Next, let's print the number of rows and columns in the array:

    ```
    print("Number of rows in array = {}"\
        .format(npArray.shape[0]))

    print("Number of columns in array = {}"\
        .format(npArray.shape[1]))
    ```

    The output is as follows:

    ```
    Number of rows in array = 5
    Number of columns in array = 6
    ```

17. Next, we will create an array filled with random numbers. Note that we cannot specify the data type while building an array filled with random numbers:

    ```
    npArray = np.random.rand(5,6)
    ```

18. As we did previously, let's print the array to find out the elements of the array:

    ```
    print(npArray)
    ```

    The output is as follows:

    ```
    [[0.19959385 0.36014215 0.8687727  0.03460717 0.66908867 0.65924373]
     [0.18098379 0.75098049 0.85627628 0.09379154 0.86269739 0.91590054]
     [0.79175856 0.24177746 0.95465331 0.34505896 0.49370488 0.06673543]
     [0.54195549 0.59927631 0.30597663 0.1569594  0.09029267 0.24362439]
     [0.01368384 0.84902871 0.02571856 0.97014665 0.38342116 0.70014051]]
    ```

19. Next, let's print the data type and shape of the random array:

    ```
    print(npArray.dtype)
    print(npArray.shape)
    ```

The output is as follows:

```
float64
(5, 6)
```

20. Finally, let's print the number of rows and columns in the array:

```
print("Number of rows in array = {}"\
      .format(npArray.shape[0]))

print("Number of columns in array = {}"\
      .format(npArray.shape[1]))
```

The output is as follows:

```
Number of rows in array = 5
Number of columns in array = 6
```

21. Finally, let's create an array that looks like the one shown in the following figure:

1	2	3	4	5	6
7	8	9	10	11	12
13	14	15	16	17	18
19	20	21	22	23	24
25	26	27	28	29	30

Figure 1.2: NumPy array

The code to create and print the array is as follows:

```
npArray = np.array([[1,2,3,4,5,6],
                    [7,8,9,10,11,12],
                    [13,14,15,16,17,18],
                    [19,20,21,22,23,24],
                    [25,26,27,28,29,30]],
                    dtype=np.uint8)
print(npArray)
```

The output is as follows:

```
[[ 1  2  3  4  5  6]
 [ 7  8  9 10 11 12]
 [13 14 15 16 17 18]
 [19 20 21 22 23 24]
 [25 26 27 28 29 30]]
```

22. Now, just like in the previous cases, we will print the data type and the shape of the array:

    ```
    print(npArray.dtype)
    print(npArray.shape)

    print("Number of rows in array = {}"\
          .format(npArray.shape[0]))

    print("Number of columns in array = {}"\
          .format(npArray.shape[1]))
    ```

    The output is as follows:

    ```
    uint8
    (5, 6)
    Number of rows in array = 5
    Number of columns in array = 6
    ```

In this exercise, we saw how to create NumPy arrays using different functions, how to specify their data types, and how to display their shape.

> **NOTE**
>
> To access the source code for this specific section, please refer to https://packt.live/3ic9R30.

Now that we are armed with the prerequisites, let's formally jump into the world of image processing by discussing the building blocks of images – pixels.

PIXELS IN IMAGES

By now, we know that images are made up of pixels. Pixels can be thought of as very small, square-like structures that, when joined, result in an image. They serve as the smallest building blocks of any image. Let's take an example of an image. The following image is made up of millions of pixels that are different colors:

Figure 1.3: An image of a lion and a girl

> **NOTE**
>
> Image source: https://www.pxfuel.com/en/free-photo-olgbr.

Let's see what pixels look like up close. What happens when we keep on zooming in on the girl's eye in the image? After a certain point, we will end up with something like the following:

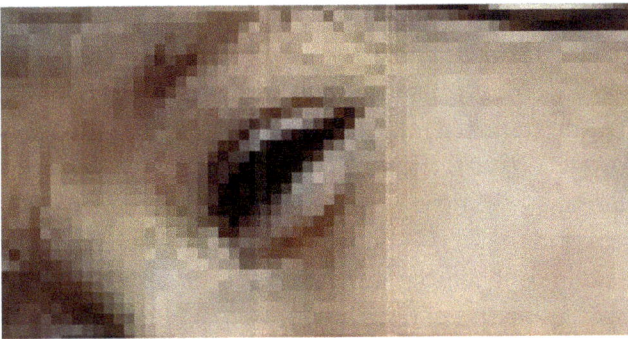

Figure 1.4: Highly zoomed-in version of the image shown in Figure 1.3

If you look carefully at the preceding image, you will be able to see some squares in the image. These are known as pixels. A pixel does not have a standard size; it differs from device to device. We frequently use the term **pixels per inches** (**PPI**) to define the resolution of an image. More pixels in an inch (or square inch) of an image means a higher resolution. So, an image from a DSLR has more pixels per inch, while an image from a laptop webcam will have fewer pixels per inch. Let's compare the image we saw in *Figure 1.4* with its higher-resolution version (*Figure 1.5*). We'll notice how, in a higher-resolution image, we can zoom in on the same region and have a much better quality image compared to the lower-resolution image (*Figure 1.4*):

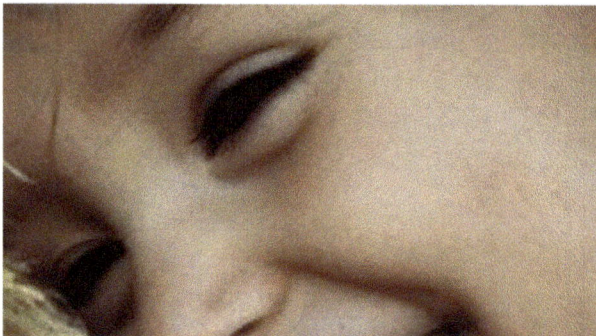

Figure 1.5: Same zoomed-in region for a higher-resolution image

Now that we have a basic idea of pixels and terms such as PPI and resolution, let's understand the properties of pixels – pixel location and the color of the pixel.

PIXEL LOCATION – IMAGE COORDINATE SYSTEM

We know that a pixel is a square and is the smallest building block of an image. A specific pixel is referenced using its location in the image. Each image has a specific coordinate system. The standard followed in OpenCV is that the top-left corner of an image acts as the origin, `(0,0)`. As we move to the right, the x-coordinate of the pixel location increases, and as we move down, the y-coordinate increases. But it's very important to understand that this is not a universally followed coordinate system. Let's try to follow a different coordinate system for the time being. We can find out the location of a pixel using this coordinate system.

Let's consider the same image (*Figure 1.3*) that we looked at previously and try to understand its coordinate system:

Figure 1.6: Image coordinate system

We have the same image as before and we have added the two axes – an *X* axis and a *Y* axis. The *X* axis is the horizontal axis, while the y-axis is the vertical axis. The origin of this coordinate system lies at the bottom-left corner of the image. Armed with this information, let's find the coordinates of the three points marked in the preceding figure – the orange point (at the bottom left), the green point (at the center of the image), and the blue point (at the top right of the image).

We know that the orange point lies at the bottom-left corner of the image, which is exactly where the origin of the image's coordinate system lies. So, the coordinates of the pixel at the bottom-left corner are `(0,0)`.

What about the blue point? Let's assume that the width of the image is `W` and the height of the image is `H`. By doing this, we can see that the x coordinate of the blue point will be the width of the image, `(W)`, while the y coordinate will be the height of the image, `(H)`. So, the coordinates of the pixel at the top-right corner are `(W, H)`.

14 | Basics of Image Processing

Now, let's think about the coordinates for the center point. The x coordinate of the center will be **W/2**, while the y coordinate of the point will be **H/2**. So, the coordinates of the pixel at the center are **(W/2, H/2)**.

Can you find the coordinates of any other pixels in the image? Try this as an additional challenge.

So, now we know how to find a pixel's location. We can use this information to extract information about a specific pixel or a group of pixels.

There is one more property associated with a pixel, and that is its color. But, before we take a look at that, let's look at the properties of an image.

IMAGE PROPERTIES

By now, we have a very good idea of images and pixels. Now, let's understand the properties of an image. From *Figure 1.1*, we can see that there are three main properties (or attributes) of an image – image size, color space, and the number of channels in an image. Let's explore each property in detail.

SIZE OF THE IMAGE

First, let's understand the size of the image. The size of an image is represented by its height and width. Let's understand this in detail.

There are quite a few ways of referring to the width and height of an image. If you have filled in any government forms, you will know that a lot of the time, they ask for images with dimensions such as **3.5 cm×4.5 cm**. This means that the width of the image will be 3.5 cm and the height of the image will be 4.5 cm.

However, when you try downloading images from websites such as Pixabay (https://pixabay.com/), you get options such as the following ones:

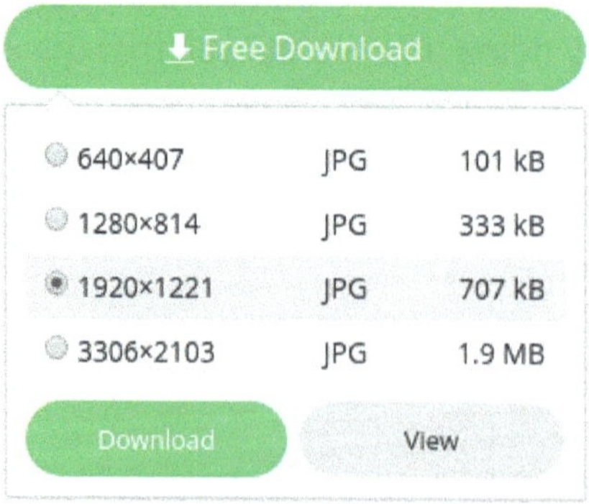

Figure 1.7: Image dimensions on Pixabay

So, what happened here? Are these numbers in centimeters or millimeters, or some other unit? Interestingly, these numbers are in pixels. These numbers represent the number of pixels that are present in the image. So, an image with size **1920×1221** will have a total of **1920×1221 = 2334420** pixels. These numbers are sometimes related to the resolution of the image as well. A higher number of pixels in an image means that the image has more detail, or in other words, we can zoom into the image more without losing its detail.

Can you try to figure out when you would need to use which image size representation? Let's try to get to the bottom of this by understanding the use cases of these different representations.

When you have your passport-size photo printed out so that you can paste it in a box in a form, you are more concerned about the size of the image in terms of units such as centimeters. Why? Because you want the image to fit in the box. Since we are talking about the physical world, the dimensions are also represented in physical units such as centimeters, inches, or millimeters. Will it matter to you how many pixels the image is made up of at that time? Consider the following form, which is where we are going to paste a passport-size photo:

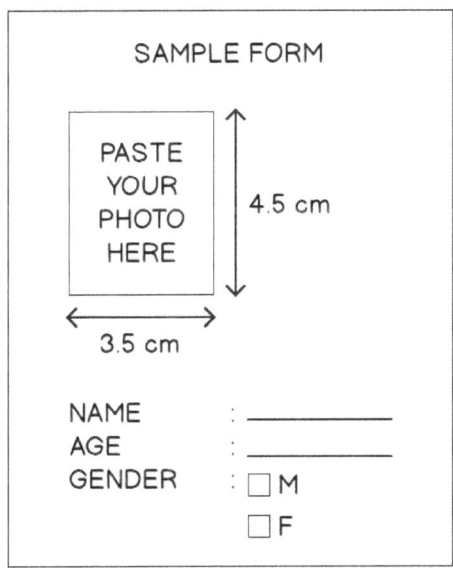

Figure 1.8: A sample form where an image must be pasted

But what about an image that's in soft copy form? Let's take the following two images as an example:

Figure 1.9: Two images with the same physical dimensions

Both of the preceding images have the same physical dimensions – a height of 2.3 inches and a width of 3.6 inches. So, are they the same images? The answer is no. The image on the left has a much higher number of pixels compared to the image on the right. This difference is evident when you are more focused on the details (or resolution) of the image rather than the physical dimensions of the image.

For example, the profile photo of every user on Facebook has the same dimensions when viewed alongside a post or a comment. But that does not mean that every image has the same sharpness/resolution/detail. Notice how we have used three words here – sharpness, resolution, and detail – to convey the same sense, that is, the quality of the image. An image with a higher number of pixels will be of far better quality compared to the same image with a lower number of pixels. Now, what kind of dimensions will we use in our book? Since we are dealing with soft copies of images, we will represent the size of images using the number of pixels in them.

Now, let's look at what we mean by color spaces and channels of an image.

COLOR SPACES AND CHANNELS

Whenever we look at a color image as humans, we are looking at three types of color, or attributes. But what three colors or attributes? Consider the two images given below. Both of them look very different but the interesting thing is that they are actually just two different versions of the same image. The difference is the color space they are represented in. Let's understand this with an analogy. We have a wooden chair. Now the wood used can be different, but the chair will still be the same. It's the same thing here. The image is the same, just the color space is different.

Let's understand this in detail. Here, we have two images:

Figure 1.10: Same images with different color spaces

While the image on the left uses **red**, **green**, and **blue** as the three attributes, thereby making its color space the **RGB color space**, the image on the left uses **hue**, **saturation**, and **value** as the three attributes, thereby making its color space the **HSV color space**.

At this point, you might be thinking, "*why do we need different color spaces?*" Well, since different color spaces use different attributes, depending on the problem we want to solve, we can use a color space that focuses on a certain attribute. Let's take a look at an example:

Figure 1.11: The red, green, and blue channels of an image

In the preceding figure, we have separated the three attributes that made the color space of the image – red, green, and blue. These attributes are also referred to as channels. So, the RGB color space has three channels – a red channel, a green channel, and a blue channel. We will understand why these images are in grayscale soon.

18 | Basics of Image Processing

Similarly, let's consider the three channels of the HSV color space – hue, saturation, and value:

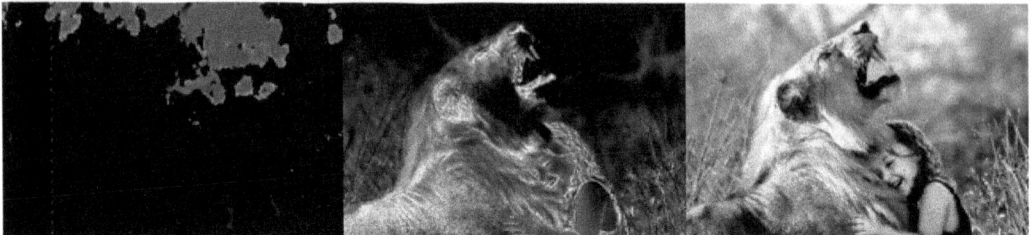

Figure 1.12: The hue, saturation, and value channels of an HSV image

Now, compare the results shown in *Figure 1.11* and *Figure 1.12*. Let's propose a problem. Let's say that we want to detect the edges present in an image. Why? Well, edges are responsible for details in an image. But we won't go into the details of that right now. Let's just assume that, for some reason, we want to detect the edges in an image. Purely based on visualization, you can see that the saturation channel of the HSV image already has a lot of edges highlighted, so even if we don't do any processing and go ahead and use the saturation channel of the HSV image, we will end up with a pretty good start for the edges.

This is exactly why we need color spaces. When we just want to see an image and praise the photographer, the RGB color space is much better than the HSV color space. But when we want to detect edges, the HSV color space is better than the RGB color space. Again, this is not a general rule and depends on the image that we are talking about.

Sometimes, the HSV color space is preferred over the RGB color space. The reason behind this is that the red, green, and blue components (or channels) in the RGB color space have a high correlation between them. The HSV color space, on the other hand, allows us to separate the value channel of the image entirely, which helps us in processing the image. Consider a case of object detection where you want to detect an object present in an image. In this problem, you will want to make sure that light invariance is present, meaning that the object can be detected irrespective of whether the image is dark or bright. Since the HSV color space allows us to separate the value or intensity channel, it's better to use it for this object detection case study.

It's also important to note that we have a large variety of color spaces; RGB and HSV are just two of them. At this point, it's not important for you to know all the color spaces. But if you are interested, you can refer to the color spaces supported by OpenCV and how an image from one color space is converted into another color space here: https://docs.opencv.org/4.2.0/de/d25/imgproc_color_conversions.html.

Let's have a look at another color space – grayscale. This will also answer your question as to why the red, green, and blue channels in *Figure 1.11* don't look red, green, and blue, respectively.

When an image has just one channel, we say that it's in grayscale mode. This is because the pixels are color in shades of gray depending on the pixel value. A pixel value of 0 represents black, whereas a pixel value of 255 represents white. You'll learn more about this in the next section.

Figure 1.13: Image in grayscale mode

When we divided the RGB and HSV images into their three channels, we were left with images that had only one channel each, and that's why they were converted into grayscale and were colored in this shade of gray.

In this section, we learned what we mean by the color space of an image and what a channel means. Now, let's look at pixel values in detail.

PIXEL VALUES

So far, we have discussed what pixels are and their properties. We learned how to represent a pixel's location using image coordinate systems. Now, we will learn how to represent a pixel's value. First, what do we mean by a pixel's value? Well, a pixel's value is nothing but the color present in that pixel. It's important to note here that a pixel can have only one color. That's why a pixel's value is a fixed value.

20 | Basics of Image Processing

If we talk about an image in grayscale, a pixel value can range between 0 and 255 (both inclusive), where 0 represents black and 255 represents white.

> **NOTE**
>
> In the following figure, there are two axes: X and Y. These axes represent the width and height of the image, respectively, and don't hold much importance in the computer vision domain. That's why they can be safely ignored. Instead, it's important to focus on the pixel values in the images.

Refer to *Figure 1.14* to understand how different pixel values decide the color present in a specific pixel:

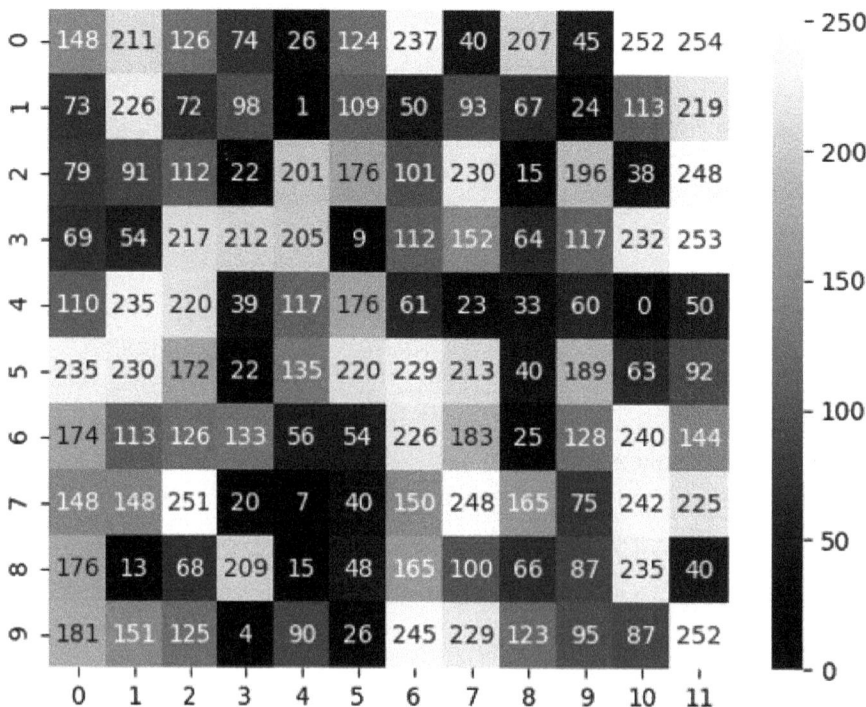

Figure 1.14: Image with pixel values annotated

Now, we know that a grayscale image has only one channel and that's why the pixel value has only one number that determines the shade of color present in that pixel. What if we are talking about an RGB image? Since the RGB image has three channels, each pixel will have three values – one value for the red channel, one for the green channel, and one for the blue channel. Consider the following image, which shows that an RGB image (on the left) is made up of three images or channels – a red channel, a green channel, and a blue channel:

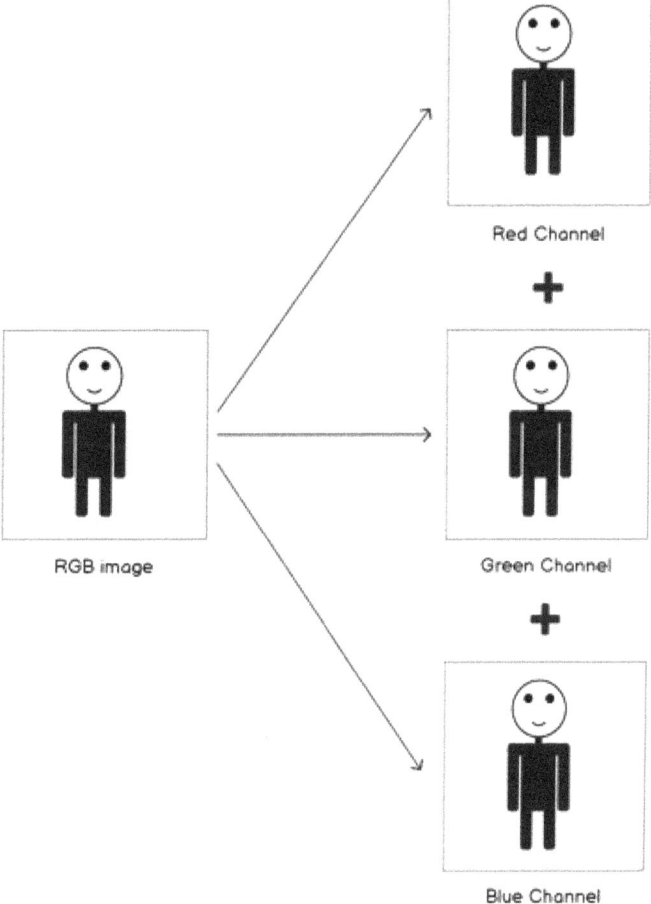

Figure 1.15: RGB image broken down into three channels

What do we know about each of these channels? In *Figure 1.11*, we saw that each channel image looks exactly like a grayscale image. That's why the pixel value for each channel will range between 0 and 255. What will happen if we assume that the following image has the same pixel values like those shown in *Figure 1.14* for the red channel, but the other two channels are zero? Let's have a look at the result:

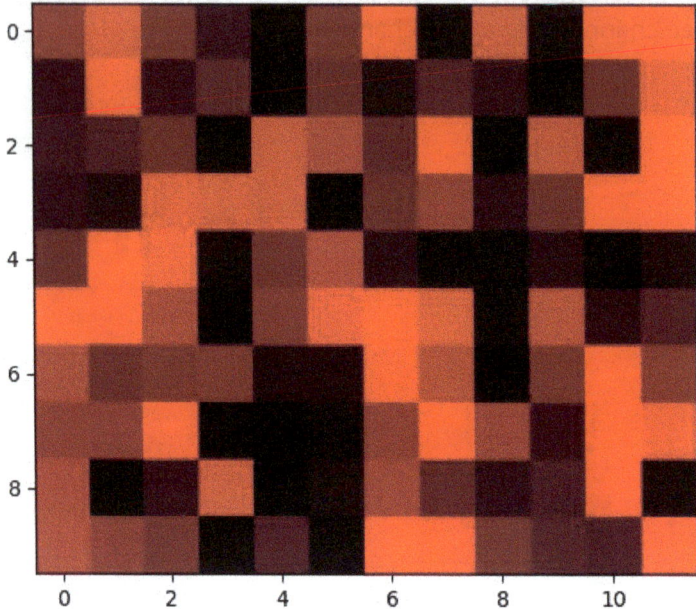

Figure 1.16: RGB image with the red channel set the same as the one used in Figure 1.13

Notice how a 0 for the red channel means that there will be no red color in that pixel. Similarly, a 255-pixel value for the red channel means that there will be a 100% red color in that pixel. By 100% red color, we mean that it won't be some darker shade of red, but the pure (lightest) red color.

Figure 1.17 and *Figure 1.18* show the RGB image with a green and a blue channel, respectively. These are the same ones that were shown in *Figure 1.14*. In each case, we are assuming that the other two channels are zero.

This way, we are highlighting the effect of only one channel:

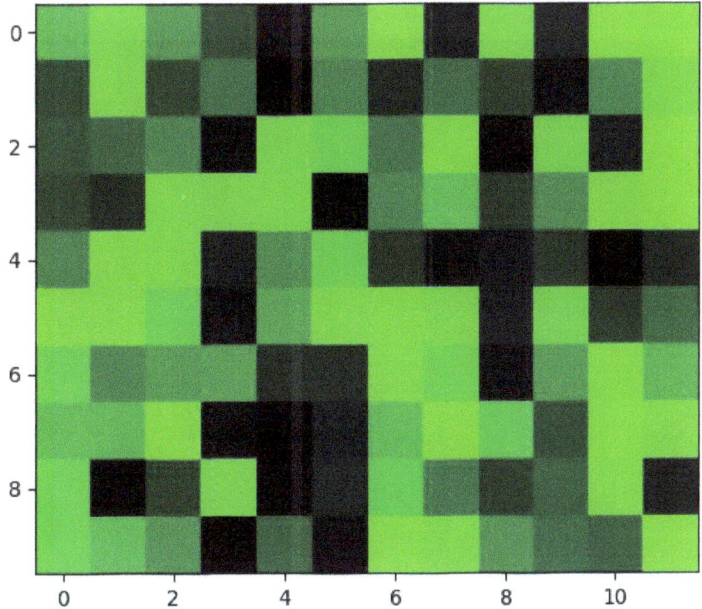

Figure 1.17: RGB image with the green channel set the same as the one used in Figure 1.14

The output for the blue channel is as follows:

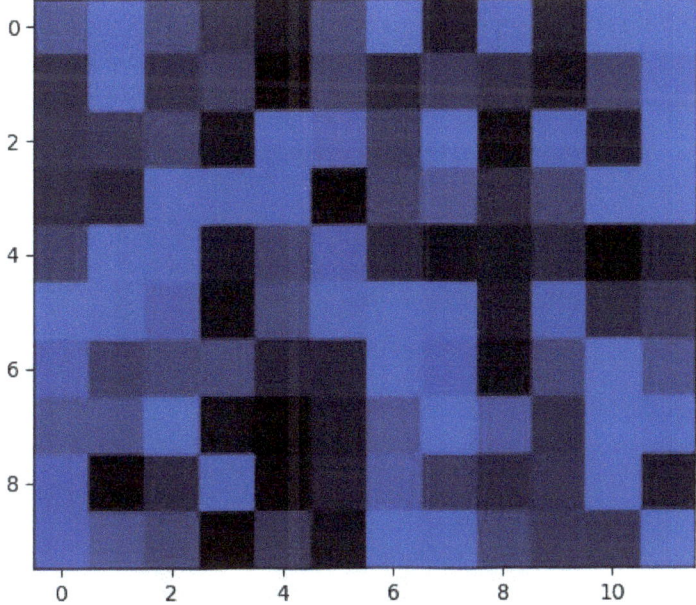

Figure 1.18: RGB image with the blue channel set the same as the one used in Figure 1.14

Now, what will happen if we combine the blue and green frames and keep the red frame set to 0?

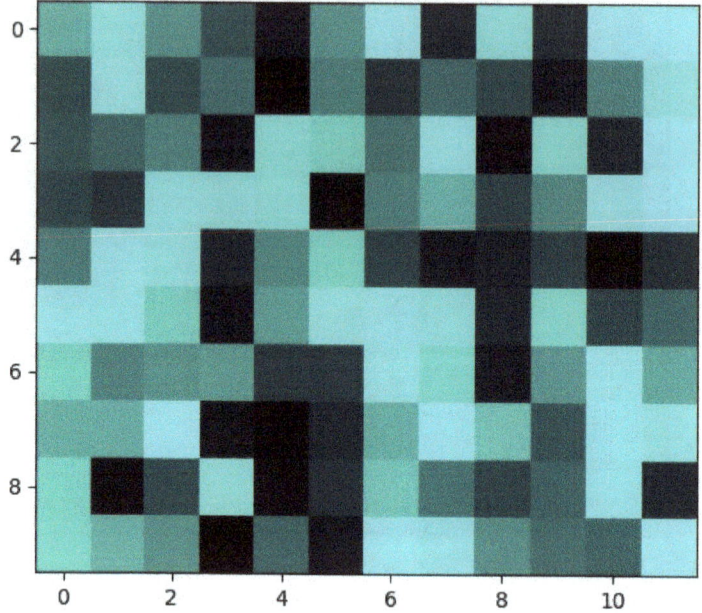

Figure 1.19: RGB image with the blue and green channels set the same as the ones used in Figure 1.14

Notice how the blue and green channels merged to create a shade of cyan. You can see the same shade being formed in the following figure when blue and green are combined:

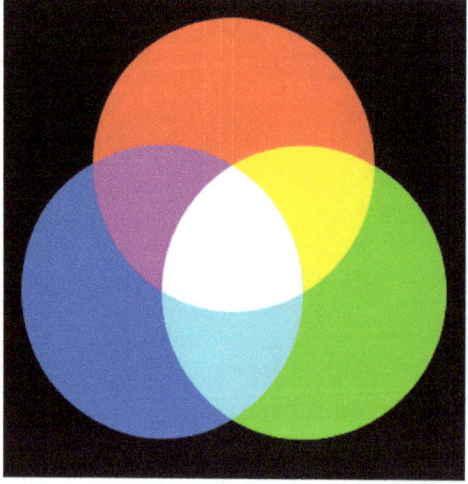

Figure 1.20: Combination of red, green, and blue

In this section, we discussed the concept of pixel values for grayscale images and images with three channels. We also saw how the pixel value affects the shade of the color present in a specific pixel. So far, we have discussed the important concepts that will be referred to throughout this book. From the next section onward, we will start with coding using various libraries such as OpenCV and Matplotlib.

INTRODUCTION TO OPENCV

OpenCV, also known as the Open Source Computer Vision library, is the most commonly used computer vision library. Primarily written in C++, it's also commonly used in Python, thanks to its Python wrappers. Over the years, OpenCV has been through multiple revisions and its current version is 4.2.0 (which is the version we are going to use in this book). What makes it different from other computer vision libraries is the fact that it's fast and easy to use, it provides support for libraries such as QT and OpenGL, and most importantly, it provides hardware acceleration for Intel processors. These powerful features/benefits make OpenCV the perfect choice for understanding the various concepts of computer vision and implementing them. Apart from OpenCV, we will also use NumPy for some basic computation and Matplotlib for visualization, wherever required.

> **NOTE**
> Refer to the *Preface* for NumPy and OpenCV installation instructions.

Let's start by understanding how images are represented in OpenCV in Python.

IMAGES IN OPENCV

OpenCV has its own class for representing images – `cv::Mat`. The "`Mat`" part comes from the term matrix. Now, this should not come as a surprise since images are nothing more than matrices. We already know that every image has three attributes specific to its dimensions – width, height, and the number of channels. We also know that every channel of an image is a collection of pixel values lying between 0 and 255. Notice how the channel of an image starts to look similar to a 2D matrix. So, an image becomes a collection of 2D matrices stacked on top of each other.

26 | Basics of Image Processing

Refer to the following diagram for more details:

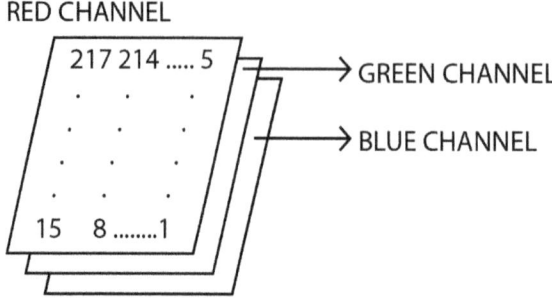

Figure 1.21: Image as 2D matrices stacked on top of each other

As a quick recap, while using OpenCV in Python, images are represented as NumPy arrays. NumPy is a Python module commonly used for numerical computation. A NumPy array looks like a 2D matrix, as we saw in *Exercise 1.01, Creating NumPy Arrays*. That's why an RGB image (which has three channels) will look like three 2D NumPy arrays stacked on top of each other.

We have restricted our discussion so far only to 2D arrays (which is good enough for grayscale images), but we know that our RGB images are not like 2D arrays. They not only have a height and a width; they also have one extra dimension – the number of channels in the image. That's why we can refer to RGB images as 3D arrays.

The only difference to the commands we discussed in the *NumPy Arrays* section is that we now have to add an extra dimension to the shape of the NumPy arrays – *the number of channels*. Since we know that RGB images have only three channels, the shape of the NumPy arrays becomes **(number of rows, number of columns, 3)**.

Also, note that the order of elements in the shape of NumPy arrays follows this format: **(number of rows, number of columns, 3)**. Here, the number of rows is equivalent to the height of the image, while the number of columns is equivalent to the width of the image. That's why the shape of the NumPy array can also be represented as **(height, width, 3)**.

Now that we know about how images are represented in OpenCV, let's go ahead and learn about some functions in OpenCV that we will commonly use.

IMPORTANT OPENCV FUNCTIONS

We can divide the OpenCV functions that we are going to use into the following categories:

- Reading an image
- Modifying an image
- Displaying an image
- Saving an image

Let's start with the function required for reading an image. The only function we will use for this is `cv2.imread`. This function takes the following arguments:

- File name of the image we want to read/load
- Flags for specifying what mode we want to read the image in

If we try to load an image that does not exist, the function returns **None**. This can be used to check whether the image was read successfully or not.

Currently, OpenCV supports formats such as `.bmp`, `.jpeg`, `.jpg`, `.png`, `.tiff`, and `.tif`. For the entire list of formats, you can refer to the documentation: https://docs.opencv.org/4.2.0/d4/da8/group__imgcodecs.html#ga288b8b3da0892bd651fce07b3bbd3a56.

The last thing that we need to focus on regarding the `cv2.imread` function is the flag. There are only three flags that are commonly used for reading an image in a specific mode:

- `cv2.IMREAD_UNCHANGED`: Reading the image as it is. This means that if an image is a PNG image with a transparent background, then it will be read as a BGRA image, where A specifies the alpha channel – which is responsible for transparency. If this flag is not used, the image will be read as a BGR image. Note that BGR refers to the blue, green, and red channels of an image. A, or the alpha channel, is responsible for transparency. That's why an image with a transparent background will be read as BGRA and not as BGR. It's also important to note here that OpenCV, by default, uses BGR mode and that's why we are discussing BGRA mode and not RGBA mode here.

- **cv2.IMREAD_GRAYSCALE**: Reading the image in grayscale format. This converts any color image into grayscale.

- **cv2.IMREAD_COLOR**: This is the default flag and it reads any image as a color image (BGR mode).

> **NOTE**
>
> Note that OpenCV reads images in BGR mode rather than RGB mode. This means that the order of channels becomes blue, green, and red. Even with the other OpenCV functions that we will use, it is assumed that the image is in BGR mode.

Next, let's have a look at some functions we can use to modify an image. We will specifically discuss the functions for the following tasks:

- Converting an image's color space
- Splitting an image into various channels
- Merging channels to form an image

Let's learn how we can convert the color space of an image. For this, we will use the **cv2.cvtColor** function. This function takes two inputs:

- The image we want to convert
- The color conversion flag, which looks as follows:

```
cv2.COLOR_{CURRENT_COLOR_SPACE}2{NEW_COLOR_SPACE}
```

For example, to convert a BGR image into an HSV image, you will use **cv2.COLOR_BGR2HSV**. For converting a BGR image into grayscale, you will use **cv2.COLOR_BGR2GRAY**, and so on. You can view the entire list of such flags here: https://docs.opencv.org/4.2.0/d8/d01/group__imgproc__color__conversions.html.

Now, let's look at splitting and merging channels. Suppose you only want to modify the red channel of an image; you can first split the three channels (blue, green, and red), modify the red channel, and then merge the three channels again. Let's see how we can use OpenCV functions to split and merge channels:

- For splitting the channels, we can use the **cv2.split** function. It takes only one argument – the image to be split – and returns the list of three channels – blue, green, and red.

- For merging the channels, we can use the `cv2.merge` function. It takes only one argument – a set consisting of the three channels (blue, green, and red) – and returns the merged image.

Next, let's look at the functions we will use for displaying an image. There are three main functions that we will be using for display purposes:

- To display an image, we will use the `cv2.imshow` function. It takes two arguments. The first argument is a string, which is the name of the window in which we are going to display the image. The second argument is the image that we want to display.

- After the `cv2.imshow` function is called, we use the `cv2.waitKey` function. This function specifies how long the control should stay on the window. If you want to move to the next piece of code after the user presses a key, you can provide 0. Otherwise, you can provide a number that specifies the number of milliseconds the program will wait before moving to the next piece of code. For example, if you want to wait for 10 milliseconds before moving to the next piece of code, you can use `cv2.waitKey(10)`.

- Without calling the `cv2.waitKey` function, the display window won't be visible properly. But after moving to the next code, the window will still stay open (but will appear as if it's not responding). To close all the display windows, we can use the `cv2.destroyAllWindows()` function. It takes no arguments. It's recommended to close the display windows once they are no longer needed.

Finally, to save an image, we will use OpenCV's `cv2.imwrite` function. It takes two arguments:

- A string that specifies the filename that we want to save the image with
- The image that we want to save

Now that we know about the OpenCV functions that we are going to use in this chapter, let's get our hands dirty by using them in the next exercise.

EXERCISE 1.02: READING, PROCESSING, AND WRITING AN IMAGE

In this exercise, we will use the OpenCV functions that we looked at in the previous section to load the image of the lion in *Figure 1.3*, separate the red, green, and blue channels, display them, and finally save the three channels to disk.

> **NOTE**
>
> The image can be found at https://packt.live/2YOyQSv.

Follow these steps to complete this exercise:

1. First of all, we will create a new notebook – **Exercise1.02.ipynb**. We will be writing our code in this notebook.

2. Let's import the OpenCV module:

```
import cv2
```

3. Next, let's read the image of the lion and the girl. The image is present at the **../data/lion.jpg** path:

> **NOTE**
>
> Before proceeding, ensure that you can change the path to the image (highlighted) based on where the image is saved in your system.

```
# Load image
img = cv2.imread("../data/lion.jpg")
```

> **NOTE**
>
> The # symbol in the preceding code snippet denotes a code comment. Comments are added into code to help explain specific bits of logic.

4. We will check whether we have read the image successfully or not by checking whether it is **None**:

```
if img is None:
    print("Image not found")
```

5. Next, let's display the image we have just read:

```
# Display the image
cv2.imshow("Lion",img)
cv2.waitKey(0)
cv2.destroyAllWindows()
```

The output is as follows:

NOTE

Please note that whenever we are going to display an image using the `cv2.imshow` function, a new display window will pop up. This output will not be visible in the Jupyter Notebook and will be displayed in a separate window, as shown in the following figure.

Figure 1.22: Lion image

6. Now comes the processing step, where we will split the image into the three channels – blue, green, and red:

```
# Split channels
blue, green, red = cv2.split(img)
```

7. Next, we can display the channels that we obtained in the preceding step. Let's start by displaying the blue channel:

```
cv2.imshow("Blue",blue)
cv2.waitKey(0)
cv2.destroyAllWindows()
```

The output is as follows:

Figure 1.23: Lion image (blue channel)

8. Next, let's display the green channel:

   ```
   cv2.imshow("Green",green)
   cv2.waitKey(0)
   cv2.destroyAllWindows()
   ```

   The output is as follows:

Figure 1.24: Lion image (green channel)

9. Similarly, we can display the red channel of the image:

   ```
   cv2.imshow("Red",red)
   cv2.waitKey(0)
   cv2.destroyAllWindows()
   ```

The output is as follows:

Figure 1.25: Lion image (red channel)

10. Finally, to save the three channels we obtained, we will use the `cv2.imwrite` function:

```
cv2.imwrite("Blue.png",blue)
cv2.imwrite("Green.png",green)
cv2.imwrite("Red.png",red)
```

This will return **True**. This indicates that the images have been successfully written/saved on the disk. At this point, you can verify whether the three channels you have obtained match the images shown here:

Figure 1.26: Blue, green, and red channels obtained using our code

> **NOTE**
>
> To access the source code for this specific section, please refer to https://packt.live/2YQlDbU.

In the next section, we will discuss another library that is commonly used in computer vision – Matplotlib.

USING MATPLOTLIB TO DISPLAY IMAGES

Matplotlib is a library that is commonly used in data science and computer vision for visualization purposes. The beauty of this library lies in the fact that it's very powerful and still very easy to use, similar to the OpenCV library.

In this section, we will have a look at how we can use Matplotlib to display images that have been read or processed using OpenCV. The only point that you need to keep in mind is that Matplotlib assumes the images will be in RGB mode, whereas OpenCV assumes the images will be in BGR mode. That's why we will be converting the image to RGB mode whenever we want to display it using Matplotlib.

There are two common ways to convert a BGR image into an RGB image:

- Using OpenCV's `cv2.cvtColor` function and passing the `cv2.COLOR_BGR2RGB` flag. Let's imagine that we have an image loaded as `img` that we want to convert into RGB mode from BGR mode. This can be done using `cv2.cvtColor(img, cv2.COLOR_BGR2RGB)`.

- The second method focuses on the fact that you are reversing the order of channels when you are converting a BGR image into an RGB image. This can be done by replacing `img` with `img[:,:,::-1]`, where `::-1` in the last position is responsible for reversing the order of channels. We will be using this approach whenever we are displaying images using Matplotlib. The only reason behind doing this is that less time is required to write this code compared to option 1.

Now, let's have a look at the functions we are going to use to display images using Matplotlib.

First, we will import the `matplotlib` library, as follows. We will be using Matplotlib's `pyplot` module to create plots and display images:

```
import matplotlib.pyplot as plt
```

36 | Basics of Image Processing

We will also be using the following magic command so that the images are displayed inside the notebook rather than in a new display window:

```
%matplotlib inline
```

Next, if we want to display a color image, we will use the following command, where we are also converting the image from BGR into RGB. We are using the same lion image as before and have loaded it as **img**:

```
plt.imshow(img[::-1])
```

> **NOTE**
>
> Execute this code in the same Jupyter Notebook where you executed *Exercise 1.02, Reading, Processing, and Writing an Image.*

The code is as follows:

```
plt.imshow(img[::-1])
```

Finally, to display the image, we will use the **plt.show()** command.

This will give us the following output:

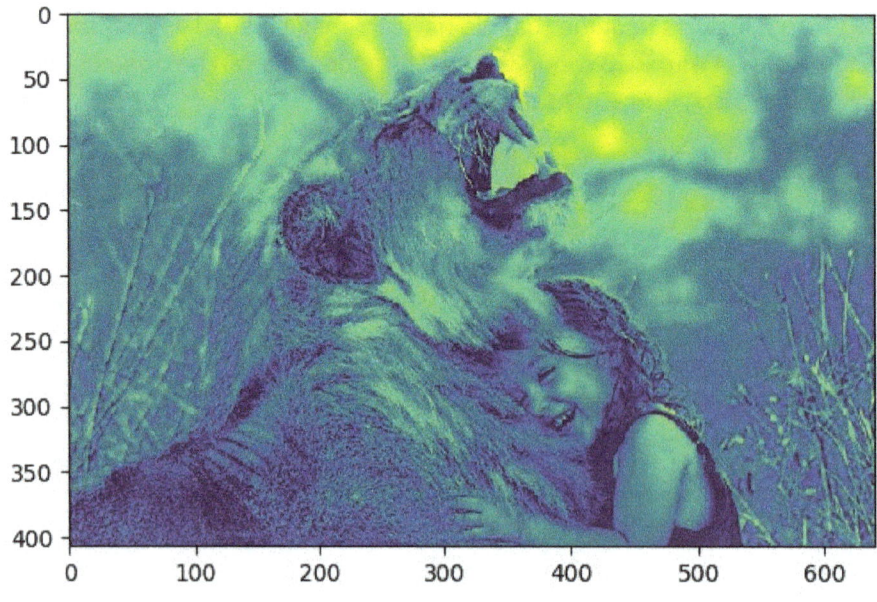

Figure 1.27: Lion image

If we want to display a grayscale image, we will also have to specify the colormap as gray. This is simply because Matplotlib displays grayscale images with a colormap of jet. You can see the difference between the two colormaps in the following plot:

```
plt.imshow(img, cmap="gray")
```

The plot looks as follows:

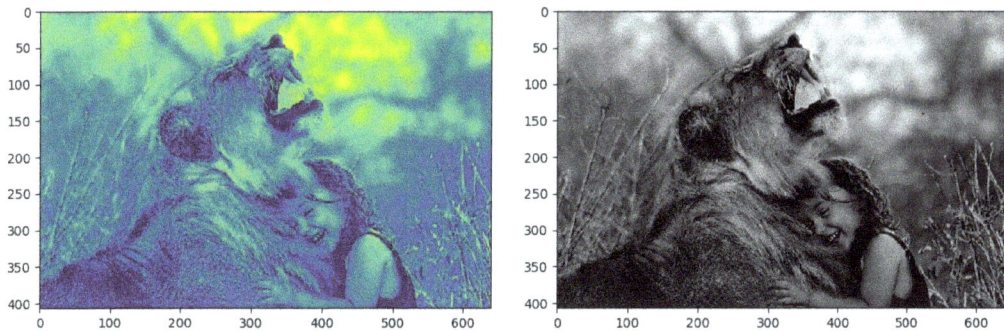

Figure 1.28: (Left) Image without colormap specified, (right) image with a gray colormap

> **NOTE**
>
> It's very important to note that when we display an image using Matplotlib, it will be displayed inside the notebook, whereas an image displayed using OpenCV's `cv2.imshow` function will be displayed in a separate window. Moreover, an image displayed using Matplotlib will have the gridlines or X and Y axes, by default. The same will not be present in an image displayed using the `cv2.imshow` function. This is because Matplotlib is actually a (graph) plotting library, and that's why it displays the axes, whereas OpenCV is a computer vision library. The axes don't hold much importance in computer vision. **Please note that irrespective of the presence or absence of the axes, the image graphic will stay the same, whether it's displayed using Matplotlib or OpenCV. That's why any and all image processing steps will also stay the same**. We will be using Matplotlib and OpenCV interchangeably in this book to display images. This means that sometimes you will find images with axes and sometimes without axes. In both cases, the axes don't hold any importance and can be ignored.

That's all it takes to display an image using Matplotlib in a Jupyter Notebook. In the next section, we will cover the final topic of this chapter – how to access and manipulate the pixels of an image.

ACCESSING AND MANIPULATING PIXELS

So far, we have discussed how to use OpenCV to read and process an image. But the image processing guide that we have covered so far was very basic and only constituted splitting and merging the channels of an image. Now, let's learn how to access and manipulate pixels, the building blocks of an image.

We can access and manipulate pixels based on their location. We'll learn how we can use the pixel locations in this section.

We already have covered how pixels are located using the coordinate system of an image. We also know that images in OpenCV in Python are represented as NumPy arrays. That's why the problem of accessing pixels becomes the general problem of accessing the elements of a NumPy array.

Let's consider a NumPy array, **A**, with **m** rows and **n** columns. If we want to access the elements present in row number **i** and column number **j**, we can do that using `A[i][j]` or `A[i,j]`.

Similarly, if we want to extract the elements of a NumPy array, **A**, within rows **a** and **b** and columns **c** and **d**, we can do that using `A[a:b][c:d]`.

What if we wanted to extract the entire i^{th} row of the array, **A**? We can do that using `A[i][:]`, where `:` is used when we want to extract the entire range of elements in that list.

Similarly, if we want to extract the entire j^{th} column, we can use `A[:][j]`.

Manipulating pixels becomes very easy once you have managed to access the pixels you want. You can either change their values to a new value or copy the values from another pixel.

Let's learn how to use the preceding operations by completing a practical exercise.

EXERCISE 1.03: CREATING A WATER EFFECT

In this exercise, we will implement a water filter that is responsible for vertically flipping an object that is floating on a body of water. You can see this effect in the following image:

Figure 1.29: Water effect

The entire problem can be broken down into the following steps:

1. Read an image.
2. Flip the image vertically.
3. Join the original image and the flipped image.
4. Display and save the final image.

In this exercise, we will create a water effect using the concepts we have studied so far. We will be applying the same water effect to the **lion.jpg** image (*Figure 1.3*) we used earlier. Follow these steps to complete this exercise:

> **NOTE**
>
> The image can be found at https://packt.live/2YOyQSv.

1. Import the required libraries – Matplotlib, NumPy, and OpenCV:

   ```
   import cv2
   import numpy as np
   import matplotlib.pyplot as plt
   ```

2. We will also use the magic command to display images using Matplotlib in the notebook:

   ```
   %matplotlib inline
   ```

3. Next, let's read the image and display it. The image is stored in the **../data/lion.jpg** path:

   > **NOTE**
   >
   > Before proceeding, ensure that you can change the path to the image (highlighted) based on where the image is saved in your system.

   ```
   # Read the image
   img = cv2.imread("../data/lion.jpg")
   # Display the image
   plt.imshow(img[:,:,::-1])
   plt.show()
   ```

The output is as follows:

Figure 1.30: Image output

4. Let's find the shape of the image to understand what we are dealing with here:

   ```
   # Find the shape of the image
   img.shape
   ```

   The shape of the image is **(407, 640, 3)**.

5. Now comes the important part. We will have to create a new image with twice the number of rows (or twice the height) but the same number of columns (or width) and the same number of channels. This is because we want to add the mirrored image to the bottom of the image:

   ```
   # Create a new array with double the size
   # Height will become twice
   # Width and number of channels will
   # stay the same
   imgNew = np.zeros((814,640,3),dtype=np.uint8)
   ```

6. Let's display this new image we created. It should be a completely black image at this point:

   ```
   # Display the image
   plt.imshow(imgNew[:,:,::-1])
   plt.show()
   ```

The output is as follows:

Figure 1.31: New black image that we have created using np.zeros

7. Next, we will copy the original image to the top half of the image. The top half of the image corresponds to the first half of the rows of the new image:

```
# Copy the original image to the
# top half of the new image
imgNew[:407][:] = img
```

8. Let's look at the new image now:

```
# Display the image
plt.imshow(imgNew[:,:,::-1])
plt.show()
```

Here's the output of the **show()** method:

Figure 1.32: Image after copying the top half of the image

44 | Basics of Image Processing

9. Next, let's vertically flip the original image. We can take some inspiration from how we reversed the channels using `::-1` in the last position. Since flipping the image vertically is equivalent to reversing the order of rows in the image, we will use `::-1` in the first position:

```
# Invert the image
imgInverted = img[::-1,:,:]
```

10. Display the inverted image, as follows:

```
# Display the image
plt.imshow(imgInverted[:,:,::-1])
plt.show()
```

The inverted image looks as follows:

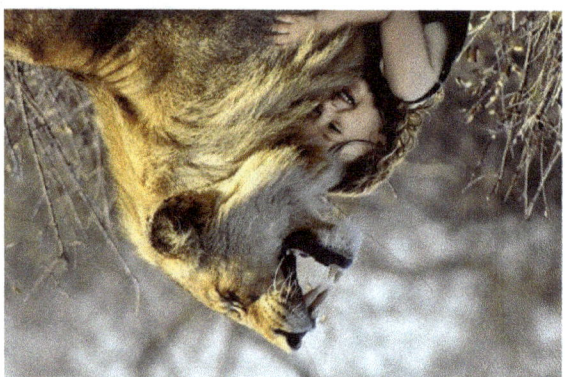

Figure 1.33: Image obtained after vertically flipping the original image

11. Now that we have the flipped image, all we have to do is copy this flipped image to the bottom half of the new image:

```
# Copy the inverted image to the
# bottom half of the new image
imgNew[407:][:] = imgInverted
```

12. Display the new image, as follows:

```
# Display the image
plt.imshow(imgNew[:,:,::-1])
plt.show()
```

The output is as follows:

Figure 1.34: Water effect

13. Let's save the image that we have just created:

```
# Save the image
cv2.imwrite("WaterEffect.png",imgNew)
```

In this exercise, you saw how the toughest-looking tasks can sometimes be completed using the very basics of a topic. Using our basic knowledge of NumPy arrays, we were able to generate a very beautiful-looking image.

> **NOTE**
>
> To access the source code for this specific section, please refer to https://packt.live/2VC7QDL.

46 | Basics of Image Processing

Let's test what we have learned so far by completing the following activity, where we will create a mirror effect. One difference between the water effect and the mirror effect image will be that the mirror effect will be laterally inverted. Moreover, we will also be introducing an additional negative effect to the mirror image. This negative effect gets its name from the image negatives that are used while processing photographs. You can see the effect of the mirror image by looking at the following figure.

Let's test what we have learned so far with the following activity.

ACTIVITY 1.01: MIRROR EFFECT WITH A TWIST

Creating a very simple mirror effect is very simple, so let's bring a twist to this. We want to replicate the effect shown in the following figure. These effects are useful when we want to create apps such as Snapchat, Instagram, and so on. For example, the water effect, the mirror effect, and so on are quite commonly used as filters. We covered the water effect in the previous exercise. Now, we will create a mirror effect:

Figure 1.35: Mirror effect

Before you read the detailed instructions, think about how you would create such an effect. Notice the symmetry in the image, that is, the mirror effect. The most important part of this activity is to generate the image on the right. Let's learn how we can do that. We will be using the same image of the lion and girl that we used in the previous exercises.

> **NOTE**
>
> The image can be found at https://packt.live/2YOyQSv.

Follow these steps to complete this activity:

1. First, load the required modules – OpenCV, Matplotlib, and NumPy.
2. Next, write the magic command to display the images in the notebook.
3. Now, load the image and display it using Matplotlib.
4. Next, obtain the shape of the image.
5. Now comes the most interesting part. Convert the image's color space from BGR into HSV and display the HSV image. The image will look as follows:

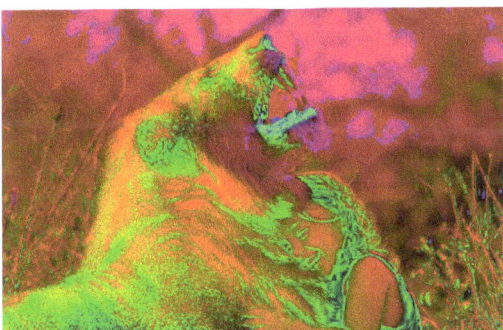

Figure 1.36: Image converted into the HSV color space

6. Next, extract the value channel from the HSV color space. Note that the value channel is the last channel of the HSV image. You can use the `cv2.split` function for this. Display the value channel. The image will look as follows:

Figure 1.37: Value channel of the HSV image

7. Now comes another interesting part. We will create a negative effect on the image. This is similar to what you see in the negatives of the images you click. To carry out this effect, all you have to do is subtract the value channel from **255**. Then, display the new value channel. The image will look as follows:

Figure 1.38: Negative of the value channel

8. Next, create a new image by merging the value channel with itself. This can be done by using `cv2.merge((value, value, value))`, where `value` refers to the negative of the value channel you obtained in the preceding step. We are doing this because we want to merge two three-channel images to create the final effect.

9. Next, flip the new image you obtained previously, horizontally. You can refer to the flipping step we did in *Exercise 1.03, Creating a Water Effect*. Note that flipping horizontally is equivalent to reversing the order of the columns of an image. The output will be as follows:

Figure 1.39: Image flipped horizontally

10. Now, create a new image with twice the width as the original image. This means that the number of rows and the number of channels will stay the same. Only the number of columns will be doubled.

11. Now, copy the original image to the left half of the new image. The image will look as follows:

Figure 1.40: Copying the original image to the left half

12. Next, copy the horizontally flipped image to the right half of the new image and display the image.

13. Finally, save the image you have obtained.

    The final image will look as follows:

Figure 1.41: Final image

> **NOTE**
>
> The solution to this activity can be found on page 466.

In this activity, you used the concepts you studied in the previous exercise to generate a really interesting result. By completing this activity, you have learned how to convert the color space of an image and use a specific channel of the new image to create a mirror effect.

SUMMARY

This brings us to the end of the first chapter of this book. We started by discussing the important core concepts of computer vision – what an image is, what pixels are, and what their attributes are. Then, we discussed the libraries that we will use throughout our computer vision journey – OpenCV, Matplotlib, and NumPy. We also learned how we can use these libraries to read, process, display, and save images. Finally, we learned how to access and manipulate pixels and used this concept to generate interesting results in the final exercise and activity.

In the next chapter, we will look deeper into image processing and how OpenCV can help us with that. The more you practice, the easier it will be for you to figure out what concept you need to employ to get a result. So, keep on practicing.

2
COMMON OPERATIONS WHEN WORKING WITH IMAGES

OVERVIEW

In this chapter, we will take a look at geometric transformations – **rotation**, **translation**, **scaling**, **affine transformation**, and **perspective transformation**. We will crop images using NumPy and OpenCV functions. Then, we will discuss binary images and how to carry out arithmetic operations on images. We will have a look at some real-life applications where these techniques can come in handy.

By the end of this chapter, you should have a fairly good idea of how to process images for your specific case study. You will be able to use affine transformations on images and carry out tasks such as motion identification using binary operations.

INTRODUCTION

If you've ever seen a "behind-the-scenes" video of your favorite Hollywood movie, you might have noticed that some of the animated scenes are shot in front of a green screen. Then, during editing, the green screen is replaced with a mind-blowing futuristic vista or a dystopian scene that's beyond your wildest imagination.

Now, think of one single frame from that scene in the movie. The original frame was shot in front of a green background. Then, the image was modified to obtain a particular result, that is, a different background.

Furthermore, sometimes, you want to process an image to directly obtain the result you seek so that you can receive an intermediate result that will make further steps easier and achievable. The following is a sample picture that was taken in front of the green screen:

Figure 2.1: A picture taken in front of a green screen

This activity of modifying or processing an image to obtain a particular result is known as **image processing**.

In the previous chapter, we went over the basics of images – what pixels are, what pixel coordinates are, how to extract pixel values using pixel coordinates, and more. We'll start this chapter by understanding what we mean by image processing and why we need it. By the end of this chapter, you will be able to process images such as the one shown previously and replace the green screen with a background of your choice using a very basic image processing technique referred to as **masking**.

This chapter can be broken down into two major parts. First, we will focus on basic techniques, such as translation, rotation, resizing, and cropping, and a more general geometric transformation called *affine transformation*. We will end the section by looking at perspective transformation. Then, we will discuss binary images and arithmetic operations that we can carry out on images. We will also talk about masking in this section.

Both sections will be accompanied by exercises and activities that you can try to complete. Now, let's start with the first section of this chapter – *Geometric Transformations*.

GEOMETRIC TRANSFORMATIONS

Often, during image processing, the geometry of the image (such as the width and height of the image) needs to be transformed. This process is called **geometric transformation**. As we saw in the *Images in OpenCV* section of *Chapter 1, Basics of Image Processing*, images are nothing but matrices, so we can use a more mathematical approach to understand these topics.

Since images are matrices, if we apply an operation to the images (matrices) and we end up with another matrix, we call this a transformation. This basic idea will be used extensively to understand and apply various kinds of geometric transformations.

Here are the geometric transformations that we are going to discuss in this chapter:

- Translation
- Rotation
- Image scaling or resizing
- Affine transformation
- Perspective transformation

While discussing image scaling, we will also discuss image cropping. which isn't actually a geometric transformation, though it is commonly used along with resizing.

First, we will go over image translation, where we will discuss moving an image in a specified direction.

IMAGE TRANSLATION

Image translation is also referred to as shifting an image. In this transformation, our basic intention is to move an image along a line. Take a look at the following figure:

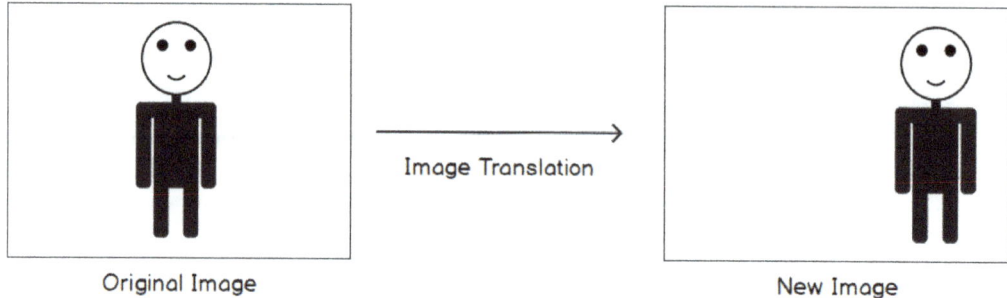

Figure 2.2: Image translation

For example, if we consider the following figure, we are shifting the human sketch to the right. We can shift an image in both the **X** and **Y** directions, as well as separately. What we are doing in this process is shifting every pixel's location in the image in a certain direction. Consider the following figure:

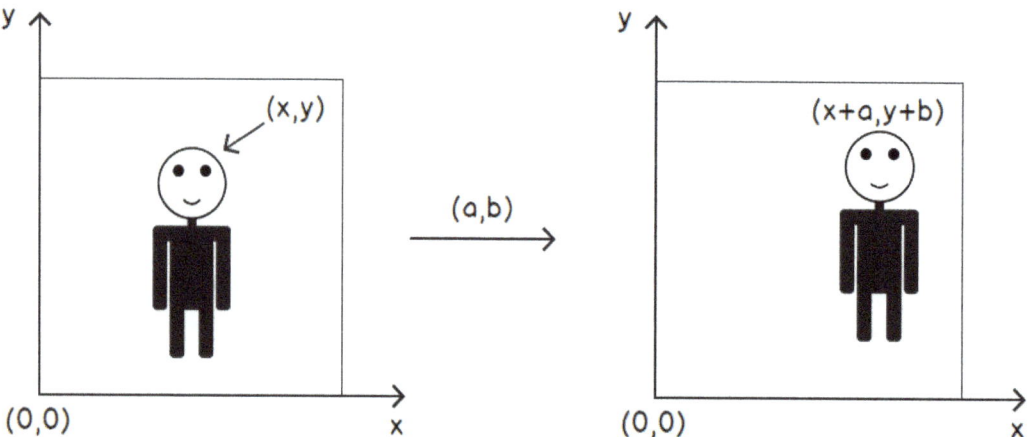

Figure 2.3: Pixel view of image translation

Using the preceding figure as reference, we can represent image translation with the following matrix equation:

$$\begin{bmatrix} x' \\ y' \end{bmatrix} = \begin{bmatrix} x+a \\ y+b \end{bmatrix} = \begin{bmatrix} x \\ y \end{bmatrix} + \begin{bmatrix} a \\ b \end{bmatrix}$$

Figure 2.4: Representation of an image matrix

In the preceding equation, `x'` and `y'` represent the new coordinates of a point/pixel after it has been shifted by **a** units in the **x** direction and **b** units in the **y** direction.

> **NOTE**
>
> Even though we have discussed image translation, it is very rarely used in image processing. It is usually combined with other transformations to create a more complex transformation. We will be studying this general class of transformations in the *Affine Transformations* section.

Now that we have discussed the theory of image translation, let's understand how translation can be performed using NumPy. We stated that images are nothing but NumPy arrays while using them in OpenCV in Python. Similarly, geometric transformations are just matrix multiplications. We also saw how image translation can be represented using a matrix equation. In terms of matrix multiplication, the same image equation can be written as a NumPy array:

```
M = np.array([[1,0,a],[0,1,b],[0,0,1])
```

If you look carefully, you will notice that the first two columns of this matrix form an identity matrix (`[[1,0],[0,1]]`). This signifies that even though we are transforming the image, the image dimensions (width and height) will stay the same. The last column is made up of **a** and **b**, which signifies that the image has been shifted by **a** units in the **x** direction and **b** units in the **y** direction. The last row, - `[0,0,1]`, is only used to make the matrix, **M**, a square matrix – a matrix with the same number of rows and columns. Now that we have the matrix for image translation with us, whenever we want to shift an image, all we need to do is multiply the image, let's say, `img` (as a NumPy array), with the array, **M**. This can be performed as follows:

```
output = M@img
```

Here, `output` is the output that's obtained after image translation.

Note that in the preceding code, it's assumed that every point that we want to shift represents a column in the matrix, that is, `img`. Moreover, usually, an extra row full of ones is also added in the `img` matrix. That's simply to make sure that that matrix multiplication can be carried out. To understand it, let's understand the dimensions of the matrix, **M**. It has three rows and three columns. For the matrix multiplication of **M** and `img` to be carried out, `img` should have the same number of rows as the number of columns in matrix **M**, that is, three.

Now, we also know that a point is represented as a column in the **img** matrix. We already have the **x** coordinate of the point in the first row and the **y** coordinate of the point in the second row of the **img** matrix. To make sure that we have three rows in the **img** matrix, we add 1 in the third row.

For example, point **[2,3]** will be presented as **[[2] , [3] , [1]]**, that is, 2 is present in the first row, 3 is present in the second row. and 1 is present in the third row.

Similarly, if we want to represent two points – **[2,3]** and **[1,0]** – they will be represented as **[[2,1] , [3,0] , [1,1]]**; that is, the **x** coordinates of both points (2 and 1) are present in the first row, the **y** coordinates (3 and 0) are present in the second row, and the third row is made up of 1s.

Let's understand this better with the help of an exercise.

EXERCISE 2.01: TRANSLATION USING NUMPY

In this exercise, we will carry out the geometric transformation known as *translation* using NumPy. We will be using the translation matrix that we looked at earlier to carry out the translation. We will be shifting 3 points by 2 units in the **x** direction and 3 units in the **y** direction. Let's get started:

1. Create a new Jupyter Notebook – **Exercise2.01.ipynb**. We will be writing our code in this notebook.

2. First, let's import the NumPy module:

    ```
    # Import modules
    import numpy as np
    ```

3. Next, we will specify the three points that we want to shift – **[2,3]**, **[0,0]**, and **[1,2]**. Just as we saw previously, the points will be represented as follows:

    ```
    # Specify the 3 points that we want to translate
    points = np.array([[2,0,1],
                       [3,0,2],
                       [1,1,1]])
    ```

    Note how the **x** coordinates (2, 0, and 1) make up the first row, the **y** coordinates (3, 0, and 2) make up the second row, and the third row has only 1s.

4. Let's print the **points** matrix:

   ```
   print(points)
   ```

   The output is as follows:

   ```
   [[2 0 1]
    [3 0 2]
    [1 1 1]]
   ```

5. Next, let's manually calculate the coordinates of these points after translation:

   ```
   # Output points - using manual calculation
   outPoints = np.array([[2 + 2, 0 + 2, 1 + 2],
               [3 + 3, 0 + 3, 2 + 3],
               [1, 1, 1]])
   ```

6. Let's also print the **outPoints** matrix:

   ```
   print(outPoints)
   ```

   The output is as follows:

   ```
   [[4 2 3]
    [6 3 5]
    [1 1 1]]
   ```

7. We need to shift these points by **2** units in the **x** direction and **3** units in the **y** direction:

   ```
   # Move by 2 units along x direction
   a = 2
   # Move by 3 units along y direction
   b = 3
   ```

8. Next, we will specify the translation matrix, as we saw previously:

   ```
   # Translation matrix
   M = np.array([[1,0,a],[0,1,b],[0,0,1]])
   ```

9. Now, we can print the translation matrix:

   ```
   print(M)
   ```

The output is as follows:

```
[[1 0 2]
 [0 1 3]
 [0 0 1]]
```

10. Now, we can perform the translation using matrix multiplication:

```
# Perform translation using NumPy
output = M@points
```

11. Finally, we can display the output points by printing them:

```
print(output)
```

The output is as follows:

```
[[4 2 3]
 [6 3 5]
 [1 1 1]]
```

Notice how the output points obtained using NumPy and using manual calculation match perfectly. In this exercise, we shifted the given points using manual calculation, as well as using matrix multiplication. We also saw how we can carry out matrix multiplication using the NumPy module in Python. The key point to focus on here is that transition is nothing but a matrix operation, just as we discussed previously at the beginning of the *Geometric Transformations* section.

> **NOTE**
>
> To access the source code for this specific section, please refer to https://packt.live/3eQdjhR.

So far, we've discussed moving an image along a direction. Next, let's discuss rotating an image around a specified point at a given angle.

IMAGE ROTATION

Image rotation, as the name suggests, involves rotating an image around a point at a specified angle. The most common notation is to specify the angle as positive if it's anti-clockwise and negative if it's specified in the clockwise direction. We will assume angles in a non-clockwise direction are positive. Let's see whether we can derive the matrix equation for this. Let's consider the case shown in the following diagram: **w** and **h** refer to the height and width of an image, respectively. The image is rotated about the origin (0,0) at an angle, alpha (α), in an anti-clockwise direction:

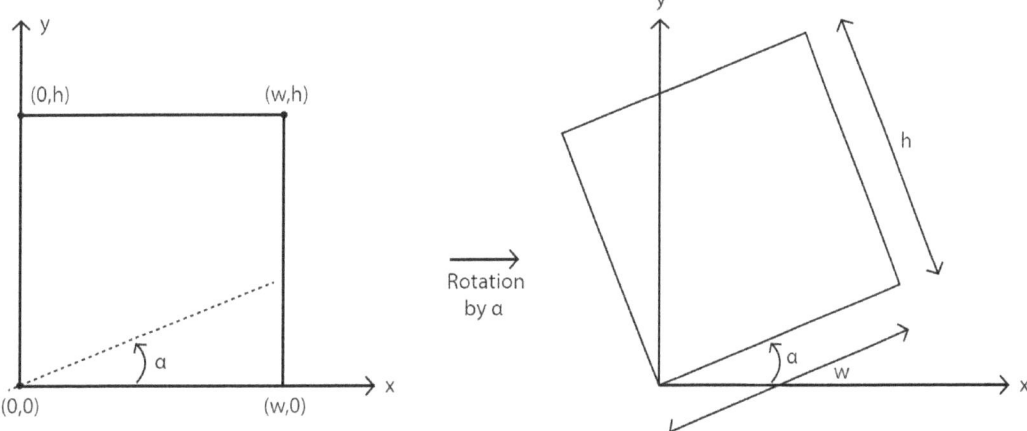

Figure 2.5: Image rotation

One thing that we can note very clearly here is that the distance between a point and the origin stays the same, even after rotation. We will be using this point when deriving the rotation matrix.

We can divide the entire problem into two parts:

- Finding the rotation matrix
- Finding the size of the new image

FINDING THE ROTATION MATRIX

Let's consider a point, **P(x,y)**, which, after rotation by angle β around the origin, **O(0,0)**, is transformed into **P'(x',y')**. Let's also assume the distance between point **P** and origin **O** is **r**:

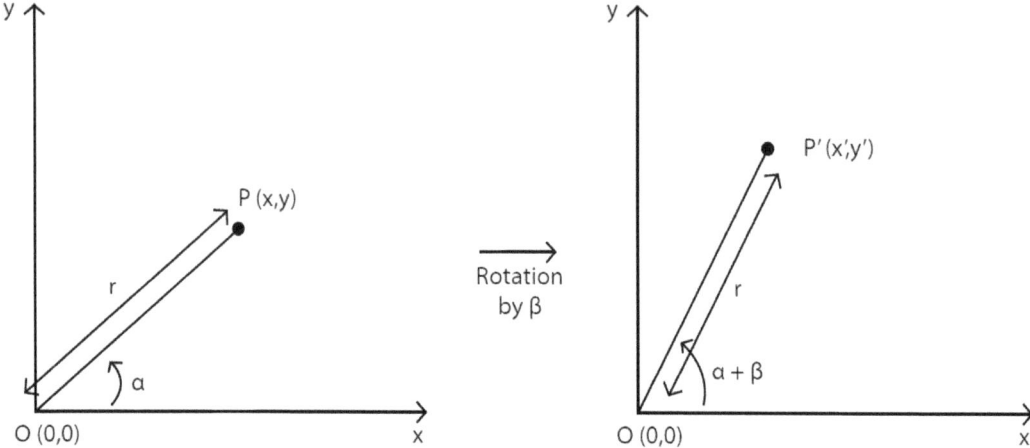

Figure 2.6: Point rotation

In the following equation, we have used the distance formula to obtain the distance between **P** and **O**:

$$r = \sqrt{(x-0)^2 + (y-0)^2} = \sqrt{x^2 + y^2}$$

Figure 2.7: Distance formula

Since we know that this distance will stay the same, even after rotation, we can write the following equation:

$$r = \sqrt{x^2 + y^2} = \sqrt{x'^2 + y'^2}$$

Figure 2.8: Distance formula after rotation

Let's assume that the **P(x,y)** point makes an angle, α, with the X-axis. Here, the formula will look as follows:

```
x = r cos(α)
y = r sin(α)
```

Similarly, the **P'(x',y')** point will make an angle, α + β, with the X-axis. Thus, the formula will look as follows:

```
x' = r cos(α + β)
y> = r sin(α + β)
```

Next, we will use the following trigonometric identity:

```
cos(α + β) = cos α cos β - sin α sin β
```

Thus, we can now simplify the equation for **x'**, as follows:

$$x' = r(\cos \alpha \cos \beta - \sin \alpha \sin \beta)$$
$$\Rightarrow x' = r(\cos \alpha \cos \beta) - r(\sin \alpha \sin \beta)$$
$$\Rightarrow x' = (r \cos \alpha) \cos \beta - (r \sin \alpha) \sin \beta$$
$$\Rightarrow x' = (x) \cos \beta - (y) \sin \beta$$
$$\Rightarrow x' = x \cos \beta - y \sin \beta$$

Figure 2.9: Simplified equation for x'

Similarly, let's simplify the equation for **y'** using the following trigonometric identity:

```
sin(α + β) = sin α cos β - cos α sin β
```

Using the preceding equation, we can simplify **y'** as follows:

$$y' = r(\sin\alpha \cos\beta + \cos\alpha \sin\beta)$$
$$\Rightarrow y' = r(\sin\alpha \cos\beta) + r(\cos\alpha \sin\beta)$$
$$\Rightarrow y' = (r \sin\alpha) \cos\beta + (r \cos\alpha) \sin\beta$$
$$\Rightarrow y' = (y) \cos\beta + (x) \sin\beta$$
$$\Rightarrow y' = y \cos\beta + x \sin\beta$$
$$\Rightarrow y' = x \sin\beta + y \cos\beta$$

Figure 2.10: Simplified equation for y'

We can now represent **x'** and **y'** using the following matrix equation:

$$\begin{bmatrix} x' \\ y' \end{bmatrix} = \begin{bmatrix} \cos\beta & -\sin\beta \\ \sin\beta & \cos\beta \end{bmatrix} \begin{bmatrix} x \\ y \end{bmatrix}$$

Figure 2.11: Representing x' and y' using the matrix equation

Now that we have obtained the preceding equation, we can transform any point into a new point if it's rotated by a given angle. The same equation can be applied to every pixel in an image to obtain the rotated image.

But have you ever seen a rotated image? Even though it's rotated, it still lies within a rectangle. This means the dimensions of the new image can change, unlike in translation, where the dimensions of the output image and the input image stayed the same. Let's see how we can find the dimensions of the image.

FINDING THE SIZE OF THE OUTPUT IMAGE

We will consider two cases here. The first case is that we keep the dimensions of the output image the same as the ones for the input image, while the second case is that we modify the dimensions of the output image. Let's understand the difference between them by looking at the following diagram. The left half shows the case when the image dimensions are kept the same, even after rotation, whereas in the right half of the diagram, we relax the dimensions to cover the entire rotated image. You can see the difference in the results obtained in both cases. In the following diagram, **L** and **H** refer to the dimensions of the original image, while **L'** and **H'** refer to the dimensions after rotation.

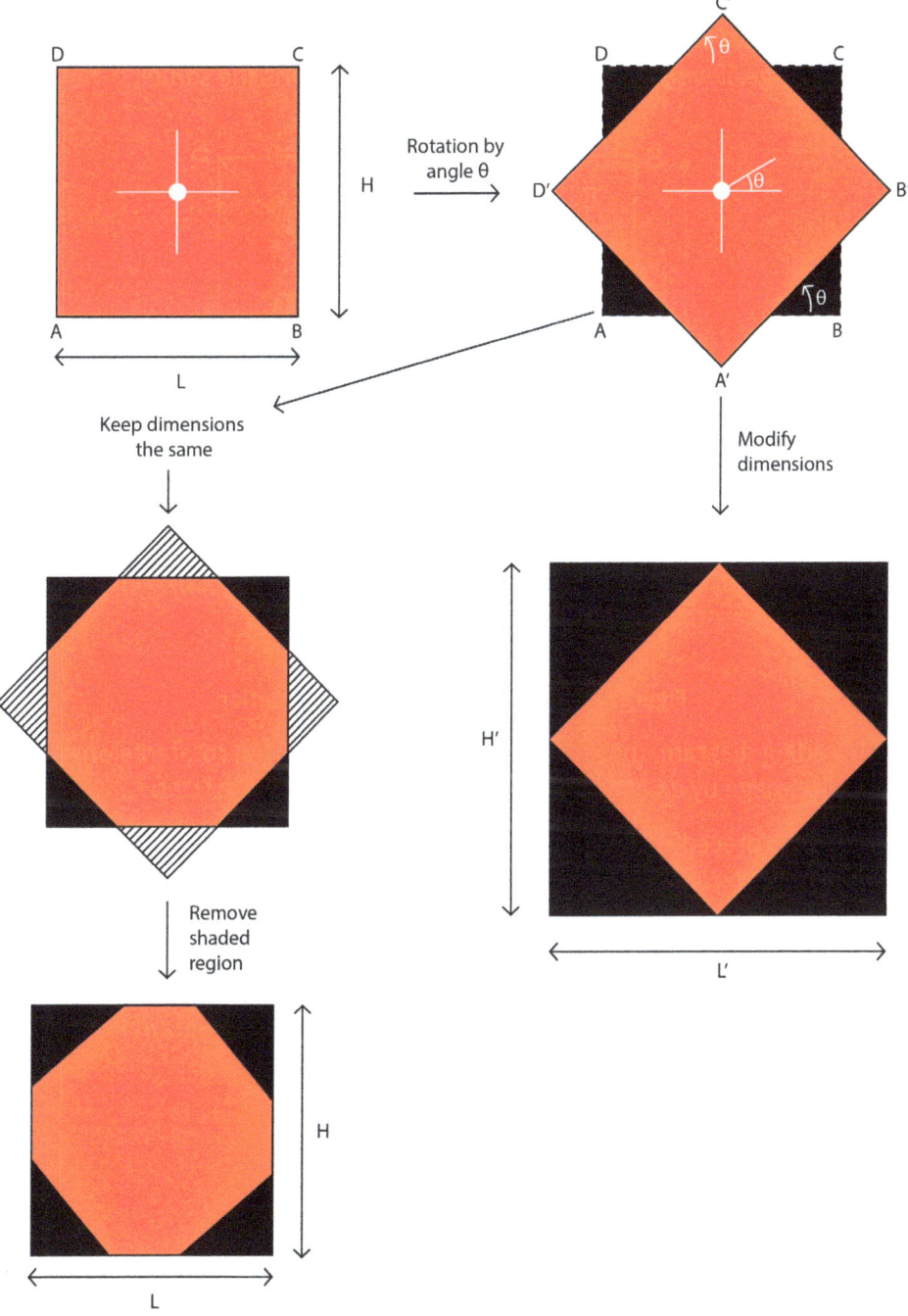

Figure 2.12: Size of the rotated image

66 | Common Operations When Working with Images

In *Figure 2.12*, the size of the rotated image is presented depending on whether the image dimensions are kept the same or are modified while rotating the image

For the case where we want to keep the size of the image the same as the initial image's size, we just crop out the extra region. Let's learn how to obtain the size of the rotated image, if we don't want to keep the dimensions the same:

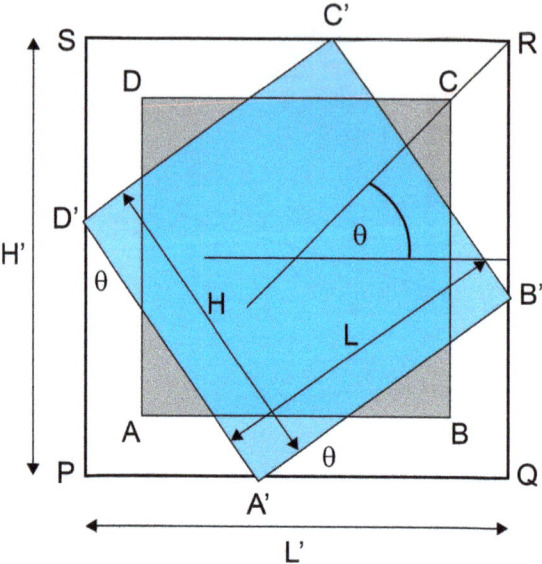

Figure 2.13: Size of the image after rotation

In the preceding diagram, the original image, ABCD, has been rotated along the center of the image by an angle, θ, to form the new image `A'B'C'D'`.

Referring to the preceding diagram, we can write the following equations:

```
A'Q = L cos θ
A'P = H sin θ
```

Thus, we can obtain `L'` in terms of `L` and `H` as follows. Note that `L'` refers to the size of the image after rotation. Refer to the preceding diagram for a visual representation:

```
L' = PQ = PA' + A'Q = H sin θ + L cos θ
```

Similarly, we can write the following equations:

```
PD' = H cos θ
D'S = L sin θ
```

Thus, we can obtain **H'** in terms of **L** and **H**, as follows:

```
H' = PS = PD' + D'S = L sin θ + H cos θ
```

Here comes a tricky part. We know the values of cosine and sine can also be negative, but the size of the rotated image cannot be smaller than the size of the input image. Thus, we will use the absolute values of sine and cosine. The equations for **L'** and **H'** will be modified so that they look as follows:

```
L' = L|cos θ| + H | sin θ|
H> = L|sin θ| + H | cos θ|
```

We can round off these values to obtain an integer value for the new width and height of the rotated image.

Now comes the question that you might be wanting to ask by now. Do we have to do all this calculation every time we want to rotate an image? Luckily, the answer is no. OpenCV has a function that we can use if we want to rotate an image by a given angle. It also gives you the option of scaling the image. The function we are talking about here is the `cv2.getRotationMatrix2D` function. It generates the rotation matrix that can be used to rotate the image. This function takes three arguments:

- The point that we want to rotate the image around. Usually, the center of the image or the bottom-left corner is chosen as the point.
- The angle in degrees that we want to rotate the image by.
- The factor that we want to change the dimensions of the image by. This is an optional argument and can be used to shrink or enlarge the image. In the preceding case, it was 1 as we didn't resize the image.

Note that we are only generating the rotation matrix here and not actually carrying out the rotation. We will cover how to use this matrix when we will look at affine transformations. For now, let's have a look at image resizing.

IMAGE RESIZING

Imagine that you are training a deep learning model or using it to carry out some prediction – for example, object detection, image classification, and so on. Most deep learning models (that are used for images) require the input image to be of a fixed size. In such situations, we resize the image to match those dimensions. Image resizing is a very simple concept where we modify an image's dimensions. This concept has its applications in almost every computer vision domain (image classification, face detection, face recognition, and so on).

Images can be resized in two ways:

- Let's say that we have the initial dimensions as **W×H**, where **W** and **H** stand for width and height, respectively. If we want to double the size (dimensions) of the image, we can resize or scale the image up to **2W×2H**. Similarly, if we want to reduce the size (dimensions) of the image by half, then we can resize or scale the image down to **W/2×H/2**. Since we just want to scale the image, we can pass the scaling factors (for length and width) while resizing. The output dimensions can be calculated based on these scaling factors.

- We might want to resize an image to a fixed dimension, let's say, **420×360** pixels. In such a situation, scaling won't work as you can't be sure that the initial dimensions are going to be a multiple (or factor) of the fixed dimension. This requires us to pass the new dimensions of the image directly while resizing.

A very interesting thing happens when we resize an image. Let's try to understand this with the help of an example:

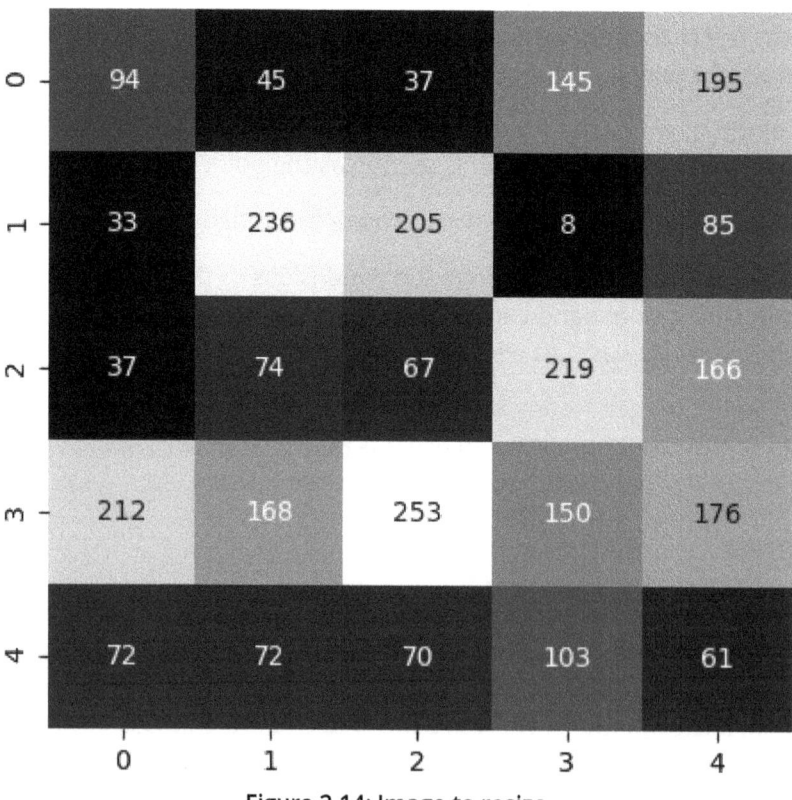

Figure 2.14: Image to resize

The preceding figure shows the image and the pixel values that we want to resize. Currently, it's of size **5×5**. Let's say we want to double the size. This results in the following output. However, we want to fill in the pixel values:

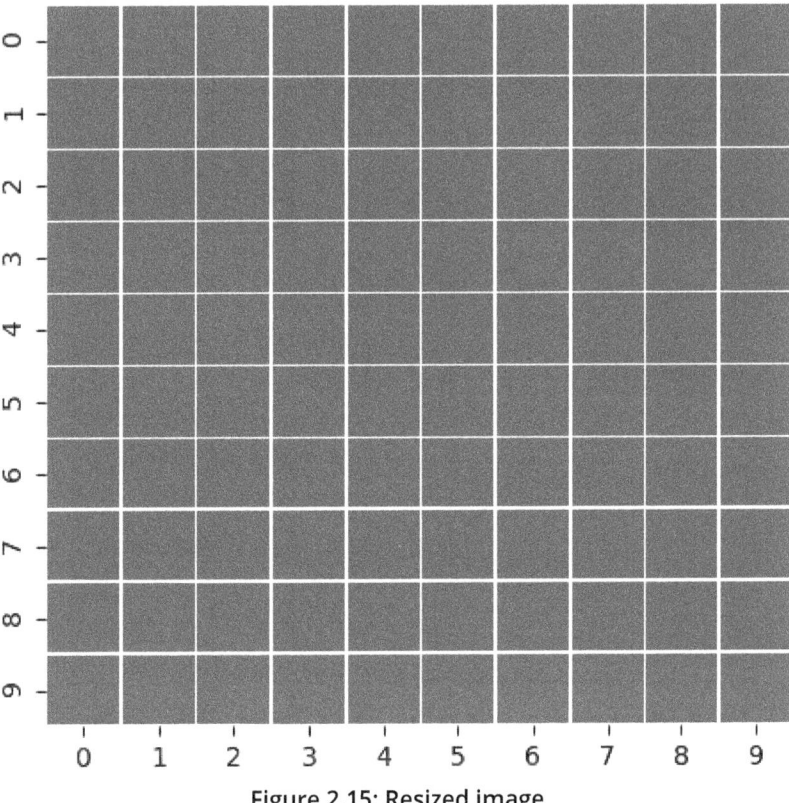

Figure 2.15: Resized image

Let's look at the various options we have. We can always replicate the pixels. This will give us the result shown in the following figure:

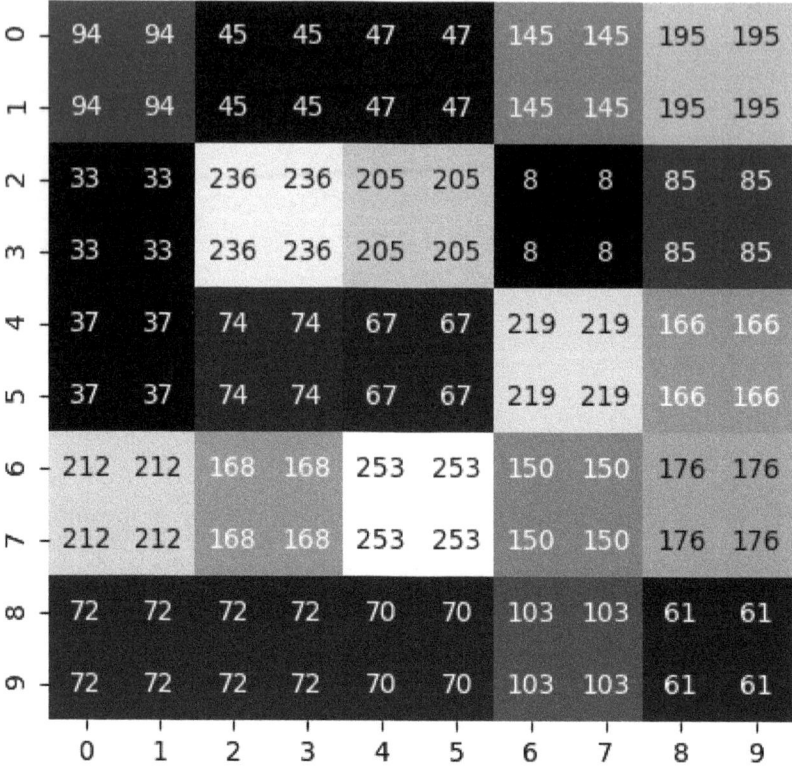

Figure 2.16: Resized image by replicating pixel values

If we remove the pixel values from the preceding image, we will obtain the image shown in the following figure. Compare this with the original image, *Figure 2.14*. Notice how similar it looks to the original image:

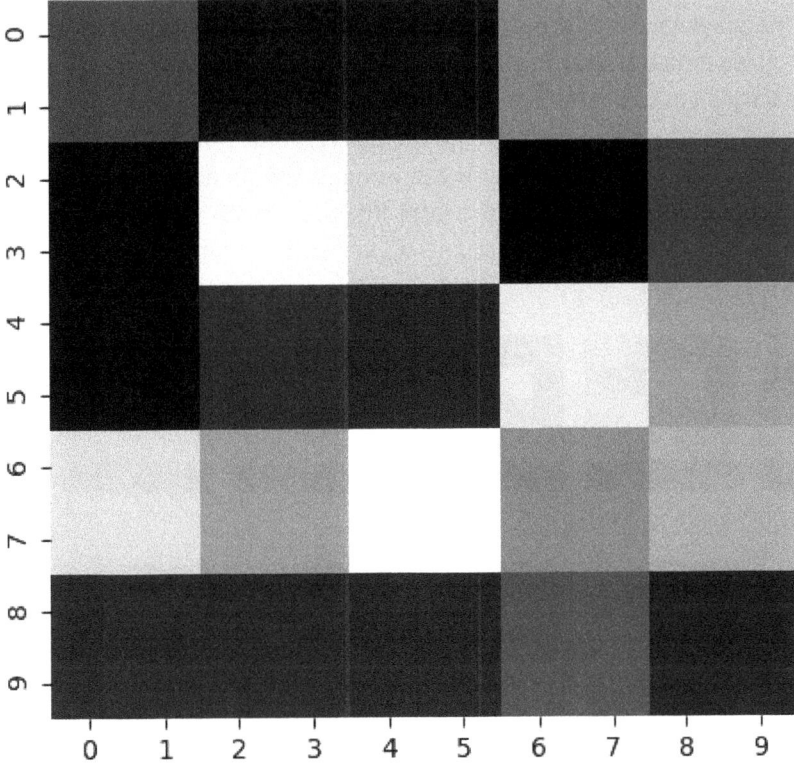

Figure 2.17: Resized image without annotation

Similarly, if we want to reduce the size of the image by half, we can drop some pixels. One thing that you would have noticed is that while resizing, we are replicating the pixels, as shown in *Figure 2.16* and *Figure 2.17*. There are some other techniques we can use as well. We can use interpolation, in which we find out the new pixel values based on the pixel values of the neighboring pixels, instead of directly replicating them. This gives a nice smooth transition in colors. The following figure shows how the results vary if we use different interpolations. From the following figure, we can see that as we progress from left to right, the pixel values of the newly created pixels are calculated differently. In the first three images, the pixels are directly copied from the neighboring pixels, whereas in the later images, the pixel values depend on all the neighboring pixels (left, right, up, and down) and, in some cases, the diagonally adjacent pixels as well:

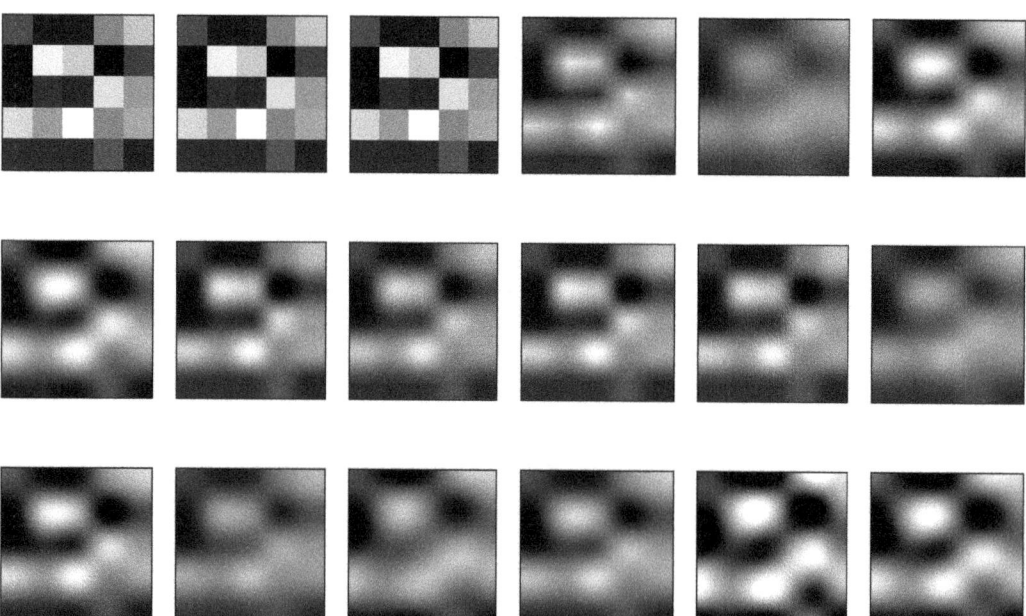

Figure 2.18: Different results obtained based on different interpolations used

Now, you don't need to worry about all the interpolations that are available. We will stick to the following three interpolations for resizing:

- If we are shrinking an image, we will use bilinear interpolation. This is represented as **cv2.INTER_AREA** in OpenCV.

- If we are increasing the size of an image, we will use either linear interpolation (**cv2.INTER_LINEAR**) or cubic interpolation (**cv2.INTER_CUBIC**).

Let's look at how we can resize an image using OpenCV's **cv2.resize** function:

```
cv2.resize(src, dsize, fx, fy, interpolation)
```

Let's have a look at the arguments of this function:

- **src**: This argument is the image we want to resize.
- **dsize**: This argument is the size of the output image (width, height). This will be used when you know the size of the output image. If we are just scaling the image, we will set this to **None**.
- **fx** and **fy**: These arguments are the scaling factors. These will only be used when we want to scale the image. If we already know the size of the output image, we will skip these arguments. These arguments are specified as **fx = 5, fy = 3**, if we want to scale the image by five in the **x** direction and three in the **y** direction.
- **interpolation**: This argument is the interpolation that we want to use. This is specified as **interpolation = cv2.INTER_LINEAR** if we want to use linear interpolation. The other interpolation flags have already been mentioned previously – **cv2.INTER_AREA** and **cv2.INTER_CUBIC**.

Next, let's have a look at a more general transformation – affine transformation.

AFFINE TRANSFORMATION

Affine transformation is one of the most important geometric transformations in computer vision. The reason for this is that an affine transformation can combine the effects of translation, rotation, and resizing into one transform. Affine transformation in OpenCV uses a 2×3 matrix and then applies the effect using the matrix using the **cv2.warpAffine** function. This function takes three arguments:

```
cv2.warpAffine(src, M, dsize)
```

Let's take a look at each argument:

- The first argument (**src**) is the image that we want to apply the transformation to.
- The second argument (**M**) is the transformation matrix.
- The third argument (**dsize**) is the shape of the output image. The order that's used is (width, height) or (number of columns, number of rows).

Let's have a look at the transformation matrices for different transformations and how to generate them:

Transformation	Transformation Matrix	Generating the Transformation Matrix
Translation	$\begin{bmatrix} 1 & 0 & t_x \\ 0 & 1 & t_y \end{bmatrix}$ Here, t_x and t_y represent translation in the **x** and **y** directions, respectively.	`M = np.float32([[1,0,tx],[0,1,ty]])` Here, t_x and t_y represent the translation.
Rotation	$\begin{bmatrix} \alpha & \beta & (1-\alpha)\cdot\text{center.x} - \beta\cdot\text{center.y} \\ -\beta & \alpha & \beta\cdot\text{center.x} + (1-\alpha)\cdot\text{center.y} \end{bmatrix}$ Here, $\alpha = \text{scale}\cdot\cos\theta$, $\beta = \text{scale}\cdot\sin\theta$ `center.x` and `center.y` represent the **x** and **y** coordinates of the point (center) around which the image has to be rotated.	`M = cv2.getRotationMatrix2D((centerX,centerY),angle,scale)` Here, `(centerX, centerY)` are the coordinates of the point around which rotation will be performed, `angle` is the angle by which the image has to be rotated, and `scale` is the factor by which the output image has to be scaled up or scaled down.

Figure 2.19: Table with transformation matrices for different transformations

To generate the transformation matrix for affine transformation, we choose any three non-collinear points on the input image and the corresponding points on the output image. Let's call the points **(in1x, in1y), (in2x, in2y), (in3x, in3y)** for the input image, and **(out1x, out1y), (out2x, out2y), (out3x, out3y)** for the output image.

Then, we can use the following code to generate the transformation matrix. We can use the following code to create a NumPy array for storing the points:

```
ptsInput = np.float32([[in1x, in1y],[in2x, in2y],\
                       [in3x, in3y]])
```

Alternatively, we can also use the following code:

```
ptsOutput = np.float32([[out1x, out1y], [out2x, out2y], \
                        [out3x, out3y]])
```

Next, we will pass these two NumPy arrays to the **cv2.getAffineTransform** function, as follows:

```
M = cv2.getAffineTransform(ptsInput, ptsOutput)
```

Now that we have seen how to generate the transformation matrix, let's see how to apply it. This can be done using **outputImage = cv2.warpAffine(inputImage, M, (outputImageWidth, outputImageHeight))**.

Now that we have discussed affine transformation and how to apply it, let's put this knowledge to use in the following exercise.

EXERCISE 2.02: WORKING WITH AFFINE TRANSFORMATION

In this exercise, we will use the OpenCV functions that we discussed in the previous sections to translate, rotate, and resize an image. We will be using the following image in this exercise.

Figure 2.20: Image of a drop

> **NOTE**
>
> This image is available at https://packt.live/3geu9Hh.

Follow these steps to complete this exercise:

1. Create a new Jupyter Notebook – **Exercise2.02.ipynb**. We will be writing our code in this notebook.

2. Start by importing the required modules:

```
# Import modules
import cv2
import numpy as np
import matplotlib.pyplot as plt
%matplotlib inline
```

3. Next, read the image.

> **NOTE**
>
> Before proceeding, be sure to change the path to the image (highlighted) based on where the image is saved in your system.

The code is as follows:

```
img = cv2.imread("../data/drip.jpg")
```

4. Display the image using Matplotlib:

```
plt.imshow(img[:,:,::-1])
plt.show()
```

The output is as follows. The *X* and *Y* axes refer to the width and height of the image, respectively:

Figure 2.21: Image with X and Y axes

5. Convert the image into grayscale for ease of use:

   ```
   img = cv2.cvtColor(img, cv2.COLOR_BGR2GRAY)
   ```

6. Store the height and width of the image:

   ```
   height,width = img.shape
   ```

7. Start with translation. Shift the image by 100 pixels to the right and 100 pixels down:

   ```
   # Translation
   tx = 100
   ty = 100
   M = np.float32([[1,0,tx],[0,1,ty]])
   dst = cv2.warpAffine(img,M,(width,height))
   plt.imshow(dst,cmap="gray")
   plt.show()
   ```

   The preceding code produces the following output. The *X* and *Y* axes refer to the width and height of the image, respectively:

   Figure 2.22: Translated image

8. Next, rotate our image 45 degrees anti-clockwise, around the center of the image, and scale it up twice:

   ```
   # Rotation
   angle = 45
   ```

```
center = (width//2, height//2)
scale = 2
M = cv2.getRotationMatrix2D(center,angle,scale)
dst = cv2.warpAffine(img,M,(width,height))
plt.imshow(dst,cmap="gray")
plt.show()
```

The output is as follows. The *X* and *Y* axes refer to the width and height of the image, respectively:

Figure 2.23: Rotated image

9. Finally, double the size of the image using the **cv2.resize** function:

```
# Resizing image

print("Width of image = {}, Height of image = {}"\
      .format(width, height))

dst = cv2.resize(img, None, fx=2, fy=2, \
                 interpolation=cv2.INTER_LINEAR)

height, width = dst.shape

print("Width of image = {}, Height of image = {}"\
      .format(width, height))
```

The output is as follows:

```
Width of image = 1280, Height of image = 849
Width of image = 2560, Height of image = 1698
```

In this exercise, we learned how to translate, rotate, and resize an image using OpenCV functions.

> **NOTE**
>
> To access the source code for this specific section, please refer to https://packt.live/38iUnFZ.

Next, let's have a look at the final geometric transformation of this chapter – **perspective transformation**.

PERSPECTIVE TRANSFORMATION

Perspective transformation, unlike all the transformations we have seen so far, requires a 3×3 transformation matrix. We won't go into the mathematics behind this matrix as it's out of the scope of this book. We will need four points in the input image and the coordinates of the same points in the output image. Note that these points should not be collinear. Next, similar to the affine transformation steps, we will carry out perspective transformation.

We will use the following code create a NumPy array for storing the points:

```
ptsInput = np.float32([[in1x, in1y],[in2x, in2y],[in3x, in3y],\
                       [in4x,in4y]])
```

Alternatively, we can use the following code:

```
ptsOutput = np.float32([[out1x, out1y], [out2x, out2y], \
                        [out3x, out3y], [out4x, out4y]])
```

Next, we will pass these two NumPy arrays to the **cv2.getPerspectiveTransform** function:

```
M = cv2.getPerspectiveTransform(ptsInput, ptsOutput)
```

The transformation matrix can then be applied using the following code:

```
outputImage = cv2.warpPerspective(inputImage, M, \
              (outputImageWidth, outputImageHeight))
```

The primary difference between perspective and affine transformation is that in perspective transformation, straight lines will still remain straight after the transformation, unlike in affine transformation, where, because of multiple transformations, the line can be converted into a curve.

Let's see how we can use perspective transformation with the help of an exercise.

EXERCISE 2.03: PERSPECTIVE TRANSFORMATION

In this exercise, we will carry out perspective transformation to obtain the front cover of the book given as the input image. We will use OpenCV's **cv2.getPerspectiveTransform** and **cv2.warpPerspective** functions. We will be using the following image of a book:

Figure 2.24: Image of a book

> **NOTE**
>
> This image can be downloaded from https://packt.live/2YM9tAA.

Follow these steps to complete this exercise:

1. Create a new Jupyter Notebook – **Exercise2.03.ipynb**. We will be writing our code in this notebook.

2. Import the modules we will be using:

```
# Import modules
import cv2
import numpy as np
import matplotlib.pyplot as plt
%matplotlib inline
```

3. Next, read the image.

> **NOTE**
>
> Before proceeding, be sure to change the path to the image (highlighted) based on where the image is saved in your system.

The code is as follows:

```
img = cv2.imread("../data/book.jpg")
```

4. Display the image using Matplotlib, as follows:

```
plt.imshow(img[:,:,::-1])
plt.show()
```

The output is as follows. The X and Y axes refer to the width and height of the image, respectively:

Figure 2.25: Output image of a book

5. Specify four points in the image so that they lie at the four corners of the front cover. Since we just want the front cover, the output points will be nothing but the corner points of the final 300×300 image. Note that the order of the points should stay the same for both the input and output points:

```
inputPts = np.float32([[4,381],
                       [266,429],
                       [329,68],
                       [68,20]])
outputPts = np.float32([[0,300],
                        [300,300],
                        [300,0],
                        [0,0]])
```

6. Next, obtain the transformation matrix:

```
M = cv2.getPerspectiveTransform(inputPts,outputPts)
```

7. Apply this transformation matrix to carry out perspective transformation:

```
dst = cv2.warpPerspective(img,M,(300,300))
```

8. Display the resultant image using Matplotlib. The *X* and *Y* axes refer to the width and height of the image, respectively:

```
plt.imshow(dst[:,:,::-1])
plt.show()
```

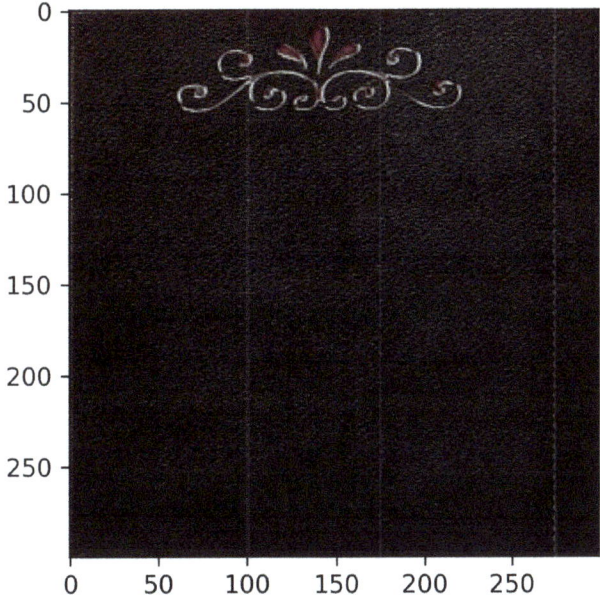

Figure 2.26: Front cover of the book

In this exercise, we saw how we can use geometric transformations to extract the front cover of the book provided in a given image. This can be interesting when we are trying to scan a document and we want to obtain a properly oriented image of the document after scanning.

> **NOTE**
>
> To access the source code for this specific section, please refer to https://packt.live/2YOqs5u.

With that, we have covered all the major geometric transformations. It's time to cover another important topic in image processing – image arithmetic.

IMAGE ARITHMETIC

We know that images are nothing but matrices. This raises an important and interesting question. If we can carry out arithmetic operations on matrices, can we carry them out on images as well? The answer is a bit tricky. We can carry out the following operations on images:

- Adding and subtracting two images
- Adding and subtracting a constant value to/from image
- Multiplying a constant value by an image

We can, of course, multiply two images if we can assume they're matrices, but in terms of images, multiplying two images does not make much sense, unless it's carried out pixel-wise.

Let's look at these operations, one by one.

IMAGE ADDITION

We know that an image is made up of pixels and that the color of a pixel is decided by the pixel value. So, when we talk about adding a constant to an image, we mean adding a constant to the pixel values of the image. While it might seem a very simple task, there is a small concept that needs to be discussed properly. We have already said that, usually, the pixel values in images are represented using unsigned 8-bit integers, and that's why the range of pixel values is from 0 to 255. Now, imagine that you want to add the value 200 to an image, that is, to all the pixels of the image. If we have a pixel value of 220 in the image, the new value will become 220 + 200 = 420. But since this value lies outside of the range (0-255), two things can happen:

- The value will be clipped to a maximum value. This means that 420 will be revised to 255.

- The value will follow a cyclic order or a modulo operation. A modulo operation means that you want to find the remainder obtained after dividing one value by another value. We are looking at division by 255 here (the maximum value in the range). That's why 420 will become 165 (the remainder that's obtained after dividing 420 by 255).

Notice how the results obtained in both approaches are different. The first approach, which is also the recommended approach, uses OpenCV's **cv2.add** function. The usage of the **cv2.add** function is as follows:

```
dst = cv2.add(src1, src2)
```

The second approach uses NumPy while adding two images. The NumPy approach is as follows:

```
dst = src1 + src2
```

`src1` refers to the first image that we are trying to add to the second image (`src2`). This gives the resultant image (`dst`) as the output. Either of these approaches can be used to add two images. The interesting part, as we will see shortly in the next exercise, is that when we add a constant value to an image using OpenCV, by default, that value gets added to the first channel of the image. Since we already know that images in OpenCV are represented in BGR mode, the value gets added to the blue channel, giving the image a more bluish tone. On the other hand, in the NumPy approach, the value is automatically broadcasted in such a manner that the value is added to every channel.

Let's compare both approaches with the help of an exercise.

EXERCISE 2.04: PERFORMING IMAGE ADDITION

In this exercise, we will learn how to add a constant value to an image and the approaches to carry out the task. We will be working on the following image. First, we will add a constant value to the image using the `cv2.add` function and then compare the output obtained with the output obtained by adding the constant value using NumPy:

Figure 2.27: Image of a Golden Retriever puppy

> **NOTE**
>
> This image can be found at https://packt.live/2ZnspEU.

Follow these steps to complete this exercise:

1. Create a new notebook – **Exercise2.04.ipynb**. We will be writing our code in this notebook.

2. Let's import the libraries we will be using and also write the magic command:

   ```
   # Import libraries
   import cv2
   import numpy as np
   import matplotlib.pyplot as plt
   %matplotlib inline
   ```

3. Read the image. The code for this is as follows:

   > **NOTE**
   >
   > Before proceeding, be sure to change the path to the image (highlighted) based on where the image is saved in your system.

   ```
   # Read image
   img = cv2.imread("../data/puppy.jpg")
   ```

4. Display the image using Matplotlib:

   ```
   # Display image
   plt.imshow(img[:,:,::-1])
   plt.show()
   ```

The output is as follows. The X and Y axes refer to the width and height of the image, respectively:

Figure 2.28: Image of a Golden Retriever puppy

5. First, try adding the value 100 to the image using NumPy:

    ```
    # Add 100 to the image
    numpyImg = img+100
    ```

6. Display the image, as follows:

    ```
    # Display image
    plt.imshow(numpyImg[:,:,::-1])
    plt.show()
    ```

The output is as follows. The X and Y axes refer to the width and height of the image, respectively:

Figure 2.29: Result obtained by adding 100 to the image using NumPy

Notice how adding the value 100 has distorted the image severely. This is because of the modulo operation being performed by NumPy on the new pixel values. This should also give you an idea of why NumPy's approach is not the recommended approach to use while dealing with adding a constant value to an image.

7. Try doing the same using OpenCV:

```
# Using OpenCV
opencvImg = cv2.add(img,100)
```

8. Display the image, as follows:

```
# Display image
plt.imshow(opencvImg[:,:,::-1])
plt.show()
```

The output is as follows. The *X* and *Y* axes refer to the width and height of the image, respectively:

Figure 2.30: Result obtained by adding 100 to the image using OpenCV

As shown in the preceding figure, there is an increased blue tone to the image. This is because the value 100 has only been added to the first channel of the image, which is the blue channel.

9. Fix this by adding an image to an image instead of adding a value to an image. Check the shape, as follows:

```
img.shape
```

The output is **(924, 1280, 3)**.

10. Now, create an image that's the same shape as the original image that has a fixed pixel value of 100. We are doing this because we want to add the value 100 to every channel of the original image:

```
nparr = np.ones((924,1280,3),dtype=np.uint8) * 100
```

90 | Common Operations When Working with Images

11. Add **nparr** to the image and visualize the result:

```
opencvImg = cv2.add(img,nparr)
# Display image
plt.imshow(opencvImg[:,:,::-1])
plt.show()
```

The output is as follows. The *X* and *Y* axes refer to the width and height of the image, respectively:

Figure 2.31: Result obtained by adding an image to another image

12. When the same image is added using OpenCV, we notice that the results that are obtained are the same as they were for NumPy:

```
npImg = img + nparr
# Display image
plt.imshow(npImg[:,:,::-1])
plt.show()
```

The output is as follows. The *X* and *Y* axes refer to the width and height of the image, respectively:

Figure 2.32: New results obtained

The most interesting point to note from the results is that adding a value to an image using OpenCV results in increased brightness of the image. You can try subtracting a value (or adding a negative value) and see whether the reverse also true.

In this exercise, we added a constant value to an image and compared the outputs obtained by using the **cv2.add** function and by using NumPy's addition operator (**+**). We also saw that while using the **cv2.add** function, the value was added to the blue channel and that it can be fixed by creating an image that's the same shape as the input image, with all the pixels of the value equal to the constant value we want to add to the image.

> **NOTE**
>
> To access the source code for this specific section, please refer to https://packt.live/2BUOUsV.

In this section, we saw how to perform image addition using NumPy and OpenCV's **cv2.add** function. Next, let's learn how to multiply an image with a constant value.

IMAGE MULTIPLICATION

Image multiplication is very similar to image addition and can be carried out using OpenCV's `cv2.multiply` function (recommended) or NumPy. OpenCV's function is recommended because of the same reason as we saw for `cv2.add` in the previous section.

We can use the `cv2.multiply` function as follows:

```
dst = cv2.Mul(src1, src2)
```

Here, `src1` and `src2` refer to the two images we are trying to multiply and `dst` refers to the output image obtained by multiplication.

We already know that multiplication is nothing but repetitive addition. Since we saw that image addition had an effect of increasing the brightness of an image, image multiplication will have the same effect. The difference here is that the effect will be manifold. Image addition is typically used when we want to make minute modifications in brightness. Let's directly see how to use these functions with the help of an exercise.

EXERCISE 2.05: IMAGE MULTIPLICATION

In this exercise, we will learn how to use OpenCV and NumPy to multiply a constant value with an image and how to multiply two images.

> **NOTE**
>
> The image used in this exercise can be found at https://packt.live/2ZnspEU.

Follow these steps to complete this exercise:

1. Create a new notebook – `Exercise2.05.ipynb`. We will be writing our code in this notebook.

2. First, let's import the necessary modules:

```
# Import libraries
import cv2
import numpy as np
import matplotlib.pyplot as plt
%matplotlib inline
```

3. Next, read the image of the puppy.

> **NOTE**
>
> Before proceeding, be sure to change the path to the image (highlighted) based on where the image is saved in your system.

The code for this is as follows:

```
# Read image
img = cv2.imread("../data/puppy.jpg")
```

4. Display the output using Matplotlib, as follows:

```
# Display image
plt.imshow(img[:,:,::-1])
plt.show()
```

The output is as follows. The *X* and *Y* axes refer to the width and height of the image, respectively:

Figure 2.33: Image of a Golden Retriever puppy

5. Multiply the image by **2** using the `cv2.multiply` function:

```
cvImg = cv2.multiply(img,2)
```

6. Display the image, as follows:

```
plt.imshow(cvImg[:,:,::-1])
plt.show()
```

The output is as follows. The *X* and *Y* axes refer to the width and height of the image, respectively:

Figure 2.34: Result obtained by multiplying the image by 2 using OpenCV

Can you think of why you obtained a high blue tone in the output image? You can use the explanation given for image addition in the previous exercise as a reference.

7. Try to multiply the image using NumPy:

```
npImg = img*2
```

8. Display the image, as follows:

```
plt.imshow(npImg[:,:,::-1])
plt.show()
```

The output is as follows. The *X* and *Y* axes refer to the width and height of the image, respectively:

Figure 2.35: Result obtained by multiplying the image by 2 using NumPy

9. Print the shape of the image, as follows:

   ```
   img.shape
   ```

   The shape is `(924, 1280, 3)`.

10. Now, let's create a new array filled with 2s that's the same shape as the original image:

    ```
    nparr = np.ones((924,1280,3),dtype=np.uint8) * 2
    ```

11. Now, we will use this new array for multiplication purposes and compare the results that are obtained:

    ```
    cvImg = cv2.multiply(img,nparr)
    plt.imshow(cvImg[:,:,::-1])
    plt.show()
    ```

The output is as follows. The X and Y axes refer to the width and height of the image, respectively:

Figure 2.36: Result obtained by multiplying two images using OpenCV

> **NOTE**
>
> To access the source code for this specific section, please refer to https://packt.live/2BRos3p.

We know that multiplication is nothing but repetitive addition, so it makes sense to obtain a brighter image using multiplication since we obtained a brighter image using addition as well.

So far, we've discussed geometric transformations and image arithmetic. Now, let's move on to a slightly more advanced topic regarding performing bitwise operations on images. Before we discuss that, let's have a look at binary images.

BINARY IMAGES

So far, we have worked with images with one channel (grayscale images) and three channels (color images). We also mentioned that, usually, pixel values in images are represented as 8-bit unsigned integers and that's why they have a range from 0 to 255. But that's not always true. Images can be represented using floating-point values and also with lesser bits, which also reduces the range. For example, an image using 6-bit unsigned integers will have a range between $0 - (2^6-1)$ or 0 to 63.

Even though it's possible to use more or fewer bits, typically, we work with only two kinds of ranges – 0 to 255 for 8-bit unsigned integers and images that have only 0 and 1. The second category of images uses only two pixel values, and that's why they are referred to as binary images. Binary images need only a single bit to represent a pixel value. These images are commonly used as masks for selecting or removing a certain region of an image. It is with these images that bitwise operations are commonly used. Can you think of a place where you have seen binary images in real life?

You can find such black-and-white images quite commonly in QR codes. Can you think of some other applications of binary images? Binary images are extensively used for document analysis and even in industrial machine vision tasks. Here is a sample binary image:

Figure 2.37: QR code as an example of a binary image

Now, let's see how we can convert an image into a binary image. This technique comes under the category of thresholding. Thresholding refers to the process of converting a color image into a binary image. There is a wide range of thresholding techniques available, but here, we will focus only on a very simple thresholding technique – binary thresholding – since we are working with binary images.

The concept behind binary thresholding is very simple. You choose a threshold value and all the pixel values below and equal to the threshold are replaced with 0, while all the pixel values above the threshold are replaced with a specified value (usually 1 or 255). This way, you end up with an image that has only two unique pixel values, which is what a binary image is.

We can convert an image into a binary image using the following code:

```
# Set threshold and maximum value
thresh = 125
maxValue = 255
# Binary threshold
th, dst = cv2.threshold(img, thresh, maxValue, \
                cv2.THRESH_BINARY)
```

In the preceding code, we first specified the threshold as 125 and then specified the maximum value. This is the value that will replace all the pixel values above the threshold. Finally, we used OpenCV's `cv2.threshold` function to perform binary thresholding. This function takes the following inputs:

- The grayscale image that we want to perform thresholding on.
- `thresh`: The threshold value.
- `maxValue`: The maximum value, which will replace all pixel values above the threshold.
- `th`, `dst`: The thresholding flag. Since we are performing binary thresholding, we will use `cv2.THRESH_BINARY`.

Let's implement what we've learned about binary thresholding.

EXERCISE 2.06: CONVERTING AN IMAGE INTO A BINARY IMAGE

In this exercise, we will use binary thresholding to convert a color image into a binary image. We will be working on the following image of zebras:

Figure 2.38: Image of zebras

> **NOTE**
>
> This image can be found at https://packt.live/2ZpQ07Z.

Follow these steps to complete this exercise:

1. Create a new Jupyter Notebook – **Exercise2.06.ipynb**. We will be writing our code in this notebook.

2. Import the necessary modules:

```
# Import modules
import cv2
import numpy as np
import matplotlib.pyplot as plt
%matplotlib inline
```

3. Read the image of the zebras and convert it into grayscale. This is necessary because we know that thresholding requires us to provide a grayscale image as an argument.

   > **NOTE**
   >
   > Before proceeding, be sure to change the path to the image (highlighted) based on where the image is saved in your system.

   The code for this is as follows:

```
img = cv2.imread("../data/zebra.jpg")
img = cv2.cvtColor(img, cv2.COLOR_BGR2GRAY)
```

4. Display the grayscale image using Matplotlib:

```
plt.imshow(img, cmap='gray')
plt.show()
```

The output is as follows. The *X* and *Y* axes refer to the width and height of the image, respectively:

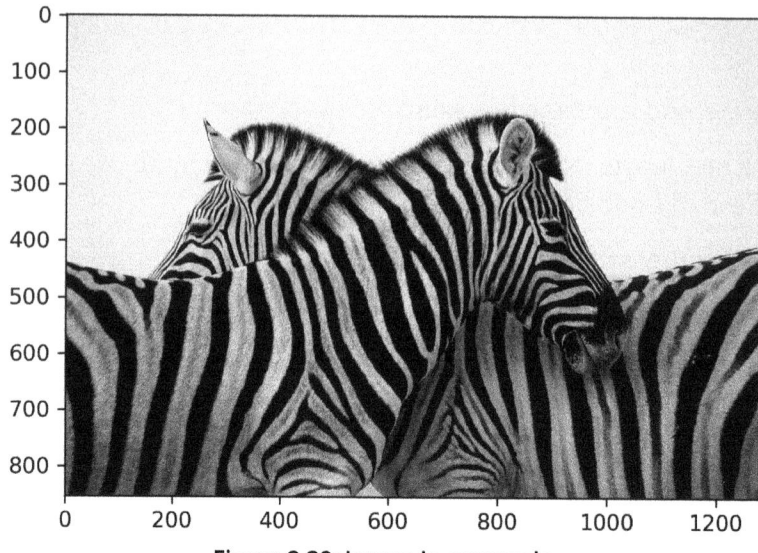

Figure 2.39: Image in grayscale

5. Use the **cv2.thresholding** function and set the threshold to **150**:

```
# Set threshold and maximum value
thresh = 150
maxValue = 255
# Binary threshold
th, dst = cv2.threshold(img, thresh, maxValue, \
                       cv2.THRESH_BINARY)
```

> **NOTE**
>
> You can try playing around with the threshold value to obtain different results.

6. Display the binary image we have obtained:

```
plt.imshow(dst, cmap='gray')
plt.show()
```

The output is as follows. The *X* and *Y* axes refer to the width and height of the image, respectively:

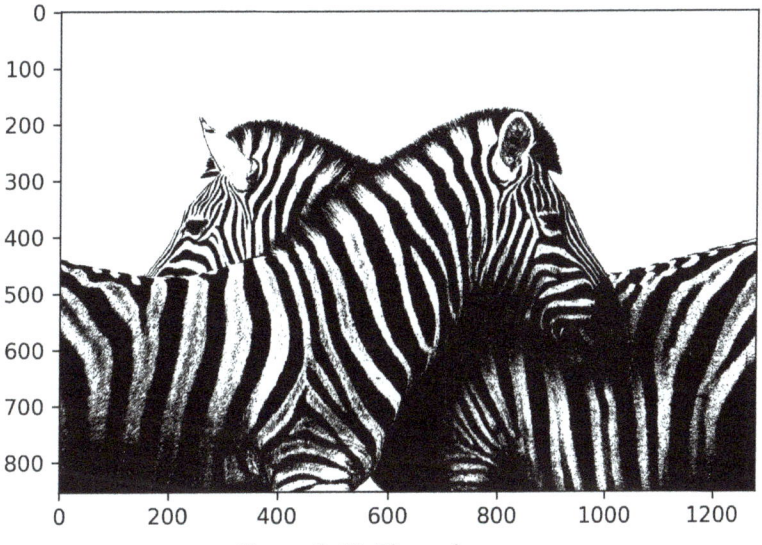

Figure 2.40: Binary image

NOTE

To access the source code for this specific section, please refer to https://packt.live/2VyYHfa.

In this exercise, we saw how to obtain a binary image using thresholding. Next, let's see how we can carry out bitwise operations on these images.

BITWISE OPERATIONS ON IMAGES

Let's start by listing the binary operations, along with their results. You must have read about these operations before, so we won't go into their details. The following table provides the truth tables for the bitwise operations as a quick refresher:

Bitwise Operation	Table		
NOT Used for generating the negative of a binary image. Function: `cv2.bitwise_not`	Input Bit	Output Bit	
	0	1	
	1	0	
OR The OR operation will return a 1 if at least one of the images has a 1 in that pixel. This can be used to generate unions of two binary images. Function: `cv2.bitwise_or`	Input Bit 1	Input Bit 2	Output Bit
	0	0	0
	0	1	1
	1	0	1
	1	1	1
AND The AND operation will return a 1, but only if both of the images have a 1 in that specific pixel. This can be used to generate the intersection of two binary images. Function: `cv2.bitwise_and`	Input Bit 1	Input Bit 2	Output Bit
	0	0	0
	0	1	0
	1	0	0
	1	1	1
XOR The XOR operation will return a 1, but only if one of the pixels is 1 for the images. This can be used to identify the moving object in two subsequent frames. Function: `cv2.bitwise_xor`	Input Bit 1	Input Bit 2	Output Bit
	0	0	0
	0	1	1
	1	0	1
	1	1	0

Figure 2.41: Bitwise operations and truth tables

Let's see how we can use these functions with the help of an exercise.

EXERCISE 2.07: CHESS PIECES

In this exercise, we will use the XOR operation to find the chess pieces that have moved using two images taken of the same chess game:

Figure 2.42: Two images of chess board

> **NOTE**
>
> These images can be found at https://packt.live/3fuxLoU.

Follow these steps to complete this exercise:

1. Create a new notebook – **Exercise2.07.ipynb**. We will be writing our code in this notebook.

2. Import the required modules:

    ```
    # Import modules
    import cv2
    import numpy as np
    import matplotlib.pyplot as plt
    %matplotlib inline
    ```

104 | Common Operations When Working with Images

3. Read the images of the board and convert them to grayscale.

> **NOTE**
>
> Before proceeding, be sure to change the path to the images (highlighted) based on where the images are saved in your system.

The code for this is as follows:

```
img1 = cv2.imread("../data/board.png")
img2 = cv2.imread("../data/board2.png")
img1 = cv2.cvtColor(img1, cv2.COLOR_BGR2GRAY)
img2 = cv2.cvtColor(img2, cv2.COLOR_BGR2GRAY)
```

4. Display these images using Matplotlib:

```
plt.imshow(img1, cmap="gray")
plt.show()
```

The output is as follows. The *X* and *Y* axes refer to the width and height of the image, respectively:

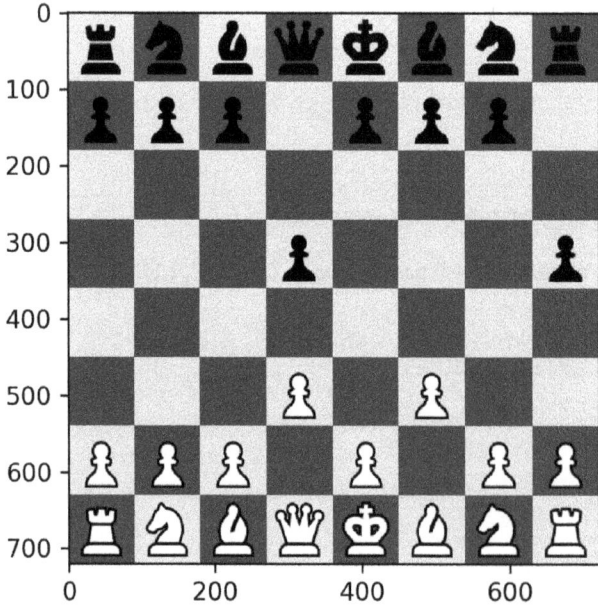

Figure 2.43: Grayscale version of the chess image

5. Plot the second image, as follows:

```
plt.imshow(img2,cmap="gray")
plt.show()
```

The output is as follows. The *X* and *Y* axes refer to the width and height of the image, respectively:

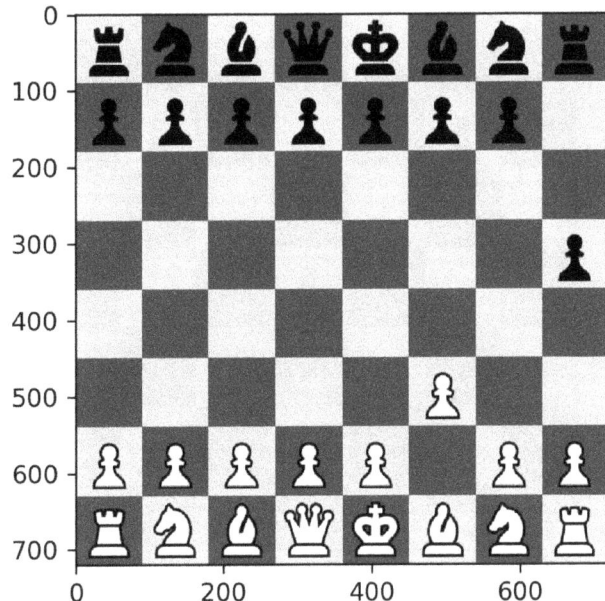

Figure 2.44: Grayscale version of the chess image

6. Threshold both the images using a threshold value of 150 and a maximum value of 255:

```
# Set threshold and maximum value
thresh = 150
maxValue = 255

# Binary threshold
th, dst1 = cv2.threshold(img1, thresh, maxValue, \
                         cv2.THRESH_BINARY)

# Binary threshold
th, dst2 = cv2.threshold(img2, thresh, maxValue, \
                         cv2.THRESH_BINARY)
```

7. Display these binary images using Matplotlib:

```
plt.imshow(dst1, cmap='gray')
plt.show()
```

The output is as follows. The *X* and *Y* axes refer to the width and height of the image, respectively:

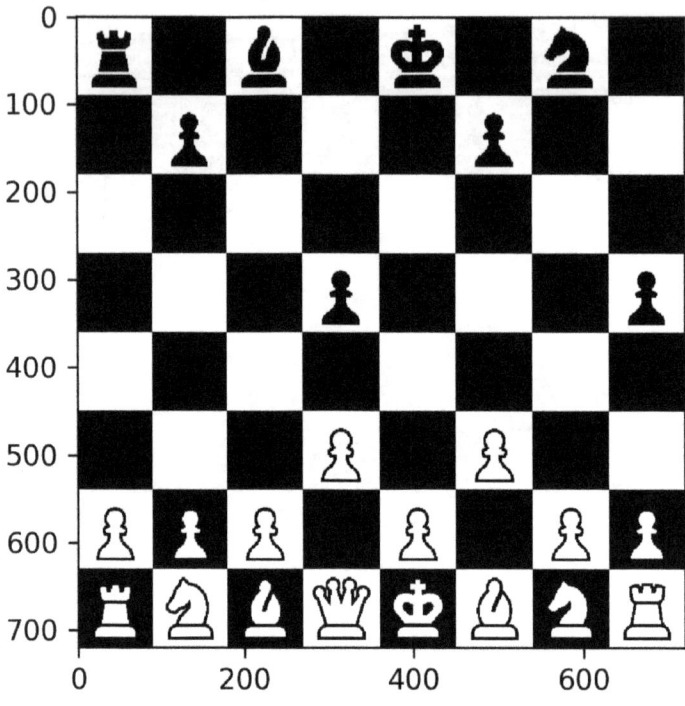

Figure 2.45: Binary image

8. Print the second image, as follows:

```
plt.imshow(dst2, cmap='gray')
plt.show()
```

The output is as follows. The X and Y axes refer to the width and height of the image, respectively:

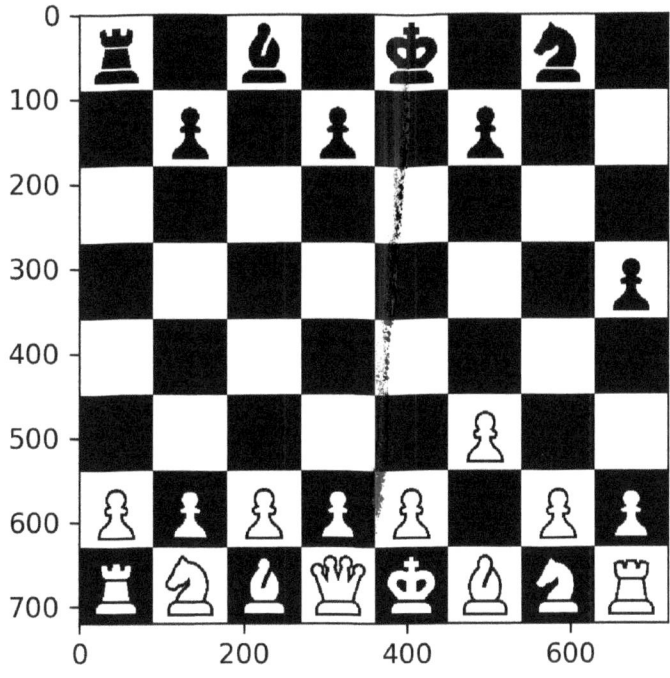

Figure 2.46: Image after thresholding

9. Use bitwise XOR to find the pieces that have moved, as follows:

```
dst = cv2.bitwise_xor(dst1,dst2)
```

10. Display the result, as follows. The X and Y axes refer to the width and height of the image, respectively:

```
plt.imshow(dst, cmap='gray')
plt.show()
```

The output is as follows:

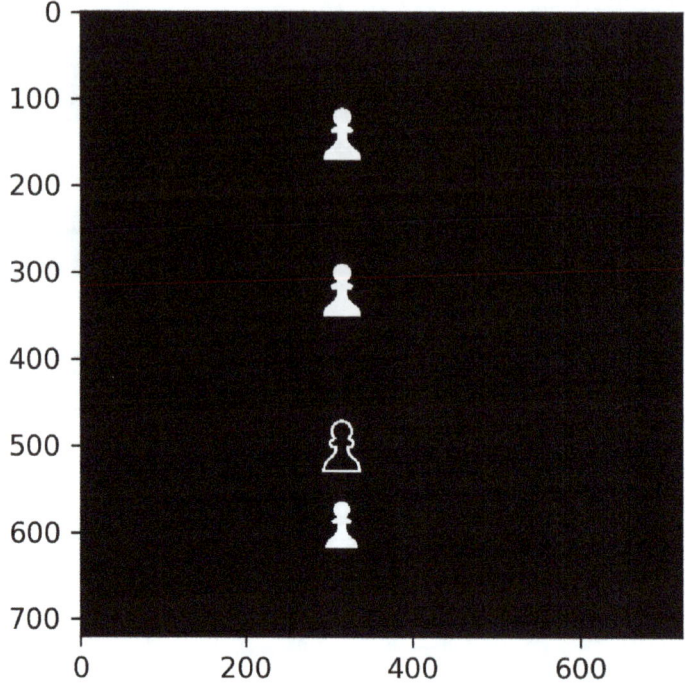

Figure 2.47: Result of the XOR operation

Notice that, in the preceding image, the four pieces that are present show the initial and final positions of the only two pieces that had changed their positions in the two images. In this exercise, we used the XOR operation to perform motion detection to detect the two chess pieces that had moved their positions after a few steps.

> **NOTE**
>
> To access the source code for this specific section, please refer to https://packt.live/2NHixQY.

MASKING

Let's discuss one last concept related to binary images. Binary images are quite frequently used to serve as a mask. For example, consider the following image. We will be using an image of a disk:

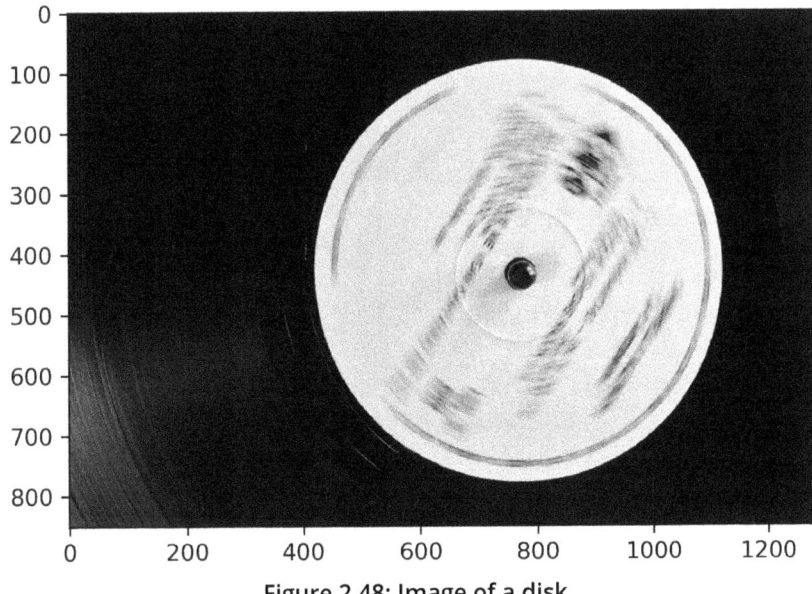

Figure 2.48: Image of a disk

After image thresholding, the mask will look as follows:

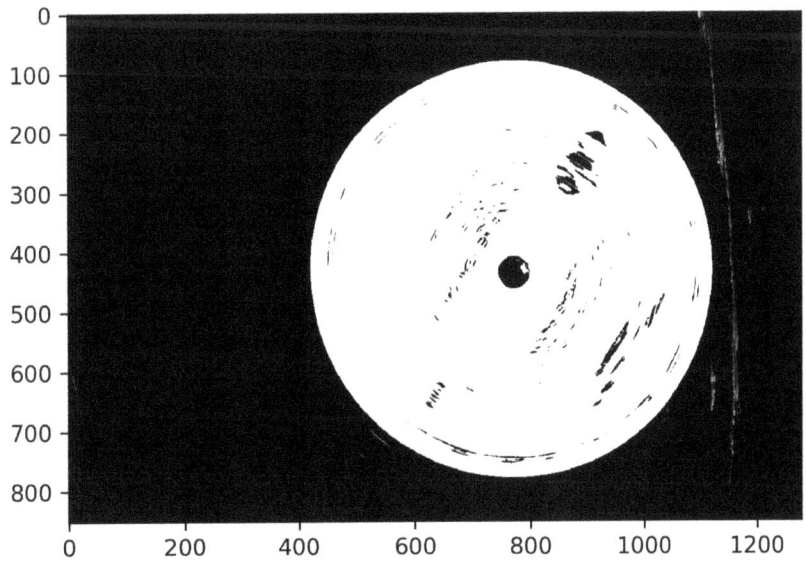

Figure 2.49: Binary mask

Let's see what happens when we apply masking to the image of the zebras that we worked with earlier:

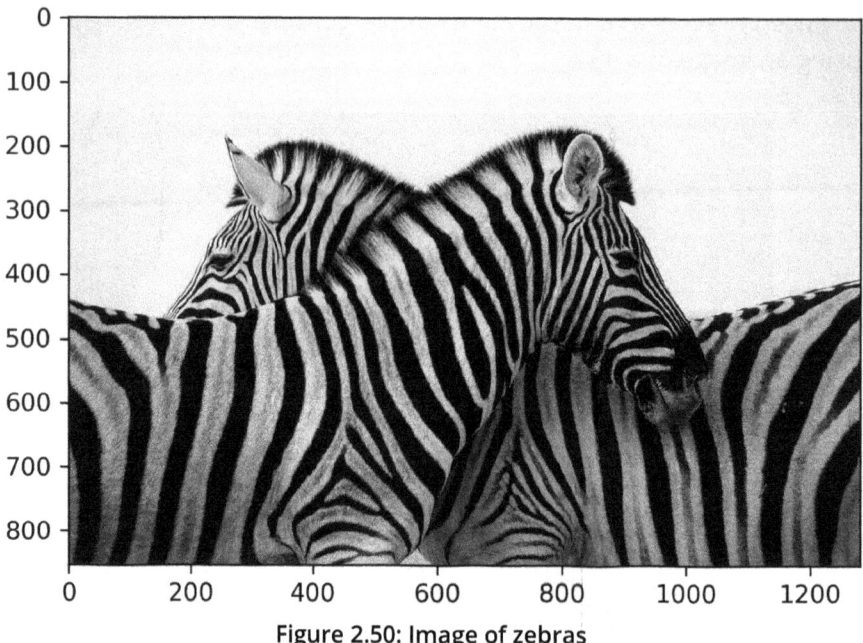

Figure 2.50: Image of zebras

The final image will look as follows:

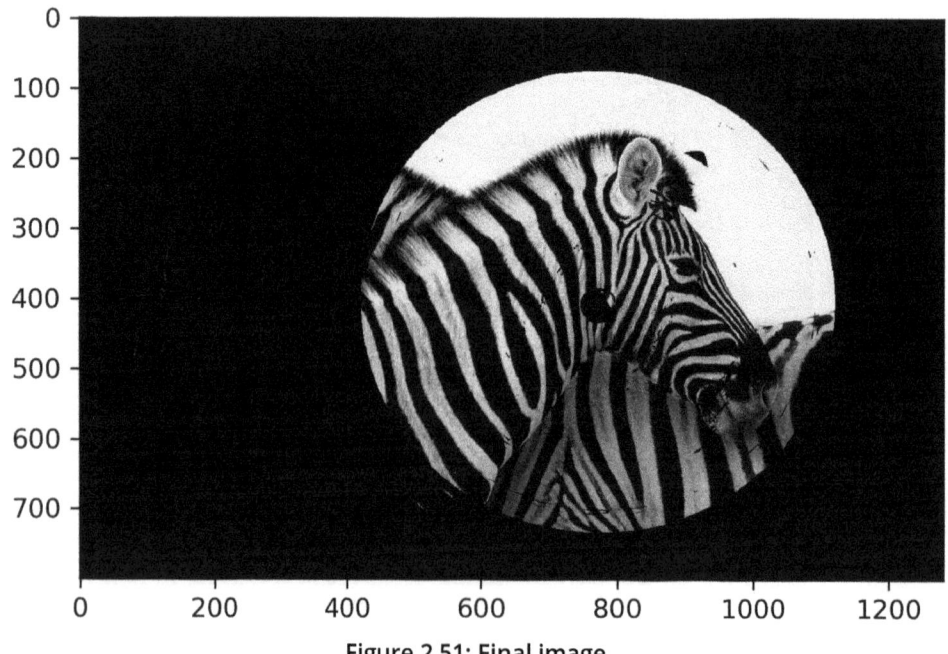

Figure 2.51: Final image

Let's start with *Figure 2.49*. This image is a binary image of a disk after thresholding. *Figure 2.50* shows the familiar grayscale image of zebras. When *Figure 2.49* is used as a mask to only keep the pixels of *Figure 2.50*, where the corresponding pixels of *Figure 2.50* are white, we end up with the result shown in *Figure 2.51*. Let's break this down. Consider a pixel, *P*, at location *(x,y)* in *Figure 2.49*. If the pixel, *P*, is white or non-zero (because zero refers to black), the pixel at location *(x,y)* in *Figure 2.50* will be left as it is. If the pixel, *P*, was black or zero, the pixel at location *(x,y)* in *Figure 2.50* will be replaced with 0. This refers to a masking operation since *Figure 2.49* is covering *Figure 2.50* as a mask and displaying only a few selected pixels. Such an operation can be easily carried out using the following code:

```
result = np.where(mask, image, 0)
```

Let's understand what is happening here. NumPy's **np.where** function says that wherever the mask (first argument) is non-zero, return the value of the image (second argument); otherwise, return 0 (third argument). This is exactly what we discussed in the previous paragraph. We will be using masks in *Chapter 5, Face Processing in Image and Video*, as well.

Now, it's time for you to try out the concepts that you have studied so far to replicate the result shown in *Figure 2.51*.

ACTIVITY 2.01: MASKING USING BINARY IMAGES

In this activity, you will be using masking and other concepts you've studied in this chapter to replicate the result shown in *Figure 2.51*. We will be using image resizing, image thresholding, and image masking concepts to display only the heads of the zebras present in *Figure 2.50*. A similar concept can be applied to create nice portraits of photos where only the face of the person is visible and the rest of the region/background is blacked out. Let's start with the steps that you need to follow to complete this activity:

1. Create a new notebook – **Activity2.01.ipynb**. You will be writing your code in this notebook.

2. Import the necessary libraries – OpenCV, NumPy, and Matplotlib. You will also need to add the magic command to display images inside the notebook.

3. Read the image titled **recording.jpg** from the disk and convert it to grayscale.

   > **NOTE**
   >
   > This image can be found at https://packt.live/32c3pDK.

4. Next, you will have to perform thresholding on this image. You can use a threshold of 150 and a maximum value of 255.

5. The thresholded image should be similar to the one shown in *Figure 2.49*.

6. Next, read the image of the zebras (titled **zebras.jpg**) and convert it to grayscale.

   > **NOTE**
   >
   > This image can be found at https://packt.live/2ZpQ07Z.

7. Before moving on to using NumPy's **where** command for masking, we need to check whether the images have the same size or not. Print the shapes of both images (zebras and disk).

8. You will notice that the images have different dimensions. Resize both images to 1,280×800 pixels. This means that the width of the resized image should be 1,280 pixels and that the height should be 800 pixels. You will have to use the **cv2.resize** function for resizing. Use linear interpolation while resizing the images.

9. Next, use NumPy's **where** command to only keep the pixels where the disk pixels are white. The other pixels should be replaced with black color.

By completing this activity, you will get an output similar to the following:

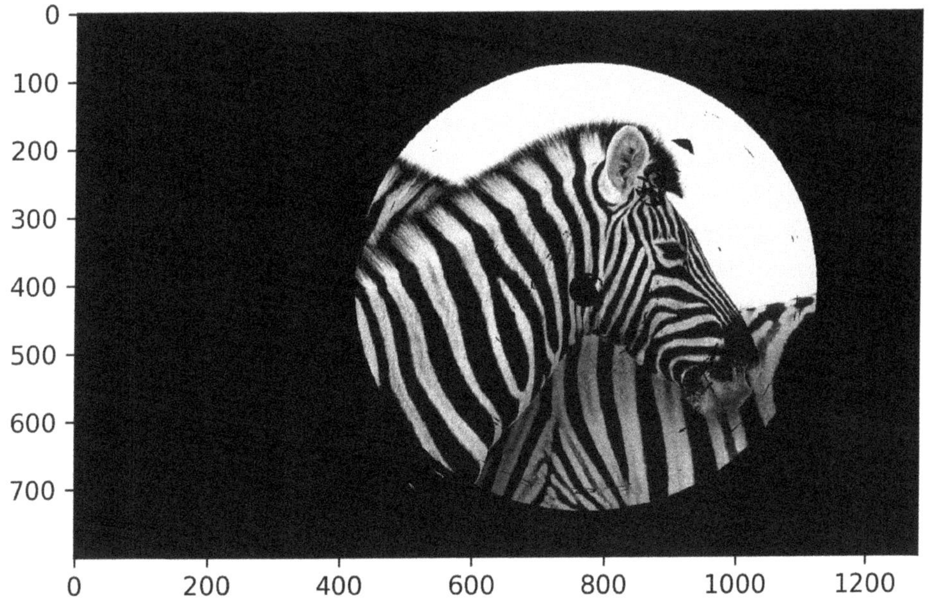

Figure 2.52: Zebra image

The result that we have obtained in this activity can be used in portrait photography, where only the subject of the image is highlighted and the background is replaced with black.

> **NOTE**
>
> The solution for this activity can be found on page 473.

By completing this activity, you have learned how to use image resizing to change the shape of an image, image thresholding to convert a color image into a binary image, and bitwise operations to perform image masking. Notice how image masking can be used to "mask" or hide certain regions of an image and display only the remaining portion of the image. This technique is used extensively in document analysis in computer vision.

SUMMARY

In this chapter, we started by understanding the various geometric transforms and their matrix representations. We also saw how we can use OpenCV's functions to carry out these transformations. Then, we performed arithmetic operations on images and saw how addition and multiplication with a constant can be understood as a brightening operation. Next, we discussed binary images and how they can be obtained using binary thresholding. Then, we discussed bitwise operations and how they can be carried out using OpenCV. Finally, we had a look at the concept of masking, where we used NumPy to obtain interesting results.

In the next chapter, we will jump into the world of histograms and discuss how we can use histogram matching and histogram equalization as more advanced image processing steps.

3

WORKING WITH HISTOGRAMS

OVERVIEW

In this chapter, you will learn about histograms and the adjustment of image histograms to enhance subtle details such as the appearance of objects hidden in shadows and the improvement of contrast in dark images. Although image histograms are simple, you will find that they are really powerful tools for enhancing an image to achieve contrast adjustment and better visualization. In almost all of your image enhancement projects, histogram equalization will be an important step. You will test your knowledge with exercises and an end-of-chapter activity, where you will draw a rectangular bounding box on an image and enhance the contrast within that box. By the end of this chapter, you will be able to plot histograms of different channels of an image and equalize histograms of grayscale and color images using different methods.

INTRODUCTION

In the previous chapter, you learned how to perform basic operations on images and how to access, split, and merge different channels of an image. You also learned how to perform geometric transformations and practiced image translation, rotation, scaling, cropping, and more. In this chapter, you will learn how to enhance the quality of a captured image in order to improve visualization and bring otherwise hidden details into focus. Histogram equalization is used for this purpose.

A histogram is a powerful tool to better understand image content. For example, many cameras display a real-time histogram of the scene that is being captured in order to adjust some parameters of the camera acquisitions, such as exposure time, brightness, or contrast, with the purpose of capturing effective images and detecting image acquisition issues. It helps the photographer see the distribution of captured pixel intensities. In this chapter, we will learn how to plot the histogram of an image channel and how to manipulate it to improve the visibility of dark objects in an image using a technique called histogram equalization. An image histogram is a graphical representation of the distribution of pixel intensities. The pixel intensity of an image has a range of **0** to **255**:

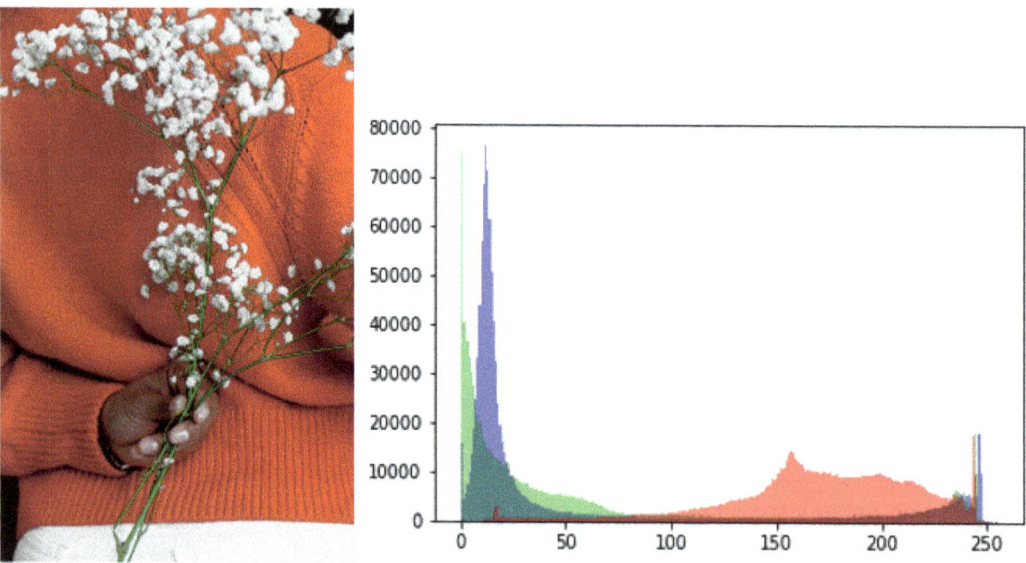

Figure 3.1: BGR image with its corresponding histogram of the three channels

The distribution of blue, green, and red intensities of a sample BGR image is shown in the preceding figure. A histogram is the most effective tool for visualizing the distribution of pixels of different intensities across an image. In this chapter, you will learn how to visualize histograms of binary, grayscale, and color images using the Matplotlib and OpenCV libraries. You will also learn how to equalize histograms to enhance the image quality of grayscale and color images.

INTRODUCTION TO MATPLOTLIB

To compute the histogram of an image, we are going to use a command from a Python library named Matplotlib. Matplotlib is a vast library, and the module you need from it to plot a histogram is called **pyplot**. The **pyplot** module gives you access to the plotting functionality of Matplotlib. This plotting functionality is what we need to display histograms.

The code you will need to import this module is as follows:

```
import matplotlib.pyplot as plt
```

Alternatively, you can also import the **pyplot** module as follows:

```
from matplotlib import pyplot as plt
```

Following is an explanation of the syntax for importing the **pyplot** module:

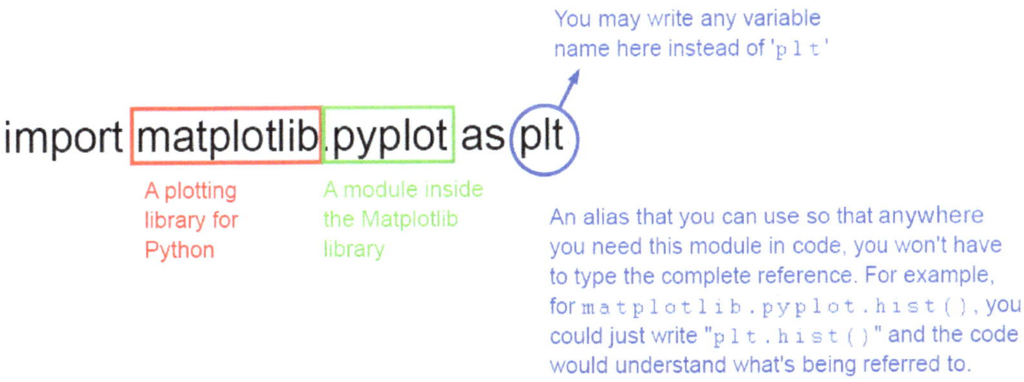

You may write any variable name here instead of 'plt'

import matplotlib.pyplot as plt

A plotting library for Python

A module inside the Matplotlib library

An alias that you can use so that anywhere you need this module in code, you won't have to type the complete reference. For example, for matplotlib.pyplot.hist(), you could just write "plt.hist()" and the code would understand what's being referred to.

Figure 3.2: Explanation of the syntax

DISPLAYING IMAGES WITH MATPLOTLIB

Using Matplotlib, you can also display images on the console itself, unlike **cv2.imshow**, which opens a new window to display the image.

To plot a grayscale image (**img**) using Matplotlib, you can use the following code:

```
imgplot = plt.imshow(img , cmap="gray")
plt.title('Original image')
plt.show()
```

When plotting a color image, Matplotlib assumes that you will always provide it with an RGB image. Python's OpenCV reads images in BGR format, so, to use Matplotlib, you will always need to convert your image to RGB first.

To convert a BGR image to RGB, use this command:

```
rgb = cv2.cvtColor(img,cv2.COLOR_BGR2RGB);
```

Similarly, to convert an LAB image to RGB, use the following command:

```
rgb = cv2.cvtColor(img,cv2.COLOR_LAB2RGB);
```

The LAB color space is more accurate than the RGB color space in terms of mimicking the human eye. It uses three values (**L**, **A**, and **B**) to specify a color. The most striking feature of this color space is its device-independence, which makes it easier to achieve exactly the same color across different media.

In LAB, the **L** channel (plane 0) has information about lightness. The **A** channel (plane 1) ranges from green to red, and the **B** channel (plane 2) ranges from blue to yellow. Therefore, the **A** and **B** channels together define the chromaticity of the color space. So, to leave chromaticity unchanged, we will not make any modifications to the **A** and **B** planes; we will only equalize the **L** plane.

To convert an HSV image to RGB, use the following command:

```
rgb = cv2.cvtColor(img, cv2.COLOR_HSV2RGB);
```

Here, **img** is the input image and **rgb** is the version of that image in RGB format.

Now, to display the RGB image, you do as follows:

```
plt.imshow(rgb)
plt.title('Converted RGB Image')
plt.show()
```

The main focus of this chapter is familiarizing you with image histograms and their equalization techniques. To compute histograms, you may use either Matplotlib or OpenCV. Both methods are described in the sections that follow.

PLOTTING HISTOGRAMS WITH MATPLOTLIB

Let's look at the command to compute a histogram. In Matplotlib, a single line of code computes a histogram and plots it. The command is as follows:

```
plt.hist(img_vec, bins = 256, color, alpha)
```

Let's take a look at the different parameters:

- **img_vec**: This is a vector containing the pixels of the image in an arrangement. If **img** is your 2D image, then the **img.ravel()** command will give you a vector containing all arranged image pixels.

- **color**: This is the color you want to plot the histogram with. By default, it is navy blue. You may specify it as **Red**, **Blue**, **Green**, **Yellow**, **Magenta**, or any other standard color.

- **bins**: This is the number of bars we want to show on the *X* axis. We know that pixel intensities range from **0** to **255**, and if we wanted to separately visualize the pixel count for each possible intensity, we would need a graph that shows the **0-255** interval as **256** separate bars. Therefore, we would keep **bins=256**. If you do not specify **bins**, Python will treat it as **10** by default (in the following example, we will see what this implies).

- **alpha**: This is the level of transparency with which you want to draw the histogram.

The only input that's mandatory for this command is **img_vec**. The rest are optional, as they stipulate the aesthetic qualities rather than the content of the graph. If the user does not provide their values, then default values are taken. There are some more optional inputs to this if you want to play around with visualization that are beyond the scope of this book. You can take a look at them in the official documentation, which can be found at https://matplotlib.org/3.1.1/api/_as_gen/matplotlib.pyplot.hist.html.

EXERCISE 3.01: PLOTTING A SAMPLE IMAGE AND ITS HISTOGRAM WITH 256 BINS

In this exercise, we will be reading an image in grayscale, plotting it, and then plotting its 2D histogram with **256** bins.

> **NOTE**
>
> The image used in this exercise can be downloaded from https://packt.live/2C9bkXM.

To complete this exercise, perform the following steps:

1. Import the required libraries:

    ```
    import cv2
    import matplotlib.pyplot as plt
    ```

2. Read the image in grayscale:

    ```
    img= cv2.imread('river_scene.jpg', \
                    cv2.IMREAD_GRAYSCALE)
    ```

3. Display the image:

    ```
    cv2.imshow('grayscale image', img)
    cv2.waitKey(0)
    cv2.destroyAllWindows()
    ```

The grayscale image looks as follows:

Figure 3.3: Sample grayscale image

124 | Working with Histograms

4. Plot its histogram:

```
ax = plt.hist(img.ravel(), bins = 256)
plt.show()
```

The X axis of the image histogram will have a range of **0-255** and the Y axis tracks the pixel count against each intensity. The plot looks as follows:

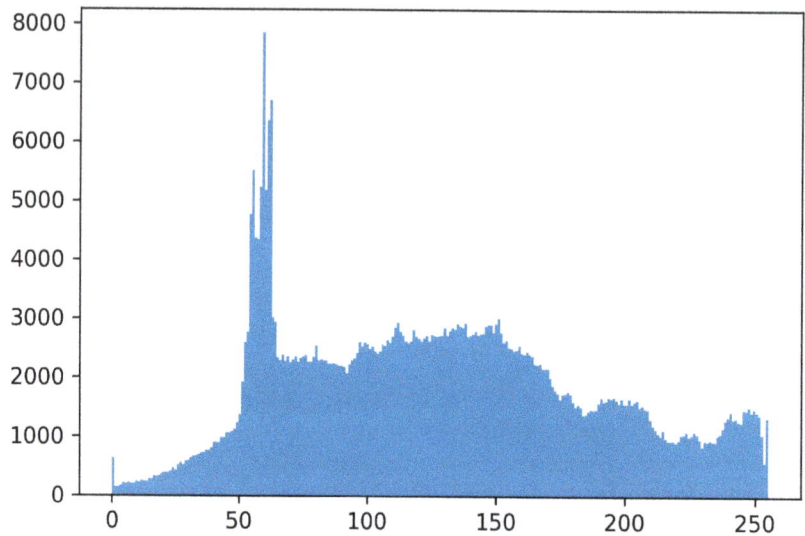

Figure 3.4: Histogram of sample grayscale image with 256 bins

From this histogram, we can see how many times each pixel intensity value (ranging from **0** to **255**) occurs in the image. This gives us an idea of the pixel distribution for this image. For example, we know that **0** is black (the darkest region) and **255** is white (the lightest region), and here we can see that the number of whiter pixels is greater than the number of darker pixels, which can be verified from the image.

> **NOTE**
>
> To access the source code for this specific section, please refer to https://packt.live/31wSYKw.

Here, we have learned how to plot a histogram for a sample grayscale image with **256** bins. Now, let's try it out with **10** bins in the upcoming exercise.

> **NOTE**
>
> It's very important to note that when we display an image using Matplotlib, it will be displayed inside our notebook, whereas an image displayed using OpenCV's `cv2.imshow` function will be displayed in a separate window. Moreover, an image displayed using Matplotlib will have the gridlines or *X* and *Y* axes, by default. These will not be present in an image displayed using the `cv2.imshow` function. This is because Matplotlib is actually a (graph) plotting library, and that's why it displays the axes, whereas OpenCV is a computer vision library and axes don't hold much importance in computer vision. Please note that irrespective of the presence or absence of the axes, the image will stay the same, whether it's displayed using Matplotlib or OpenCV. That's why any and all image processing steps will also stay the same between Matplotlib and OpenCV. We will be using Matplotlib and OpenCV interchangeably in the entire book to display images. This means that sometimes you will find images with axes and sometimes without axes. In both cases, the axes don't hold any importance and can be ignored.

EXERCISE 3.02: PLOTTING A SAMPLE IMAGE AND ITS HISTOGRAM WITH 10 BINS

In this exercise, we will be reading an image in grayscale, plotting it, and then plotting its 2D histogram with **10** bins.

> **NOTE**
>
> The image used in this exercise can be downloaded from https://packt.live/2C9bkXM.

126 | Working with Histograms

To complete this exercise, perform the following steps:

1. Import the required libraries:

    ```
    import cv2
    import matplotlib.pyplot as plt
    ```

2. Read the image in grayscale:

    ```
    img= cv2.imread('river_scene.jpg', \
                    cv2.IMREAD_GRAYSCALE)
    ```

3. Display the image:

    ```
    cv2.imshow('grayscale image', img)
    cv2.waitKey(0)
    cv2.destroyAllWindows()
    ```

    The grayscale image looks as follows:

Figure 3.5: Sample grayscale image

4. Plot the histogram of the grayscale image:

```
plt.hist(img.ravel())
plt.show()
```

The plot will look as follows:

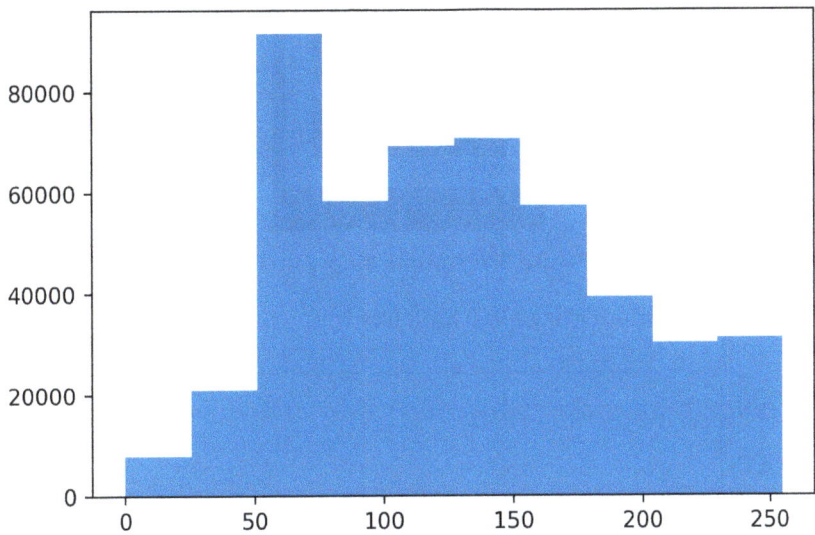

Figure 3.6: Histogram of sample grayscale image with the default 10 bins

If you count the number of bars in the preceding figure, you will see that it is **10**, which is the default number for bins. This means we can know how many pixels of each intensity range (this range consists of **25.6** pixels here because **256/10=25.6**) lie within each bar segment. The first bar, for example, tells us how many total pixels with intensities between **0** to **25.6** lie in this image, but it cannot tell us exactly how many pixels of one particular value of intensity lie in this image.

> **NOTE**
>
> To access the source code for this specific section, please refer to https://packt.live/2Zu9FDX.

Similarly, we can observe the histogram of a sample binary image with **256** bins, using the following code:

```
ax = plt.hist(img.ravel(), bins = 256)
plt.show()
```

128 | Working with Histograms

Let's try to see this code in action. We will use the binary image shown in *Figure 3.7*. When we plot the histogram for the image using **256** bins, we will see the histogram shown in *Figure 3.8*:

Figure 3.7: Sample binary image

The histogram of the binary image will look like this:

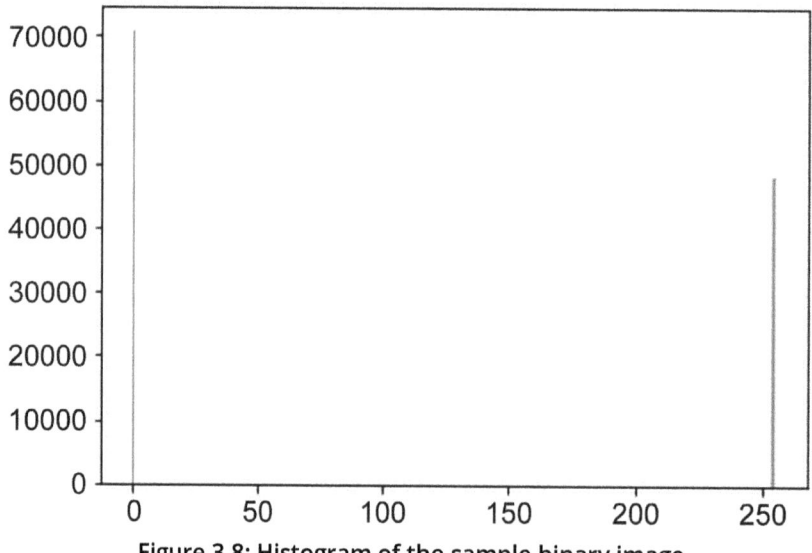

Figure 3.8: Histogram of the sample binary image

Note that there are only two possible intensities on a binary image (**0** and **255**), hence there are only two lines on the histogram. The histogram in *Figure 3.8* tells us that there are more zeros (the black region) than ones (the white region) on the binary image, as shown in *Figure 3.7*.

A BGR image, on the other hand, has three channels, so it will have three 2D histograms – one for each channel:

Figure 3.9: Sample color image

> **NOTE**
>
> The image used in this example can be downloaded from https://packt.live/3emQUrb.

To plot the three 2D histograms of the three channels (**B**, **G**, and **R**) on a single plot but in different colors, the `plt.show()` command can be called after using the `plt.hist()` command on each individual channel.

130 | Working with Histograms

Let's use a transparency value of **0.5** to show the three channels in their respective colors on the histogram plot:

```
img= cv2.imread('sunflower.jpg')
im=  img[:, :, 2]
plt.hist(im.ravel(), bins = 256, color = 'Red', \
         alpha = 0.5)

im=  img[:, :, 1]
plt.hist(im.ravel(), bins = 256, color = 'Green', \
         alpha = 0.5)

im=  img[:, :, 0]
plt.hist(im.ravel(), bins = 256, color = 'Blue', \
         alpha = 0.5)

plt.show()
```

The output is as follows:

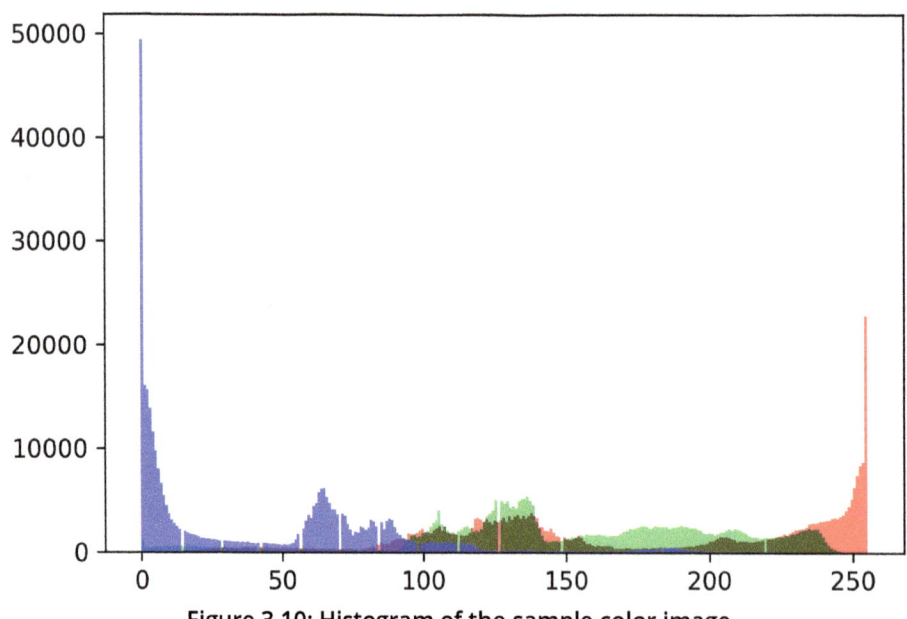

Figure 3.10: Histogram of the sample color image

In the next section, we will learn about histograms with OpenCV.

HISTOGRAMS WITH OPENCV

OpenCV also gives you a command to generate histograms. To plot, you will need Matplotlib as well. The core strength of this method is that it allows you to find the histogram of your selected **region of interest** (**ROI**) in an image. This ROI is communicated to the command with the use of an appropriate mask. This mask is a binary image. It has white pixels on the ROI and black pixels elsewhere.

The command used to compute the histogram of a particular channel of an image is as follows:

```
cv2.calcHist([images], [channels], mask, [histSize], \
             [ranges])
```

> **NOTE**
>
> The inputs with square brackets must always be given as lists.

Let's take a look at the different parameters:

- `[images]`: This is the list containing your input image(s). It may be 2D or 3D.
- `[channels]`: This is the index of the plane of the image whose histogram you want to compute. For a grayscale or binary image, it can only be 0. It is also given as a list, `[]`, so that for a BGR image, you can provide either individual channel's index, as in `[0]`, `[1]`, or `[2]`, or when computing histograms of multiple planes together, you can provide values accordingly: for example, `[0, 1]` for the **B** and **G** planes combined or `[0, 1, 2]` for all three planes combined.
- `mask`: This is a binary image. It has a white region inside the corresponding ROI segment and black elsewhere. If you want to find a histogram of the complete image (no mask), then in its place, simply write **None**.
- `[histSize]`: This is the number of bins you want to use. If you have a single channel, then this list has a single number (for example, `[10]`). When computing a histogram of two channels, this list will have two numbers (for example, `[10,10]`), and when computing a histogram of three channels together, this list will have three numbers (for example, `[10,10,10]`).

- **[ranges]**: This is the range of histogram bin boundaries in each dimension. It is also given as a list, **[]**. For a single channel, you can give a single value such as **[256]**, but for more than one channels together you have to provide complete ranges for each channel (for example **[0, 255, 0, 255]** for two channels and **[0, 255, 0, 255, 0, 255]** for three channels).

Let's see an example where we apply a given mask to an RGB image to see the distribution of the red color plane within an ROI.

Consider the following image:

Figure 3.11: Sample image of the cherries

> **NOTE**
>
> The image used in this example can be downloaded from https://packt.live/32aOytc.

Suppose we have a mask over the red cherry highlighted in the blue box, as shown in the following figure:

Figure 3.12: Masked cherry

The mask would look as follows:

Figure 3.13: Displaying the mask

134 | Working with Histograms

Let's plot the histogram of the image segment's red plane that lies inside the ROI shown in this mask. For this, we will first read the image in BGR format and its mask in grayscale format:

```
im = cv2.imread('activity im.jpeg')
mymask= cv2.imread('mask cherry.png', 0)
```

The **cv2.calcHist()** command will be used to compute a histogram for the red plane (channel = 2) of the BGR image:

```
hist = cv2.calcHist([im],[2], mymask,[256], [0,255])
```

We will use the Matplotlib library to display the plot of the histogram and to display a title at the top:

```
plt.plot(hist)
plt.title('Red plane of ROI cherry', color='r')
```

The output is as follows:

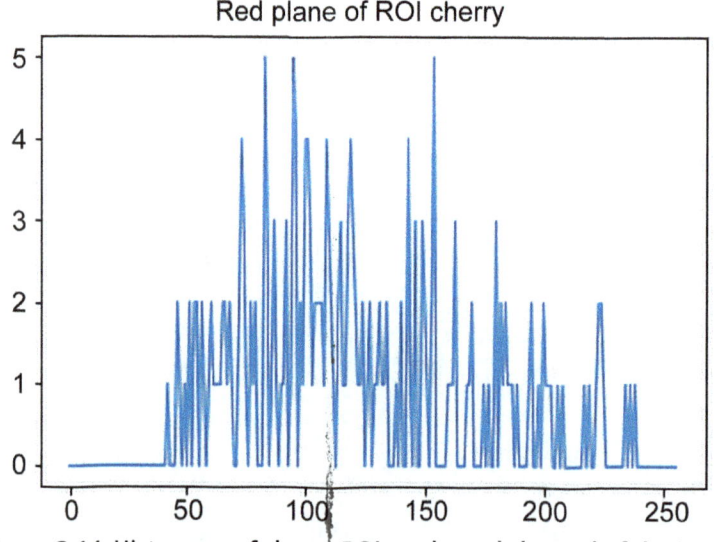

Figure 3.14: Histogram of cherry ROI on the red channel of the image

In this section, we learned how to plot the histogram of a region of an image that is enclosed in a corresponding mask image.

Readers should note that RGB is a highly correlated color space. By computing individual channels' histograms and combining them (by overlaying), we won't be able to capture their correlations. The correct way is to compute a multi-dimensional histogram, that is, to compute a histogram of the RGB color space. Multi-dimensional histograms are out of the scope of this book, so when we compute a histogram of a color image, it will entail first separately computing each individual channel's histogram and then combining them by overlaying all the histograms:

- For an image with three planes (RGB), the combined histogram will be 4D (three dimensions for the three planes and one dimension for the computed frequencies). The fourth dimension (frequencies) will tell you, for a given combination of RGB pixel intensities (say, **(5, 5, 5)**), how many pixels can have this value. You can bin them using small 3D cubes (say, **8x8x8**), dividing a space of **256x256x256** into **N** number of **8x8x8**.

- Similarly, for any two channels combined (say, **R** and **G**), the histogram will be 3D (two dimensions for the **R** and **G** channels and one for frequencies), and, of course, you can bin them as well, for example, into small 2D cubes (say, **8x8**).

For the image shown in *Figure 3.11*, if we want to compute the histograms of the **B** and **G** planes together, then the command would be as follows:

```
hist= cv2.calcHist(im, channels=[0, 1], mask=None, \
                histSize=[10, 10], ranges=[0, 255, 0, 255])
```

The output would be a 3D histogram with a **10x10** single bin size. Let's show the shape of the histogram:

```
print("3D histogram shape: %s, with %d values" \
      % (hist.shape, hist.flatten().shape[0]))
```

That will give the following output:

```
3D histogram shape: (10, 10), with 100 values
```

If we wanted to compute the **B**, **G**, and **R** histograms all together, then we would use this:

```
hist= cv2.calcHist(im, channels=[0, 1, 2], mask=None, \
      histSize=[10, 10, 10], ranges=[0, 255, 0, 255, 0, 255])

print ("4D histogram shape: %s, with %d values" \
% (hist.shape, hist.flatten().shape[0]))
```

136 | Working with Histograms

That would give the following output:

```
4D histogram shape: (10, 10, 10), with 1000 values
```

USER-SELECTED ROI

Next, let's see how a user can select an ROI in an image by drawing a bounding box over it. If the color image you have read into the code is represented by the `im` variable, then you can use the following command to prompt the user to draw a bounding box on the image:

```
x,y,w,h = cv2.selectROI(im, fromCenter=0)
```

This command opens an image interface and expects the user to first click the mouse at the top-left corner of the bounding box and then drag it to the bottom-right corner to form a rectangular bounding box, and then hit *Enter*:

Figure 3.15: User-selected ROI

`x` and `y` are the starting coordinates of this ROI box and `w` and `h` are the `width` and `height`, respectively:

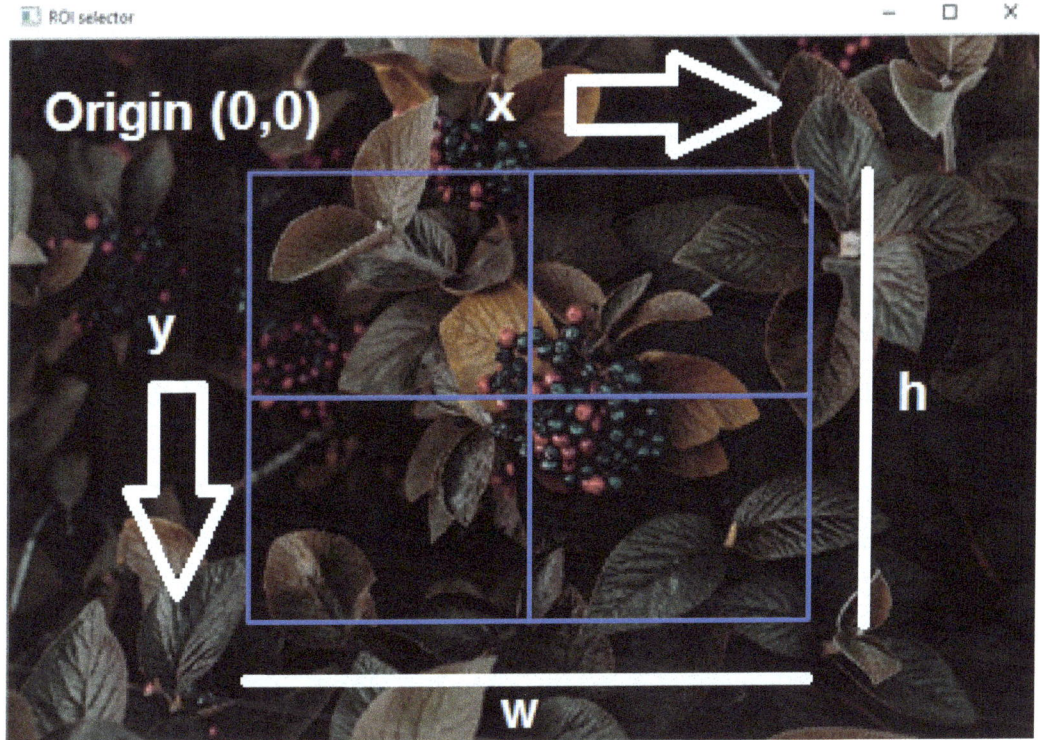

Figure 3.16: x, y, w, and h coordinates of the ROI box

EXERCISE 3.03: CREATING A MASK IMAGE USING A USER-SELECTED ROI

Consider the image shown in *Figure 3.11*. Your task is to write a program that enables the user to draw a bounding box on this image around an ROI. Your program should then automatically create a binary image of a mask corresponding to the selected ROI.

> **NOTE**
>
> The image used in this exercise can be downloaded from https://packt.live/2VxrBfq.

138 | Working with Histograms

To complete the exercise, perform the following steps:

1. Import OpenCV for image processing operations and Matplotlib for plotting. You will also need the NumPy library to create the binary mask, so you can also import that at the start of the code:

   ```
   import cv2
   import numpy as np
   ```

2. Read the image:

   ```
   im = cv2.imread('flowers.jpeg')
   ```

   There is no need to display the image because, in our next step, it will be automatically displayed by the ROI selector window.

3. Prompt the user to draw a rectangular bounding box on the image to mark the ROI:

   ```
   x,y,w,h = cv2.selectROI(im, fromCenter=0)
   ```

   The ROI selector looks as follows:

Figure 3.17: ROI selector

Suppose the box is drawn like this:

Figure 3.18: User-drawn bounding box to select the ROI

4. Use NumPy to create a totally blank image with all pixels being black (**0**). This new image will have the same dimensions for the width and height as the original image:

```
mymask= np.zeros(im.shape[:2], dtype = "uint8")
```

5. Within the mask's ROI, make all the pixels white (**255**):

```
mymask[int(y):int(y+h) , int(x):int(x+w)]= 255
```

This is now the mask image.

6. Display the mask:

```
cv2.imshow('Created mask', mymask)
cv2.waitKey(0)
cv2.destroyAllWindows()
```

140 | Working with Histograms

The mask should look as follows:

Figure 3.19: Displaying the created mask

7. Save the mask to a **.png** file:

```
cv2.imwrite('mask cherry.png' , mymask)
```

The above command will output **True** showing that the image has been saved successfully. You can open up the saved PNG to find the mask similar to Figure 3.19.

> **NOTE**
>
> To access the source code for this specific section, please refer to https://packt.live/2NLngkA.

This exercise has equipped you with the skills necessary to create customized masks for your images if you want to visualize histograms of selected regions using the `cv2.calcHist()` command.

A COMPARISON OF SOME SAMPLE HISTOGRAMS

To understand histogram equalization, it is first important to know what the histograms of images with different illumination levels look like.

Take a close look at the following grayscale images and their corresponding histograms:

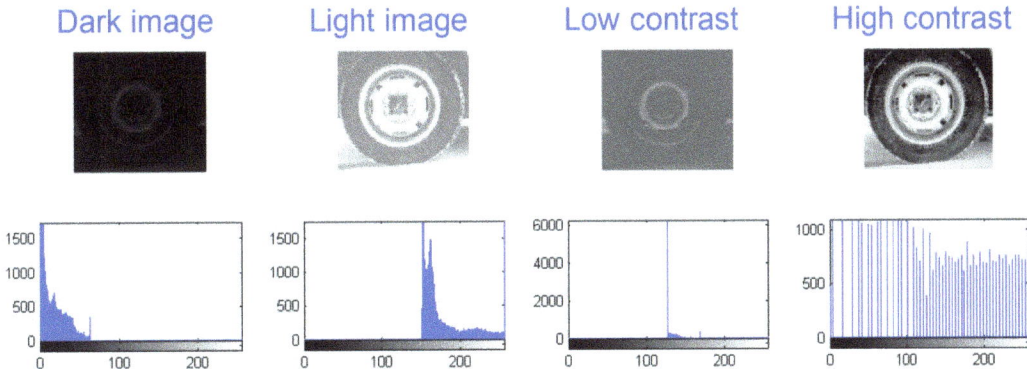

Figure 3.20: Histograms of differently illuminated images, ranging from dark to well-lit images

You will notice the following:

- For a dark image, the majority of the pixels lie in the low-pixel-intensity areas, so the histogram is located around pixel intensity **0**.

- For a light image, the majority of the pixels lie in the high-pixel-intensity areas, so the histogram is located around pixel intensity **255**.

- For an image with low contrast, the histogram is not evenly spread and is concentrated around some particular value.

- For an image with high contrast where details are clearly distinguishable, the histogram weight is spread throughout the range of **0-255**.

WHAT IS HISTOGRAM EQUALIZATION?

Keeping in mind the analysis at the end of the previous section on histogram comparison, your next objective is to improve the contrast of an image by modifying its histogram. To do this, you must make sure that the histogram is not concentrated around any particular intensity. This is achieved by spreading the histogram graph across all intensity values from **0-255** – this will give you a high-contrast image and is called **histogram equalization**. Python's OpenCV library provides us with a very handy command that lets us achieve this in a single line:

```
imgOut = cv2.equalizeHist(imgIn)
```

Here, `imgIn` is the input grayscale image whose histogram you want to equalize, and `imgOut` is the output (histogram equalized) image.

EXERCISE 3.04: HISTOGRAM EQUALIZATION OF A GRAYSCALE IMAGE

In this exercise, we will be reading the image shown in *Figure 3.21* in grayscale, plotting it, and then plotting its 2D histogram with **256** bins. After that, we will apply histogram equalization to it and then plot the histogram of the histogram equalized image. To complete this exercise, take the following steps.

> **NOTE**
>
> The image used in this exercise can be downloaded from https://packt.live/3f0Klvu.

Compare the histograms of the image before and after equalization. Analyze the output image and see whether you can find any limitations:

1. Import the OpenCV library:

   ```
   import cv2
   ```

2. Read the image as grayscale:

   ```
   img= cv2.imread('dark_image1.png', 0)
   ```

3. Display the image:

```
cv2.imshow('Original Image', img)
cv2.waitKey(0)
cv2.destroyAllWindows()
```

The output grayscale image looks as follows:

Figure 3.21: Grayscale image

4. Import the Matplotlib library:

```
import matplotlib.pyplot as plt
```

5. Draw the histogram of the original image with **256** bins:

```
ax = plt.hist(img.ravel(), bins= 256)
plt.title('Histogram of Original Image')
plt.show()
```

The output is as follows:

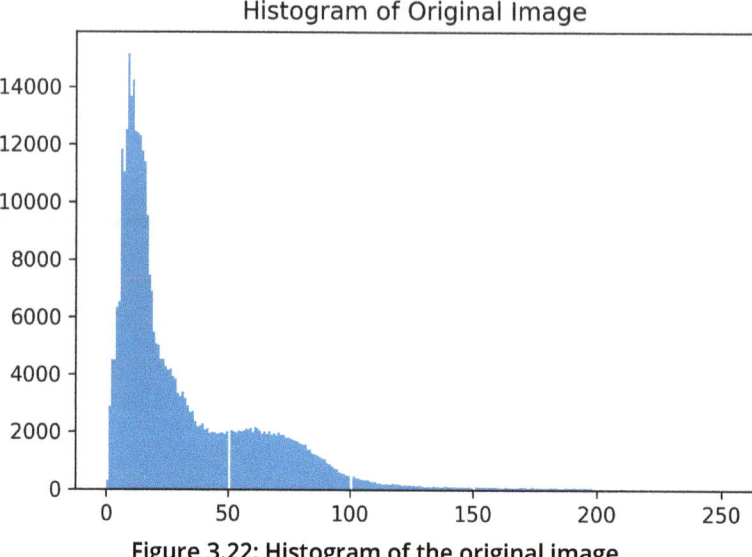

Figure 3.22: Histogram of the original image

6. Apply histogram equalization to the original image:

```
histequ = cv2.equalizeHist(img)
```

7. Display the histogram equalized image:

```
cv2.imshow('Histogram Equalized Image', histequ)
cv2.waitKey(0)
cv2.destroyAllWindows()
```

The histogram equalized image looks as follows:

Figure 3.23: Histogram equalized image

8. Draw the histogram of the histogram equalized image with **256** bins:

   ```
   ax = plt.hist(histequ.ravel(), bins= 256)
   plt.title('Histogram of Histogram Equalized Image')
   plt.show()
   ```

   The histogram of the histogram equalized image looks as follows:

   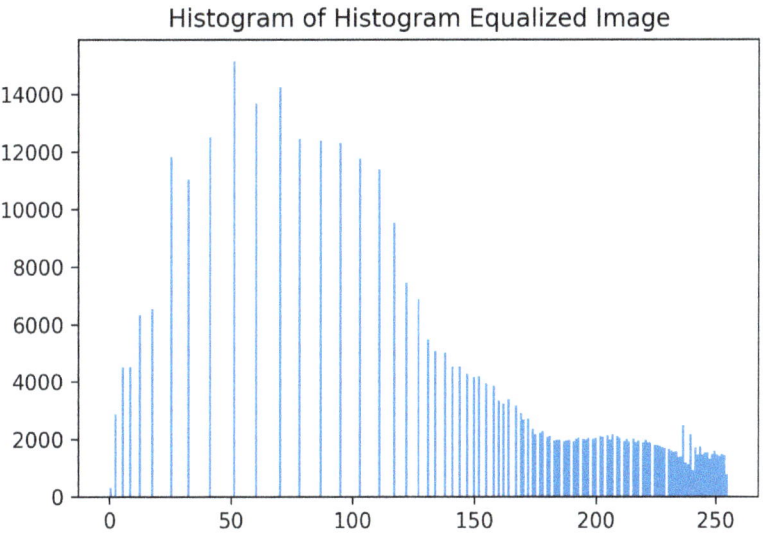

   Figure 3.24: Histogram of the histogram equalized image

9. To have a better view of the differences, let's make another image in which the original and histogram equalized images are stacked side by side. The simplest way to do this is by using the **hstack** command from the **numpy** library like this:

   ```
   import numpy as np
   img_with_histequ = np.hstack((img,histequ))
   ```

10. Now display the resultant image:

    ```
    cv2.imshow('Comparison', img_with_histequ)
    cv2.waitKey(0)
    cv2.destroyAllWindows()
    ```

The image looks as follows:

Figure 3.25: Comparison of the original image with its histogram equalized version

> **NOTE**
>
> To access the source code for this specific section, please refer to https://packt.live/31zvKmT.

A comparison of the histograms is given here:

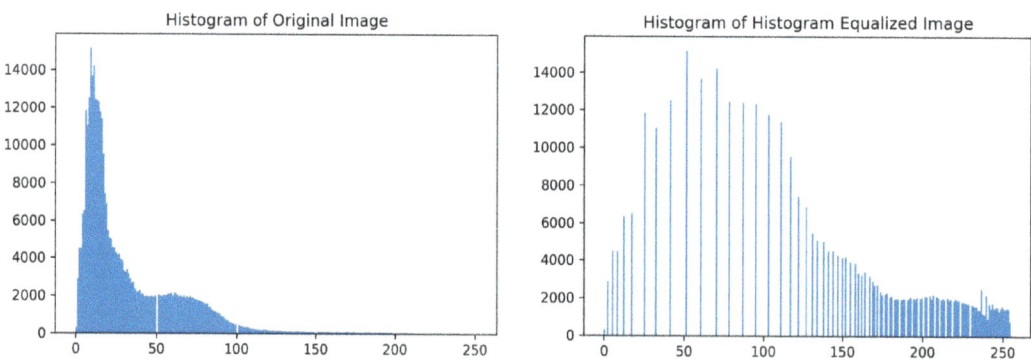

Figure 3.26: Comparison of histograms of the original and the histogram equalized image

So, what histogram equalization did here is as follows:

- It stretched the histogram of the original image and forced it to cover the entire `0-255` range.
- It changed the shape of the histogram – it essentially "flattened" it so there are no uneven or sharp changes in its shape.

- The number of light pixels (close to **255**) is now much more than the number of dark pixels (close to **0**). This can be seen from the histograms and also verified using the original and resultant images.

Visually, it made the image more pleasant to look at and easier to comprehend. Many hidden details that could easily have been missed from the original image are now in plain view:

- We can now clearly see that this person is carrying a bag.
- There is a pattern of cracks on the wall behind this person.
- Now we can see into the alley on the right side of the image, which was not possible to do in the original dark image.

However, there are still problems with the histogram equalized image:

Figure 3.27: Regions brightened too much by histogram equalization

See *Figure 3.27* and you will notice that there are some regions where the brightness is a little too much – for example, the bags behind this person. They have been rendered so brightly that we cannot make out what exactly their surface looks like. Furthermore, the entire ground region appears to have a uniformly bright and clean surface. Additionally, we cannot clearly see the texture of the jeans that this person is wearing. These regions of excessive brightness are called artifacts and were introduced by our enhancement attempt. In the next section, we will see how to remedy this to retain those missed details.

CONTRAST LIMITED ADAPTIVE HISTOGRAM EQUALIZATION (CLAHE)

Contrast Limited Adaptive Histogram Equalization (**CLAHE**), although a simple algorithm, is a very powerful tool for retrieving hidden details from images without the over-amplification of noise. What it does is divide the entire image into small segments and equalize the histogram of each segment. To make the image smooth, the visibility of the block regions is reduced by bilinear interpolation. This algorithm linearly interpolates in one direction (for example, horizontally) and then performs linear interpolation in the other direction (for example, vertically). More about this can be found at https://packt.live/3gTFsoP.

> **NOTE**
>
> For more details related to bilinear interpolation, you can refer to the paper *Application of the CLAHE Algorithm Based on Optimized Bilinear Interpolation in Near Infrared Vein Image Enhancement*. You can also refer to https://packt.live/2OoSmPq and https://packt.live/2OqIAvW.

Let's see the basic steps of how **CLAHE** works. Take a look at the following grayscale image:

Figure 3.28: A sample grayscale image

We have to first divide it into blocks to limit the dynamic range of the pixel intensity. According to the selected block size, clipping and redistributing gray peaks is done later on. Let's choose a block size of **6x6** so that each block has **6*6= 36** pixels:

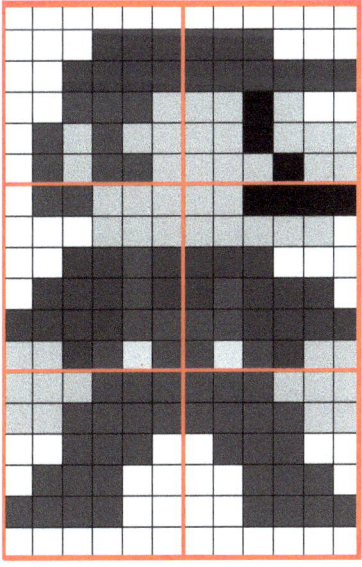

Figure 3.29: Image divided into blocks

Now we need to find the central point of each block. We will call this the grid point from here onward. Grid points are shown as yellow dots in the following figure:

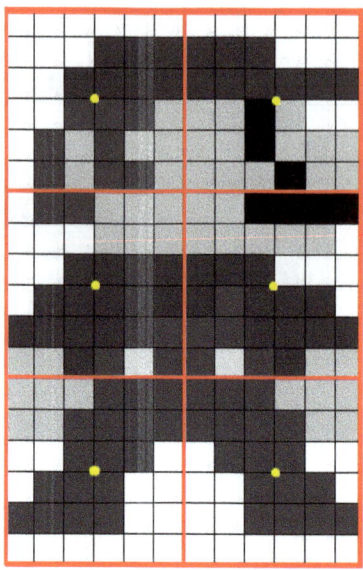

Figure 3.30: Central grid points for each block

First, the original histogram of each block is clipped from the top if it exceeds a predefined peak (clip limit). The portion of the histogram that was above the clip limit is uniformly distributed throughout the pixel range of the histogram:

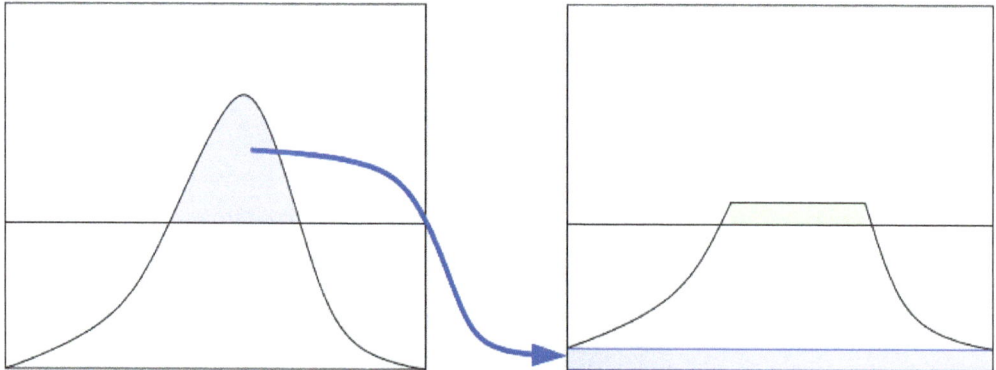

Figure 3.31: Uniform re-distribution of the histogram

This uniform distribution would make the clipped region again somewhat surpass the clip limit. To handle this, the clipping process is iteratively repeated until only a negligible portion remains above this limit.

Each pixel within a block is then transformed depending on the clipped original histogram of that block region. The slope of this transformation function controls how much contrast amplification is to be made. Computing the transformation function for each pixel on the block would be a computationally expensive procedure, so it is only done for the grid point (central pixel), and the rest of the pixels of the block are transformed according to that central pixel through interpolation:

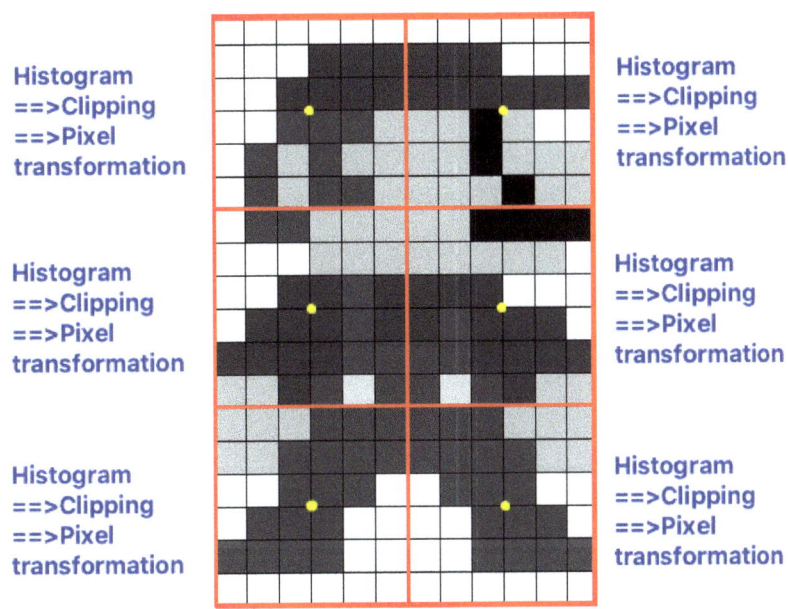

Figure 3.32: Operations applied to each block separately

To apply CLAHE to an image in Python using OpenCV, firstly, a CLAHE object needs to be created. The command for that is as follows:

```
clahe = cv2.createCLAHE(clip_limit, block_size)
```

clip_limit: This tells the function how much of the histogram peak should be clipped off. If its value is **1**, then no clipping is done. Usually, it is kept at **2** or **3**. The higher the clip limit, the more the local contrast is amplified.

block_size: This is the size of the blocks you want to divide the image into.

After your CLAHE object is created, you can use it to apply CLAHE on any grayscale image with the following command:

```
img_out = clahe.apply(img_in)
```

Here **img_in** is the input image and **img_out** is the transformed image after the CLAHE operation.

EXERCISE 3.05: APPLICATION OF CLAHE ON A GRAYSCALE IMAGE

In this exercise, we will be reading the image shown in *Figure 3.21* in grayscale, plotting it, and then plotting its 2D histogram with **256** bins. After that, we will apply CLAHE to it and then plot the histogram of the resultant image.

Finally, we will compare how the image transformed by CLAHE varies from the image that is enhanced by simple histogram equalization.

> **NOTE**
>
> The image used in this exercise can be downloaded from https://packt.live/3f0Klvu. To complete this exercise, perform the following steps:

1. Import the OpenCV library:

   ```
   import cv2
   ```

2. Read the image as grayscale:

   ```
   img= cv2.imread('dark_image1.png', 0)
   ```

3. Display the image:

   ```
   cv2.imshow('Original Image', img)
   cv2.waitKey(0)
   cv2.destroyAllWindows()
   ```

The output grayscale image looks as follows:

Figure 3.33: Grayscale image

4. Import the Matplotlib library:

```
import matplotlib.pyplot as plt
```

5. Draw the histogram of the original image with **256** bins:

```
ax = plt.hist(img.ravel(), bins= 256)
plt.title('Histogram of Original Image')
plt.show()
```

The output is as follows:

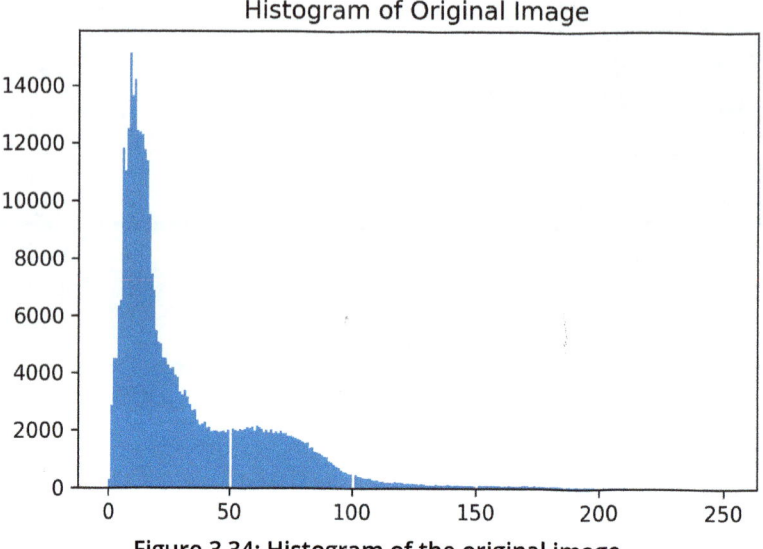

Figure 3.34: Histogram of the original image

6. Create a CLAHE object. Let's set the clip limit as **3** and the grid size as **8x8**. You can change these parameters if you want:

```
clahe = cv2.createCLAHE(clipLimit=3.0, tileGridSize=(8,8))
```

7. Then apply it to the image:

```
out = clahe.apply(img)
```

8. Display the resultant image:

```
cv2.imshow('CLAHE image', out)
cv2.waitKey(0)
cv2.destroyAllWindows()
```

The resultant image looks as follows:

Figure 3.35: Image after CLAHE

9. Plot the histogram for the CLAHE image:

```
ax = plt.hist(out.ravel(), bins= 256)
plt.title('Histogram after CLAHE')
plt.show()
```

The output is as follows:

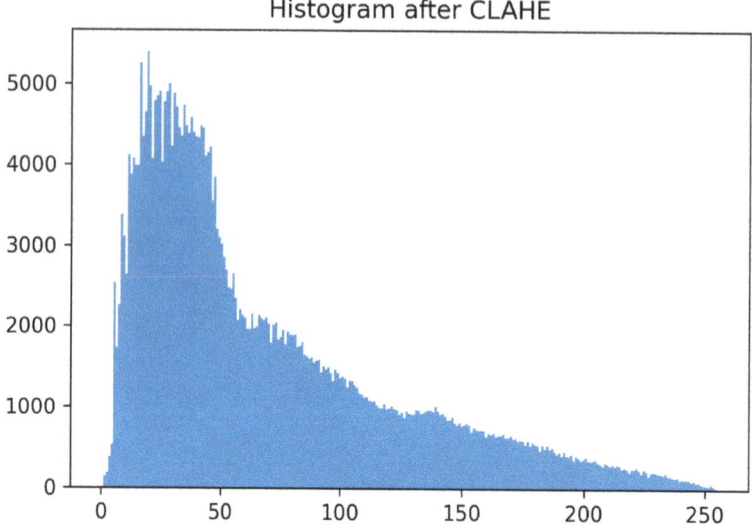

Figure 3.36: Histogram of the image after the application of CLAHE

In this exercise, we applied CLAHE on an image and saw how its results are different from simple histogram equalization. It can be seen that after CLAHE, the basic shape of the histogram is preserved – it is just stretched throughout the **0-255** range of possible pixel intensities. It is sharply in contrast with the shape of the histogram gained from simple histogram equalization, where not only was the histogram stretched to the **0-255** range but also the sharp peaks present in it were lost.

Let's see what happens if we play around with the clip limit and block size.

If we fix the block size at **8x8** and increase the clip limit gradually from **3** to **12**, the results can be seen in *Figures 3.37* to *3.39*:

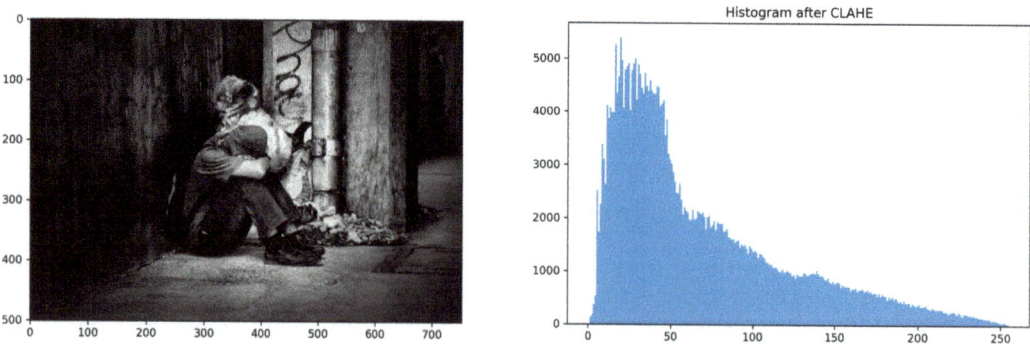

Figure 3.37: Result for clip limit = 3 and block size = 8x8

Following is the result for block size at **8x8** and increase the clip limit of **6**:

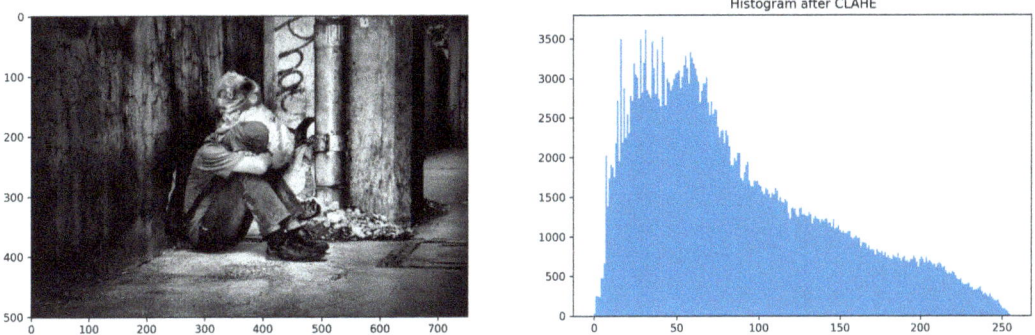

Figure 3.38: Result for clip limit = 6 and block size = 8x8

Following is the result for block size at **8x8** and increase the clip limit of **12**:

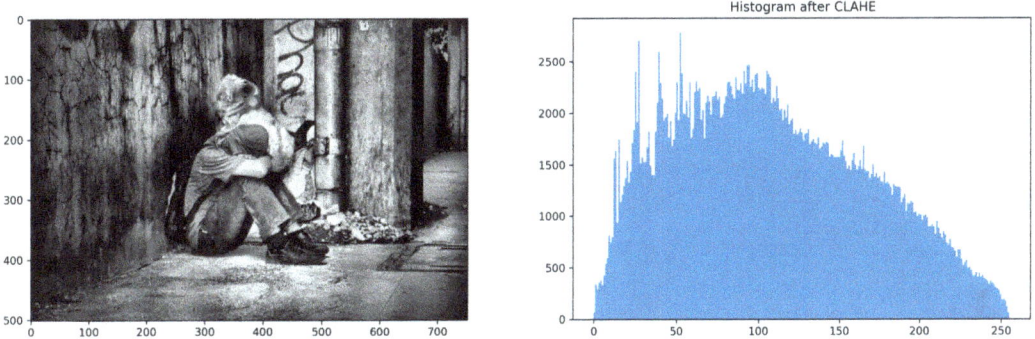

Figure 3.39: Result for clip limit = 12 and block size = 8x8

If we fix the clip limit at **3** and increase the block size gradually from **16x16** to **128x128**, the results can be seen in *Figures 3.40* to *3.42*:

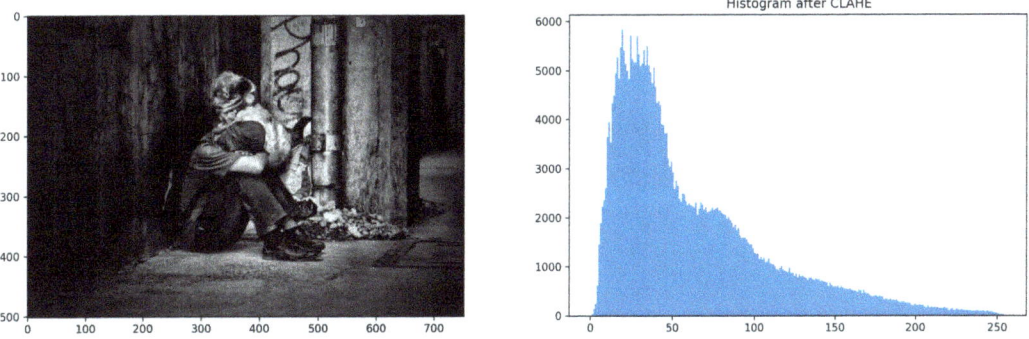

Figure 3.40: Result for clip limit = 3 and block size = 16x16

158 | Working with Histograms

Following is the result for block size at **32x32** and increase the clip limit of **3**:

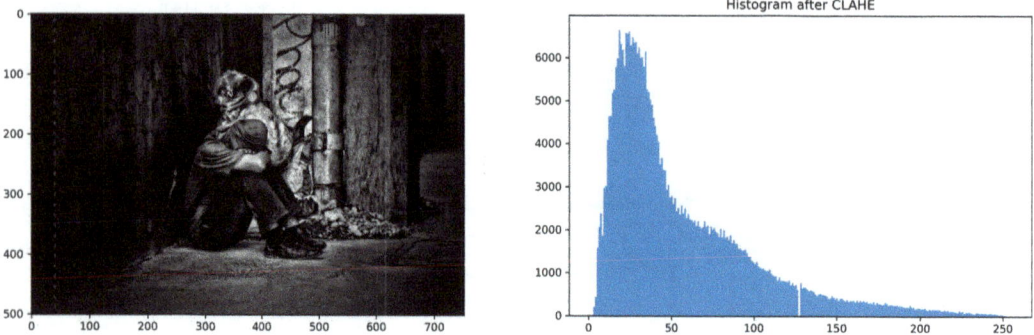

Figure 3.41: Result for clip limit = 3 and block size = 32x32

Following is the result for block size at **128x128** and increase the clip limit of **3**:

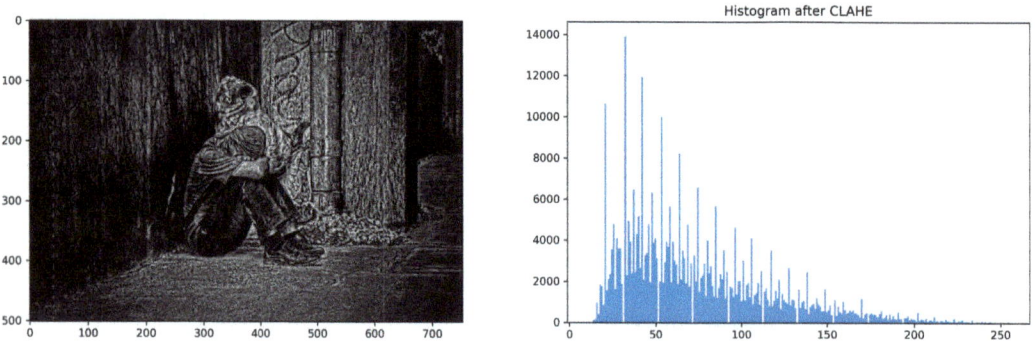

Figure 3.42: Result for clip limit = 3 and block size = 128x128

From this, we can conclude that increasing the clip limit introduces a higher level of contrast, making the regions of the image brighter, and increasing the block size would bring more focus to the edges.

> **NOTE**
>
> To access the source code for this specific section, please refer to https://packt.live/3ggRCI4.

Whether you go with histogram equalization or CLAHE completely depends on what sort of data you have and what kind of information you want to bring into focus. CLAHE is usually a good place to start. After applying CLAHE, you may apply histogram equalization on the result. What is most important is that you take the time to play around with your data. Apply different enhancement procedures to it and then decide for yourself what works best. Blindly following the technique of any one research method will not do you any good, as what works for one kind of data might not work on another kind of data.

ACTIVITY 3.01: ENHANCING IMAGES USING HISTOGRAM EQUALIZATION AND CLAHE

You will all be familiar with face recognition technology, where a person's identity is recognized using an image of their face. Similarly, fingerprint recognition is where a person scans their finger, and then a computer analyzes the fingerprint and recognizes the person's identity. One currently emerging technology is finger vein recognition. It recognizes a person based on the pattern of veins that are spread beneath the skin of the finger:

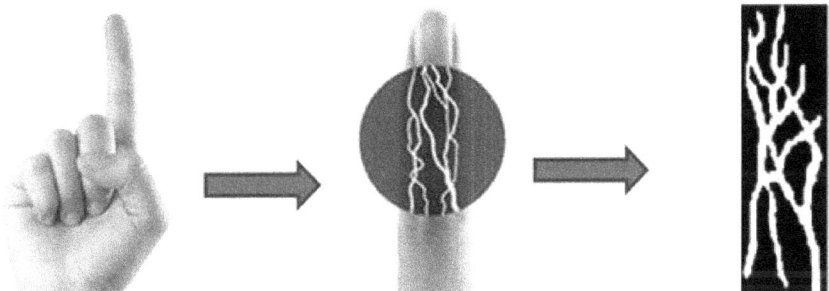

Figure 3.43: Segmentation (or extraction) of finger veins

Suppose you are to design a system for the recognition of finger vein images of different people. Images are acquired by throwing infrared light on a finger enclosed in a dark box. The acquired finger image looks like this:

Figure 3.44: Cropped finger vein image

You can see that this image is quite dark, and no vein pattern is visible. So, the first step toward designing this recognition system (and any other, for that matter) is to perform image enhancement.

Your task is to enhance this image using the following:

- Histogram equalization
- CLAHE

Plot the resultant histograms for each of the two aforementioned tasks.

You can keep the clip limit as **4** and the block size as **16x16** (or any other combination that you think is suitable). Perform the following steps to complete the activity:

1. Read the image in grayscale and view its histogram.
2. Perform histogram equalization on the image and view its histogram.
3. Perform CLAHE on the image and view the histogram of its output.

> **NOTE**
>
> The image used in this activity can be downloaded from https://packt.live/3ilbmw3.

In this activity, you will see how histogram equalization and CLAHE are important in preprocessing biometric data.

> **NOTE**
>
> The solution for this activity can be found on page 479.

We can see from this activity how powerful a tool histogram equalization is in the preprocessing stage when working with image data, and that CLAHE outperforms histogram equalization in bringing subtle details into focus.

EXERCISE 3.06: HISTOGRAM EQUALIZATION IN THE BGR COLOR SPACE

Python reads an image in BGR format. You can separate out the three planes of **b**, **g**, and **r** using the following command:

```
b,g,r = cv2.split(img)
```

Now you can individually read each plane. It will be like a grayscale image. Your task is to read a BGR image and split its planes. Then, perform histogram equalization on each plane. After that, combine the planes again using the following command:

```
bgr = cv2.merge([b,g,r]);
```

Based on your observations, consider whether this is a good approach to perform histogram equalization for color images.

Do this first with histogram equalization and then with CLAHE:

> **NOTE**
>
> The image used in this exercise can be downloaded from
> https://packt.live/3ijvmPt.

1. Import OpenCV and Matplotlib. OpenCV will be used for image processing and Matplotlib will be used for plotting:

   ```
   import cv2
   import matplotlib.pyplot as plt
   ```

2. Read the image:

   ```
   img = cv2.imread(bloom.jpg')
   ```

3. Display the image. You can either use OpenCV's **cv2.imshow()** to display it or you can use Matplotlib. In this example, we are going to use Matplotlib to familiarize you with its syntax. Matplotlib takes an image in RGB format. So, to use it, you should convert your BGR image to RGB first:

   ```
   origrgb = cv2.cvtColor(img,cv2.COLOR_BGR2RGB);
   imgplot = plt.imshow(origrgb)
   plt.title('Original')
   plt.show()
   ```

The preceding code produces the following output:

Figure 3.45: Image for histogram equalization

The values on the X and Y axes show the width and height of the displayed image.

4. Split the BGR image into three channels. The first channel in a BGR image is the **b** (blue) channel. The second one is the **g** (green) channel, and the third one is the **r** (red) channel:

```
b,g,r = cv2.split(img);
```

5. Apply histogram equalization to each channel:

```
b = cv2.equalizeHist(b);
g = cv2.equalizeHist(g);
r = cv2.equalizeHist(r);
```

6. Merge the channels again:

```
bgr = cv2.merge([b,g,r]);
```

7. Convert the image to RGB for display using Matplotlib:

```
rgb = cv2.cvtColor(bgr,cv2.COLOR_BGR2RGB);
```

8. Display the enhanced image:

```
imgplot = plt.imshow(rgb)
plt.title(' Histogram equalization in rgb')
plt.show()
```

The plot looks as follows:

Figure 3.46: Histogram equalization in RGB

9. To repeat the same process with CLAHE, split the original image into three channels again:

```
b,g,r = cv2.split(img);
```

10. Create a CLAHE object and then apply it to each channel:

```
clahe = cv2.createCLAHE(clipLimit=4.0, tileGridSize=(16,16))

b = clahe.apply(b)
g = clahe.apply(g)
r = clahe.apply(r)
```

11. Merge the channels again:

```
bgr = cv2.merge([b,g,r]);
```

12. Convert the image to RGB for display using Matplotlib:

```
rgb = cv2.cvtColor(bgr,cv2.COLOR_BGR2RGB);
```

13. Display the enhanced image:

```
imgplot = plt.imshow(rgb)
plt.title('CLAHE in rgb')
plt.show()
```

The preceding code produces the following output:

Figure 3.47: The resultant color image after applying CLAHE to the individual R, G, and B channels

Based on our observations, we can conclude that applying histogram equalization to all three color channels (R, G, and B) is definitely not a good idea because of the artifacts introduced in the enhanced image. It looks quite unnatural. Applying CLAHE, on the other hand, produced a better result because no artifacts were introduced and the colors seemed more natural.

> **NOTE**
>
> To access the source code for this specific section, please refer to https://packt.live/2NJkTih.

In the next section, we will see some better ways to perform histogram equalization of color images.

THE HISTOGRAM EQUALIZATION OF COLOR IMAGES USING THE HSV AND LAB COLOR SPACES

In the previous exercise, we saw that applying histogram equalization to the R, G, and B planes led to modifications in the color channels, which results in an image with distorted colors. The reason for this is that for each pixel, there is a correlation between its three color channels, and changing them independently changes the color chromaticity, making it appear unnatural.

CLAHE produced relatively good results, but it could still have been better. The best practice here would be to process the image in either the HSV or LAB color space. In both of these approaches, color information is separated out from intensity information and the color planes are not changed.

> **NOTE**
>
> The HSV color space has been extensively discussed in *Chapter 1, Basics of Image Processing*. In HSV, the hue plane (**H** or plane 0) has information about basic color, the saturation plane (**S** or plane 1) has information about the vibrance of that color, and the value plane (**V** or plane 2) has information about the brightness of that color. So, here, we will only equalize the **V** plane.

The steps for processing in the HSV space are as follows:

1. Convert the image from RGB/BGR to the HSV color space.
2. Split the image into the **h**, **s**, and **v** planes.
3. Equalize the histogram of the **v** plane.
4. Merge the **h**, **s**, and **v** planes back together.
5. Convert back to RGB/BGR.

Now, here are the steps for processing in the LAB space:

1. Convert the image from RGB/BGR to the LAB color space.
2. Split the image into the **l**, **a**, and **b** planes.
3. Equalize the histogram of the **l** plane.
4. Merge the **l**, **a**, and **b** planes back together.
5. Convert back to RGB/BGR.

EXERCISE 3.07: HISTOGRAM EQUALIZATION IN THE HSV COLOR SPACE

In this exercise, we will apply histogram equalization to a color image in the HSV plane. We will use the same image we used in the previous exercise to have a fair comparison. Do this first with histogram equalization and then with CLAHE:

> **NOTE**
>
> The image used in this exercise can be downloaded from https://packt.live/3ijvmPt.

1. Import OpenCV and Matplotlib. OpenCV is for image processing and Matplotlib is for plotting:

```
import cv2
import matplotlib.pyplot as plt
```

2. Read the image using the following command:

```
img = cv2.imread(bloom.jpg')
```

3. We will use Matplotlib to display the image. Matplotlib takes an image in RGB format. So, to use it, we will convert our BGR image to RGB first:

```
origrgb = cv2.cvtColor(img,cv2.COLOR_BGR2RGB);
imgplot = plt.imshow(origrgb)
plt.title('Original')
plt.show()
```

The image looks as follows:

4. cConvert the image to the HSV space using the following code:

   ```
   imgHSV = cv2.cvtColor(img,cv2.COLOR_BGR2HSV);
   ```

5. Split the image into the **h**, **s**, and **v** planes as follows:

   ```
   h,s,v = cv2.split(imgHSV);
   ```

6. Apply histogram equalization to the **v** channel:

   ```
   v = cv2.equalizeHist(v);
   ```

168 | Working with Histograms

7. Stack the three planes back together to get an updated HSV image:

```
hsv = cv2.merge([h,s,v]);
```

8. Finally, transform it back to RGB so that it can be visualized:

```
rgb = cv2.cvtColor(hsv,cv2.COLOR_HSV2RGB);
```

> **NOTE**
>
> If you want to use `cv2.imshow()` to display the image or `cv2.imwrite()` to save the image – or do anything else to do with OpenCV – then instead of using the RGB color space, you need to use the BGR color space with the `cv2.cvtColor(hsv,cv2.COLOR_HSV2BGR)` command.

9. Display the image in Matplotlib:

```
plt.imshow(rgb)
plt.title('HE in HSV')
plt.show()
```

The resultant image after applying histogram equalization to the individual channels and then merging them will look as follows:

Figure 3.49: The resultant image

10. To repeat the process with CLAHE, again split the image into the **h**, **s**, and **v** planes as follows:

    ```
    h,s,v = cv2.split(imgHSV);
    ```

11. Create a CLAHE object:

    ```
    clahe = cv2.createCLAHE(clipLimit=4.0, tileGridSize=(16,16))
    ```

12. Apply CLAHE on the **v** channel:

    ```
    v = clahe.apply(v)
    ```

13. Stack the three planes back together to get an updated HSV image:

    ```
    hsv = cv2.merge([h,s,v]);
    ```

14. Finally, transform it back to RGB so that it can be visualized by Matplotlib:

    ```
    rgb = cv2.cvtColor(hsv,cv2.COLOR_HSV2RGB);
    ```

15. Display the image with Matplotlib:

    ```
    plt.imshow(rgb)
    plt.title('CLAHE in HSV')
    plt.show()
    ```

The output is as follows:

Figure 3.50: Resultant image after applying CLAHE to the individual planes and then merging them

From this, we can infer that CLAHE outperforms histogram equalization in HSV as well.

> **NOTE**
>
> To access the source code for this specific section, please refer to https://packt.live/2BUEfhl.

Please note that the **cv2.split** command takes a toll on your computational resources and can slow down your code. A faster approach is to directly access the required plane and modify it in a single line using NumPy indexing.

The cv2.split() approach	Equivalent NumPy indexing approach

h,s,v = cv2.split(imgHSV);
v = cv2.equalizeHist(v);
imgHSV = cv2.merge([h,s,v]);
	imgHSV[:, :, 2] = cv2.equalizeHist(imgHSV[:, :, 2]);

Figure 3.51: Comparison of two similar approaches

EXERCISE 3.08: HISTOGRAM EQUALIZATION IN THE LAB COLOR SPACE

To go from BGR to LAB, use the following command:

```
cv2.cvtColor(img,cv2.COLOR_BGR2LAB)
```

For an image in the LAB color space, the first plane is **L**, the second plane is **A**, and the third one is **B**.

The aim of this exercise is to enhance the contrast of the image provided to you after transforming it to the LAB color space. After this, you will be able to compare the results of the histogram equalization of color images in the RGB, HSV, and LAB color spaces.

First, do it with simple histogram equalization and then with CLAHE. Do not use **cv2. split()**:

> **NOTE**
>
> The image used in this exercise can be downloaded from https://packt.live/3ijvmPt.

1. Import OpenCV for image processing and Matplotlib for plotting:

   ```
   import cv2
   import matplotlib.pyplot as plt
   ```

2. Read the image as follows:

   ```
   img = cv2.imread('bloom.jpg')
   ```

3. Convert the image from BGR to RGB and then display it with Matplotlib:

   ```
   origrgb = cv2.cvtColor(img,cv2.COLOR_BGR2RGB);
   imgplot = plt.imshow(origrgb)
   plt.title('Original')
   plt.show()
   ```

   The converted image looks as follows:

   Figure 3.52: Original image

172 | Working with Histograms

4. Convert the image to the LAB space:

   ```
   imgLAB = cv2.cvtColor(img,cv2.COLOR_BGR2LAB);
   ```

5. Perform histogram equalization on the **L** channel (plane **0** of the LAB image):

   ```
   imgLAB[:,:,0]= cv2.equalizeHist(imgLAB[:,:,0])
   ```

6. Transform it back to RGB so that it can be visualized using Matplotlib:

   ```
   rgb = cv2.cvtColor(imgLAB,cv2.COLOR_LAB2RGB);
   ```

7. Display the final image as follows:

   ```
   plt.imshow(rgb)
   plt.title('HE in LAB')
   plt.show()
   ```

The image looks as follows:

Figure 3.53: Resultant image after applying histogram equalization in the LAB color space

8. To repeat the process with CLAHE, make a CLAHE object:

   ```
   clahe = cv2.createCLAHE(clipLimit=4.0, \
           tileGridSize=(16,16))
   ```

9. Apply CLAHE to the **L** channel (plane **0** of the original LAB image):

   ```
   imgLAB = cv2.cvtColor(img,cv2.COLOR_BGR2LAB);
   imgLAB[:,:,0] = clahe.apply(imgLAB[:,:,0])
   ```

10. Transform it back to RGB so that it can be visualized using Matplotlib:

    ```
    rgb = cv2.cvtColor(imgLAB,cv2.COLOR_LAB2RGB);
    ```

11. Display the final image:

    ```
    plt.imshow(rgb)
    plt.title('CLAHE in LAB)
    plt.show()
    ```

    The final image will look as follows:

Figure 3.54: Resultant image after applying CLAHE in the LAB color space

174 | Working with Histograms

In this exercise, we had a walk-through of how we can enhance the contrast of an image in the LAB color space.

> **NOTE**
>
> To access the source code for this specific section, please refer to https://packt.live/3dV957q.

Now you have seen how histogram equalization and CLAHE can be applied to the RGB, HSV, and LAB color spaces.

In *Figure 3.56*, you'll find a quick summary of the results of histogram equalization and CLAHE in different color spaces for the image shown in *Figure 3.55*:

Figure 3.55: Original image

Following table represents a summary of the histogram equalization and CLAHE outputs in the RGB, HSV, and LAB color spaces:

Figure 3.56: Summary of the histogram equalization and CLAHE outputs

From this table, we can observe that for color images, CLAHE always outperforms simple histogram equalization in terms of image clarity. Furthermore, simple histogram equalization in all these color spaces has introduced artifacts that have distorted the contrast. For the best results, it is suggested to use the HSV or LAB color spaces for the histogram equalization of color images.

ACTIVITY 3.02: IMAGE ENHANCEMENT IN A USER-DEFINED ROI

Suppose you need to make an application where the user will select a rectangular region of an image and the image will be enhanced only within that region.

Take a look at the image provided. The user will select an ROI on it (a rectangular shape) by dragging the mouse to form a rectangle. Your task is to perform CLAHE image enhancement on that image but only within the selected ROI.

Here is the image:

Figure 3.57: Activity image

> **NOTE**
>
> The image used in this activity can be downloaded from https://packt.live/3itjACm.

Here are the steps to follow:

1. Read the BGR image.

2. Prompt the user to select an ROI by drawing a rectangular box on it.

3. Create a binary mask according to that ROI.

4. Bring the BGR image into the HSV space.
5. Plot the histogram of the V plane of the image portion that lies inside the ROI.
6. Equalize the histogram of the V plane of the HSV image.
7. Plot the equalized histogram of the V plane of the image portion that lies inside the ROI.
8. Bring the image back to the RGB space and visualize it.

By completing this activity, you will learn how to manipulate the histogram of a particular ROI in an image. This will enable you to see a sharp contrast between the histogram equalized region of the image and other regions.

The BGR image will look as follows:

Figure 3.58: Image enhancement of only a certain part (a user-selected ROI) of an image

> **NOTE**
>
> The solution for this activity can be found on page 484.

By completing this activity, you have now gained hands-on experience in equalizing the histogram of a user-selected ROI of a color image to highlight the details of that selected portion.

SUMMARY

In this chapter, we had a walk-through of the methods of histogram equalization and CLAHE in grayscale and color images. For color images, we implemented these methods in the RGB, LAB, and HSV color spaces. We concluded that CLAHE outperforms simple histogram equalization for both grayscale and color images. We also saw that for color images, the histogram equalization of the R, G, and B planes in the BGR color space introduces artifacts because color information is independently changed in all these planes. Hence, it is always best to do histogram equalization for color images in the HSV or LAB color spaces, because the color information is separated from the intensity information and the color planes are not changed.

In the next chapter, you will learn about contours and how contour detection is a handy technique for fetching objects of interest from an image. You will learn how you can select certain objects from an image according to their shape and hierarchy.

4
WORKING WITH CONTOURS

OVERVIEW

This chapter will familiarize you with the concept of contours. You will learn how to detect nested contours and access specific contours with a given shape or moment distribution. You will gain the ability to play with contours. This ability will come in handy during object detection tasks, especially when you are searching and selecting your object of interest. You will also learn how to access contours according to their hierarchy. Moreover, you'll learn how to fetch contours of a particular shape from an image with multiple contours. By the end of this chapter, you will be able to detect and handle contours of different shapes and sizes.

INTRODUCTION

In the previous chapter, we learned about histogram equalization, which was used to enhance an image by bringing out the hidden details in a dark image. In this chapter, we will learn how to fetch objects of interest from an image. We will start with a gentle introduction to contours and then move ahead to see some interesting ways in which you can apply this concept. A contour is the boundary of an object – it is a closed shape along the portion of an image that has the same color or intensity. Speaking in terms of OpenCV, a contour is *the boundary of a group of white pixels on a black background*. Yes, in OpenCV, contours can be extracted only from binary images.

In practical terms, contours can help you to count the number of objects in an image. You can also use contours to identify your object(s) of interest in a given image, for example, to detect a basketball net in an image (*Exercise 4.05, Detecting a Basketball Net in an Image*). Furthermore, you will find that contour detection can be used to identify objects with a particular height or width ratio (you will find a use case for this in *Exercise 4.06, Detecting Fruits in an Image*).

This chapter provides a walkthrough of how contours can be detected, plotted, and counted. You will see what the difference is between contours and edges. You will learn about contour hierarchy, how to access different contours, and how to find contours of a given reference shape in an image with multiple contours. It will be an interesting journey and will equip you with a solid skill set before you venture out on your image processing adventures.

CONTOURS – BASIC DETECTION AND PLOTTING

A contour can only be detected in a binary image using OpenCV. To detect contours in colored (BGR) or grayscale images, you will first need to convert them to binary.

> **NOTE**
>
> In OpenCV, a colored image has its channels in the order of BGR (`blue`, `green`, and `red`) instead of RGB.

To find the contours in a colored image, first, you will need to convert it to grayscale. After that, you will segment it to convert it to binary. This segmentation can be done either with thresholding based on a fixed grayscale value of your choice or by using Otsu's method (or any other method that suits your data). Following is a flowchart of contour detection:

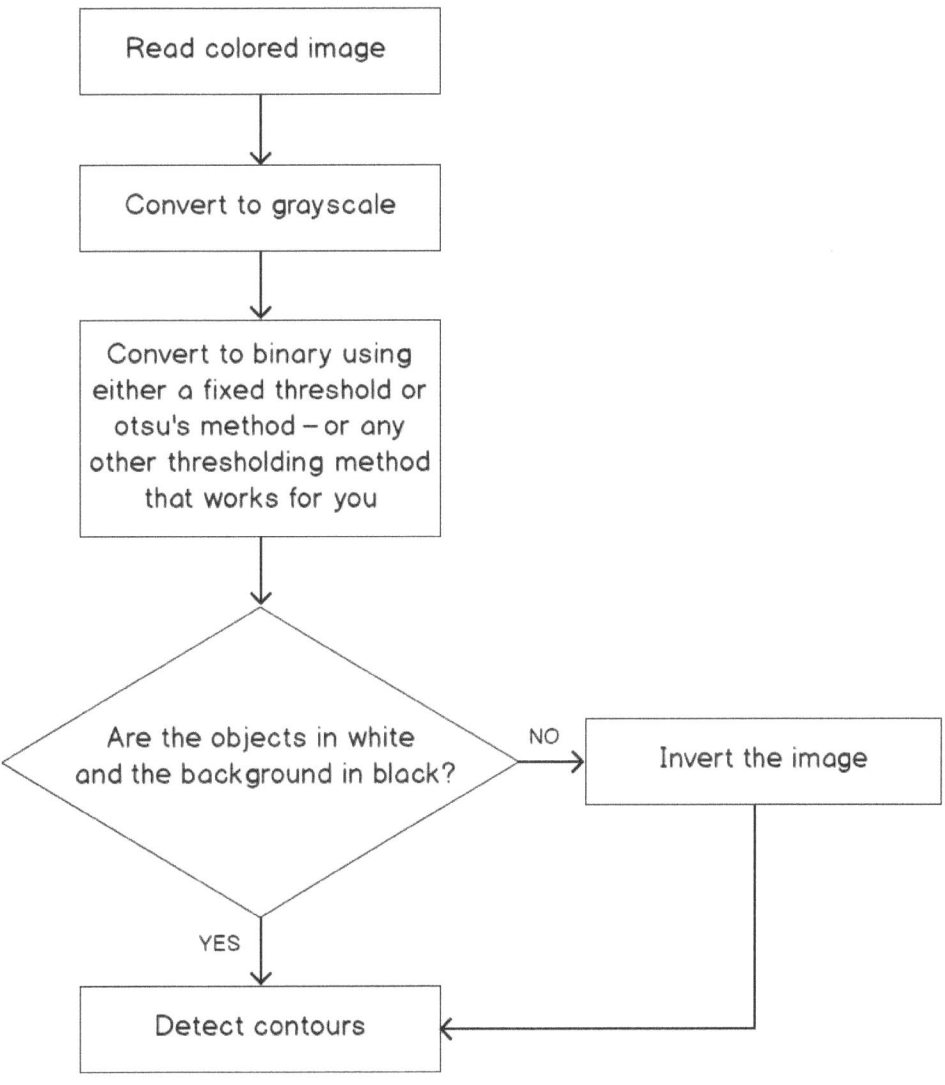

Figure 4.1: Flowchart of contour detection

The command to detect contours in an image is as follows:

```
contours, hierarchy = cv2.findContours(source_image, \
                                retrieval_mode, approx_method)
```

There are two outputs of this function:

- **contours** is a list containing all detected contours in the source image.
- **hierarchy** is a variable that tells you the relationship between contours. For example, is a contour enclosed within another? If yes, then is that larger contour located within a still larger contour?

In this section, we will just explore the **contours** output variable. We will explore the **hierarchy** variable in detail in the *Hierarchy* section.

Here is a brief explanation of the inputs of this function:

- **source_image** is your input binary image. To detect contours, the background of this binary image must be black. Contours are extracted from white blobs.
- **retrieval_mode** is a flag that instructs the **cv2.findContours** function on how to fetch all contours. (Should all contours be fetched independently? If a contour lies inside another larger contour, should that information be returned? If many contours lie inside a larger contour, should the internal contours be returned, or will just the outer contour suffice?)

    If you only want to fetch all the extreme outer contours, then you should keep it to be **cv2.RETR_EXTERNAL**. This will ignore all those contours that lie inside other contours.

    If you simply want to retrieve all contours independently and do not care which lies inside which one, then **cv2.RETR_LIST** would be your go-to choice. This does not create parent-child relationships and so all contours are at the same hierarchical level.

    If you are interested in finding the outer and inner boundaries of objects (two levels only: the border of the outer shape and the boundary of the inner hole), then you would go for **cv2.RETR_CCOMP**.

    If you want to make a detailed family tree covering all generations (parent, child, grandchild, and great-grandchild), then you need to set this input as **cv2.RETR_TREE**.

- **approx_method** is a flag that tells this function how to store the boundary points of a detected contour. (Should each coordinate point be saved? If so, keep this flag's value as **cv2.CHAIN_APPROX_NONE**. Or should only those points be saved that are strictly needed to draw the contour? In that case, set the flag to **cv2.CHAIN_APPROX_SIMPLE**. This will save you a lot of memory.) If you want to dig into this further, do check out https://opencv-python-tutroals.readthedocs.io/en/latest/py_tutorials/py_imgproc/py_contours/py_contours_begin/py_contours_begin.html.

Suppose you have an image of silhouettes (as shown in the following screenshot) and you need to detect contours on it:

Figure 4.2: Sample binary image

186 | Working with contours

We can store the preceding image in an `img` variable. Then to detect contours, we can apply the `cv2.findContours` command. For the sake of this introductory example, we will use external contour detection and the `CHAIN_APPROX_NONE` method:

```
contours, hierarchy = cv2.findContours(img, cv2.RETR_EXTERNAL, \
                                      cv2.CHAIN_APPROX_NONE)
```

To plot the detected contours on a BGR image, the command is as follows:

```
marked_img = cv2.drawContours(img, contours, contourIdx, color, \
                              thickness, lineType = cv.LINE_8, \
                              hierarchy = new cv.Mat(), \
                              maxLevel = INT_MAX, \
                              offset = new cv.Point(0, 0)))
```

The output of this function is the image, `img`, with the contours drawn on it.

Here is a brief explanation of the inputs of this function:

- `img` is the BGR version of the image on which you want the contours to be marked. It should be BGR because you will need to draw the contours with some color other than black or white and the color will have three values in its BGR code. So, you will need three channels in the image.

- `contours` is the Python list of detected contours.

- `contourIdx` is the contour you want to draw from the list of contours. If you want to draw all of them, then its input value will be **-1**.

- `color` is the BGR color code of the color you want to use for plotting. For example, for red it will be **(0, 0, 255)**. To see the RGB color codes for some commonly used colors, visit https://www.rapidtables.com/web/color/RGB_Color.html. To convert these RGB color codes to BGR, simply reverse the order of the three values.

- `thickness` is the width of the line used for plotting the contours. This is an optional input. If you specify it to be negative, then the drawn contours will be filled with color.

- `lineType` is an optional argument that specifies line connectivity. For further details, visit https://docs.opencv.org/master/d6/d6e/group__imgproc__draw.html#gaf076ef45de481ac96e0ab3dc2c29a777.

- **hierarchy** is an optional input containing information about the hierarchy levels present so that if you want to draw up to a particular hierarchy level, you can specify that in the next input.

- **maxLevel** corresponds to the depth level of the hierarchy you want to draw. If **maxLevel** is **0**, only the outer contours will be drawn. If it is **1**, all contours and their nested (up to level **1**) contours will be drawn. If it is **2**, all contours, their nested contours, and nested-to-nested contours (up to level **2**) will be drawn.

- **offset** is an optional contour shift parameter.

To apply this command to the preceding example, let's use a red color **(BGR code: 0, 0, 255)** and a thickness of **2**:

```
with_contours = cv2.drawContours(im_3chan, contours,-1,(0,0,255),2)
```

Here, **im_3chan** is the BGR version of the image on which we want to draw the contours. Now, if we visualize the output image, we will see the following:

Figure 4.3: Contours drawn

In the next exercise, we will see how we can detect contours in a colored image.

EXERCISE 4.01: DETECTING SHAPES AND DISPLAYING THEM ON BGR IMAGES

Let's say you have the following colored image with different shapes:

Figure 4.4: Image with shapes

> **NOTE**
>
> The image is also available at https://packt.live/2ZtOSQP.

Your task is to count all the shapes and detect their outer boundaries as follows:

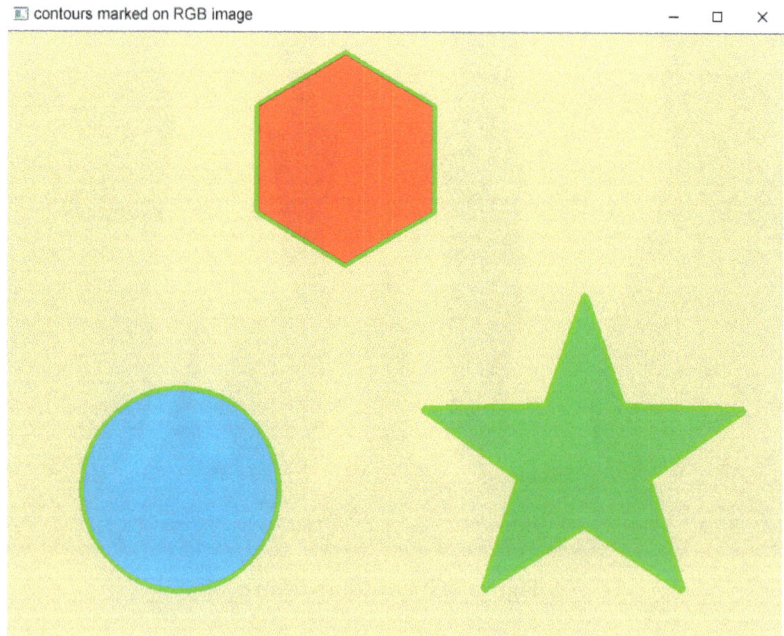

Figure 4.5: Required output

Perform the following steps:

1. Open a new Jupyter notebook. Your working directory must contain the image file provided for this exercise. Click **File | New File**. A new file will open up in the editor. In the following steps, you will begin to write code in this file. Name the file **Exercise4.01**.

   > **NOTE**
   >
   > The filename must start with an alphabetical character and must not have any spaces or special characters.

2. Import OpenCV as follows:

   ```
   import cv2
   ```

3. Read the image as a BGR image:

   ```
   image = cv2.imread('sample shapes.png')
   ```

4. Convert it to grayscale:

   ```
   gray_image = cv2.cvtColor(image, cv2.COLOR_BGR2GRAY)
   ```

5. Now, display our grayscale image to see what it looks like. To display an image, we use the **cv2.imshow** command as in the following code. The first input to it is the text label that we want to caption our image with. The next input is the variable name that we have used for that image:

   ```
   cv2.imshow('gray' , gray_image)
   ```

The output is as follows:

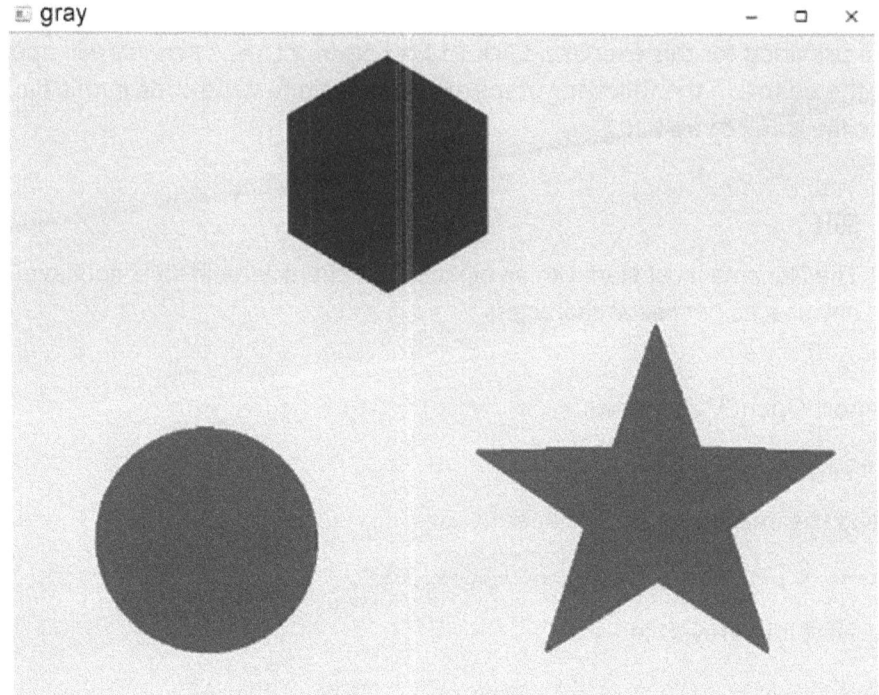

Figure 4.6: Grayscale image

The preceding command will display the image stored in the `gray_image` variable, with `gray` as the caption. However, the image will be displayed and all subsequent steps of the code will be brought to a standstill until you instruct your code on what to do next: that is, how long to wait before the program closes the image window and moves on.

6. Add some code to have the image window wait until the user presses any key on the keyboard. For this instruction, we have the following command:

```
cv2.waitKey(0)
```

If you want to wait for a fixed length of time before closing the window, instead of having to wait for user input, you can use the `cv2.waitKey(time_ms)` command, where `time_ms` is the number of milliseconds you want the program to wait before automatically closing the window.

Now, our program will wait for the user to press any key on the keyboard and once our user presses it, the program execution will pass this line, so we need to give it the next instruction.

7. Next, let's tell our program to close all open image windows before moving on to the next steps, using the following code:

```
cv2.destroyAllWindows()
```

8. After this, we will now convert our image to binary. We are going to use Otsu's method to do this segmentation because it provides acceptable results for this image. In case Otsu's method does not give good results on your image, then you can always select a fixed threshold:

```
ret,binary_im = cv2.threshold(gray_image,0,255,\
                            cv2.THRESH_OTSU)
cv2.imshow('binary image', binary_im)
cv2.waitKey(0)
cv2.destroyAllWindows()
```

The output is as follows:

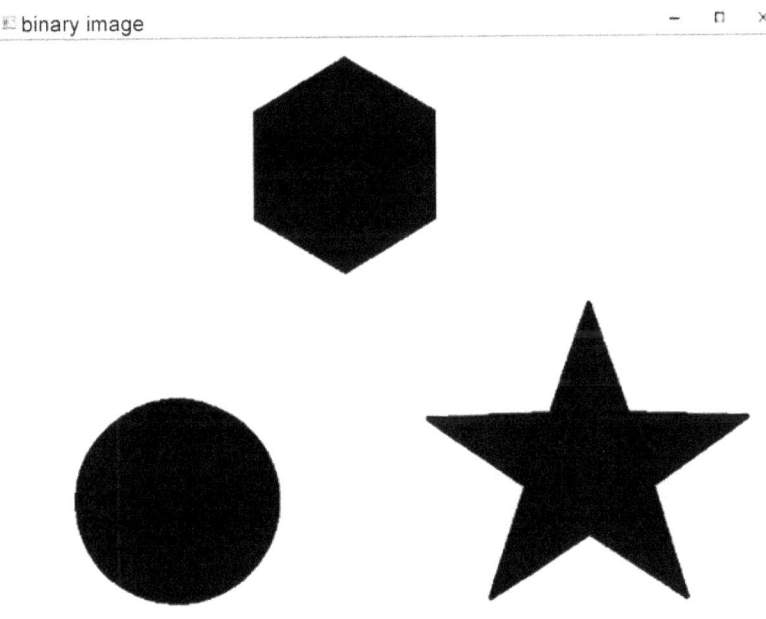

Figure 4.7: Binary image

Now, before we apply contour detection to this image, let's analyze the objects we want to detect. We want to detect these black shapes on a white background. To apply OpenCV's implementation of contour detection here, first we need to make the background black and the foreground white. This is called inverting an image.

> **NOTE**
>
> In colored images, black is represented by 0 and white by 255. So, to invert it, we will simply apply the following formula: `pixel= 255 - pixel`. This is applied to the whole image as `image= 255 - image`.
>
> In Python, an OpenCV image is stored as a NumPy array. Therefore, we can simply invert it as `image= ~image`.
>
> Both commands give identical results.

9. Next, invert the image and display it as follows:

```
inverted_binary_im= ~binary_im
cv2.imshow('inverse of binary image', inverted_binary_im)
cv2.waitKey(0)
cv2.destroyAllWindows()
```

The output is as follows:

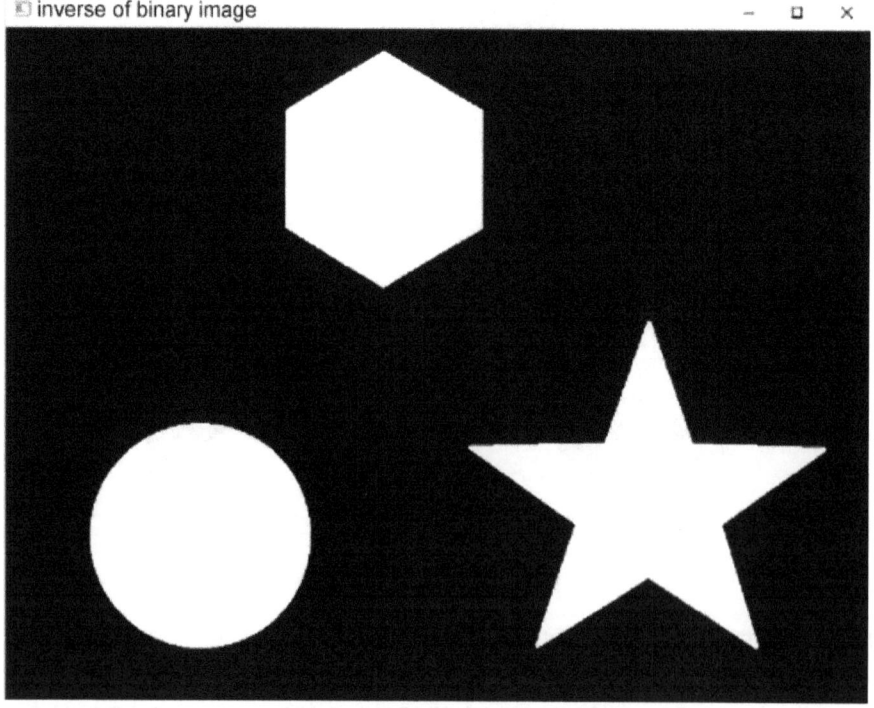

Figure 4.8: Inverted binary image

10. Find the contours in the binary image:

```
contours,hierarchy = cv2.findContours(inverted_binary_im,\
                                 cv2.RETR_TREE,\
                                 cv2.CHAIN_APPROX_SIMPLE)
```

11. Now, mark all the detected contours on the original BGR image in any color (say, **green**). We will set the thickness to **3**:

```
with_contours = cv2.drawContours(image, contours, -1,(0,255,0),3)
cv2.imshow('Detected contours on RGB image', with_contours)
cv2.waitKey(0)
cv2.destroyAllWindows()
```

The output is as follows:

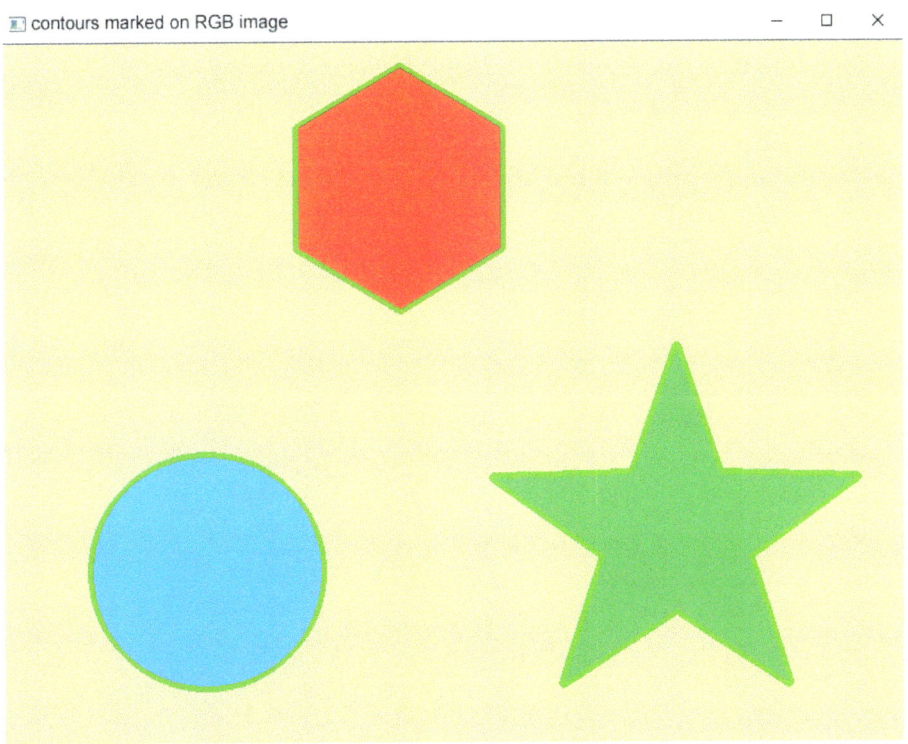

Figure 4.9: Detected contours on an RGB image

> **NOTE**
> You can experiment with the value for thickness.

12. Finally, display the total count of the detected contours:

    ```
    print('Total number of detected contours is:')
    print(len(contours))
    ```

    The output is as follows:

    ```
    Total number of detected contours is:
    3
    ```

In this exercise, we practiced how to detect contours on a colored image. First, the image must be converted to grayscale and then to binary with a black background and a white foreground. After this, contours are detected and visualized on a BGR image using OpenCV functions.

> **NOTE**
>
> To access the source code for this specific section, please refer to https://packt.live/2CVdTg1.

EXERCISE 4.02: DETECTING SHAPES AND DISPLAYING THEM ON BLACK AND WHITE IMAGES

In the preceding exercise, you plotted the contours on a colored image. Your next task is to plot the detected contours in a blue **(BGR code: 255, 0, 0)** color, but on the following black and white image:

Figure 4.10: Required output

> **NOTE**
>
> Complete *steps 1* to *10* of *Exercise 4.01, Detecting Shapes and Displaying Them on BGR Images*, before you begin this exercise. Since this exercise is built on top of the previous one, it should be executed in the same Jupyter notebook.

1. To draw contours with a BGR color code, the image must have three channels. So, we will replicate the single plane of the binary image three times and then merge the three planes to extend it into BGR color space. To merge the channels, refer the **cv2.merge** command you learned in the *Important OpenCV Functions* section of *Chapter 1, Basics of Image Processing*:

   ```
   bgr = cv2.merge([inverted_binary_im, \
                   inverted_binary_im, inverted_binary_im]);
   ```

2. Now, mark all the detected contours on this generated BGR image in any color (say, **blue**). We will keep the thickness set to **3**:

   ```
   with_contours = cv2.drawContours(bgr, contours, \
                   -1, (255,0, 0),3)
   cv2.imshow('Detected contours', with_contours)
   cv2.waitKey(0)
   cv2.destroyAllWindows()
   ```

The output is as follows:

Figure 4.11: Detected contours on black and white images

> **NOTE**
>
> You can experiment with the value for line thickness.

In this exercise, we saw how contours can be drawn on a black and white image. Binary images have a single plane, whereas colored contours can only be drawn on an image with three planes, so, we will replicate the binary image to get three similar images (planes). These planes are merged to get an image in BGR format. After this, the detected contours are drawn on it.

> **NOTE**
>
> To access the source code for this specific section, please refer to https://packt.live/3eQyobZ.

EXERCISE 4.03: DISPLAYING DIFFERENT CONTOURS WITH DIFFERENT COLORS AND THICKNESSES

This is an extension of the previous exercise. Having completed all the steps under *Exercise 4.02, Detecting Shapes and Displaying Them on Black and White Images*, we will do the following:

1. Mark contour number 1 in red color with a thickness of **10**.
2. Mark contour number 2 in green color with a thickness of **20**.
3. Mark contour number 3 in blue color with a thickness of **30**.

> **NOTE**
>
> This exercise is built on top of *Exercise 4.02, Detecting Shapes and Displaying Them on Black and White Images* and should be executed in the same Jupyter Notebook.

Perform the following steps:

1. Take the BGR image stored in the **bgr** variable created in *Step 1* of the previous exercise and draw contour **#0** on it (that is, index **0** in the list of contours) in red **(BGR code: 0, 0, 255)**. The thickness value will be **10**. Let's call the output image (with the drawn contour) **with_contours**:

   ```
   with_contours = cv2.drawContours(bgr, contours, 0, (0,0,255),10)
   ```

2. Take the **with_contours** image and draw contour number **1** on it (that is, index **1** in the list of contours) in green **(BGR code: 0, 255, 0)**. The thickness value will be **20**. The **with_contours** image is updated in this step (the output image is given the same name as the input image):

   ```
   with_contours = cv2.drawContours(with_contours, contours, \
                                    1, (0, 255, 0),20)
   ```

3. Take the **with_contours** image and draw contour number **2** on it (that is, index **2** in the list of contours) in blue **(BGR code: 255, 0, 0)**. The thickness value will be 30. The **with_contours** image is updated in this step as well:

   ```
   with_contours = cv2.drawContours(with_contours, contours, \
                                    2, (255,0, 0), 30)
   ```

4. Display the result using the following code:

   ```
   cv2.imshow('Detected contours', with_contours)
   cv2.waitKey(0)
   cv2.destroyAllWindows()
   ```

The output is as follows:

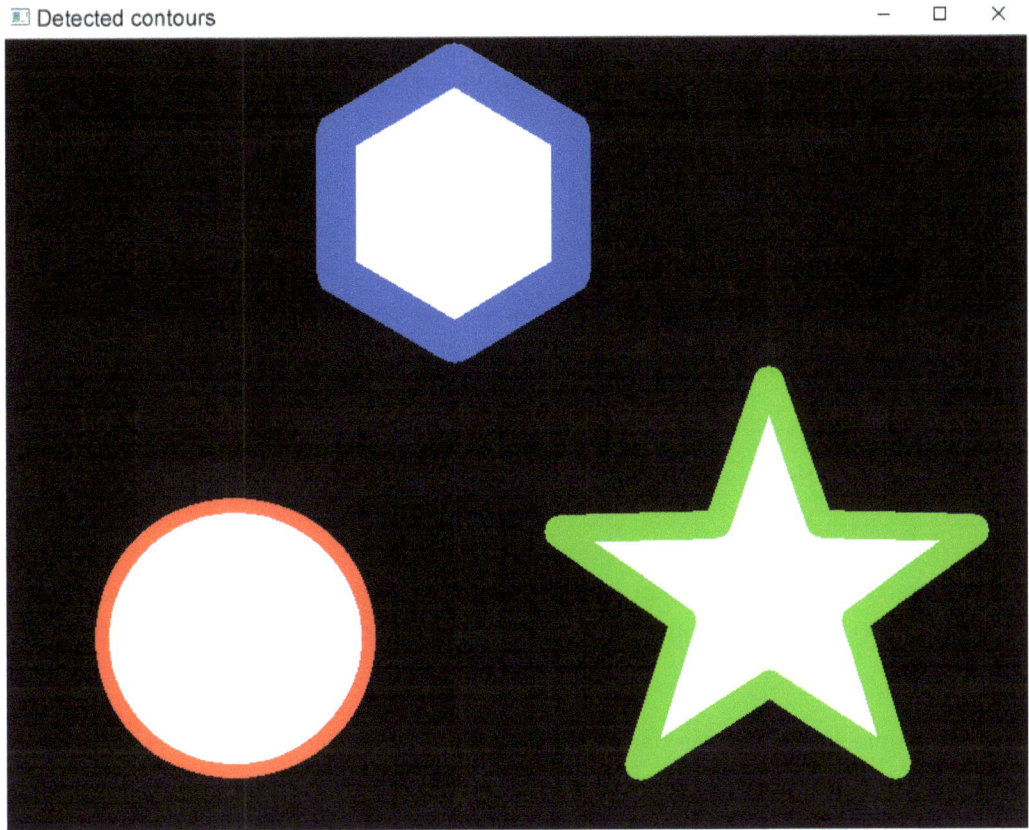

Figure 4.12: Final result

In this exercise, we practiced how to access individual contours from the list of detected contours. We also practiced giving different thicknesses and colors to the individual contours.

> **NOTE**
>
> To access the source code for this specific section, please refer to https://packt.live/2Zqfr9t.

DRAWING A BOUNDING BOX AROUND A CONTOUR

So far, you have learned how to draw the exact shape of the detected contours. Another important thing you might need to do in your projects is to draw an upright rectangular bounding box around your contours of interest.

For example, say you are asked to track a moving car in a video. The first thing you would do is to draw a bounding box around that car in all frames of the video. Similarly, face detection algorithms also draw bounding boxes around faces present in an image.

To draw a bounding box on any region of an image, you need the following:

- The starting **x** coordinate of the box
- The starting **y** coordinate of the box
- The width of the box (**w**)
- The height of the box (**h**)

You can get these parameters for a contour (around which you want to draw the bounding box) by passing that contour into the **cv2.boundingRect** function:

```
x, y, w, h= cv2.boundingRect(my_contour)
```

As an example, let's get these parameters for the first detected contour (which is the contour at **index=0** in the list of contours) on the image in *Figure 4.3*:

```
x,y,w,h = cv2.boundingRect(contours[0])
```

The next step is to plot these values on the image. For this, we will use OpenCV's **cv2.rectangle** command:

```
cv2.rectangle(img,(x,y), (x+w,y+h), color_code, thickness)
```

This command takes as input the **x** and **y** coordinates in the top left of the image and the **x** and **y** coordinates in the bottom right of the image (with which we can compute the width and height). The color code and thickness values are also inputted according to the user's choice. Let's plot our box in blue with a thickness of **5**:

```
cv2.rectangle(with_contours,(x,y), (x+w,y+h), \
              (255,0,0), 5)
cv2.imshow('contour#1', with_contours)
cv2.waitKey(0)
cv2.destroyAllWindows()
```

This will give you an output along the lines of the following:

Contours – Basic Detection and Plotting | 201

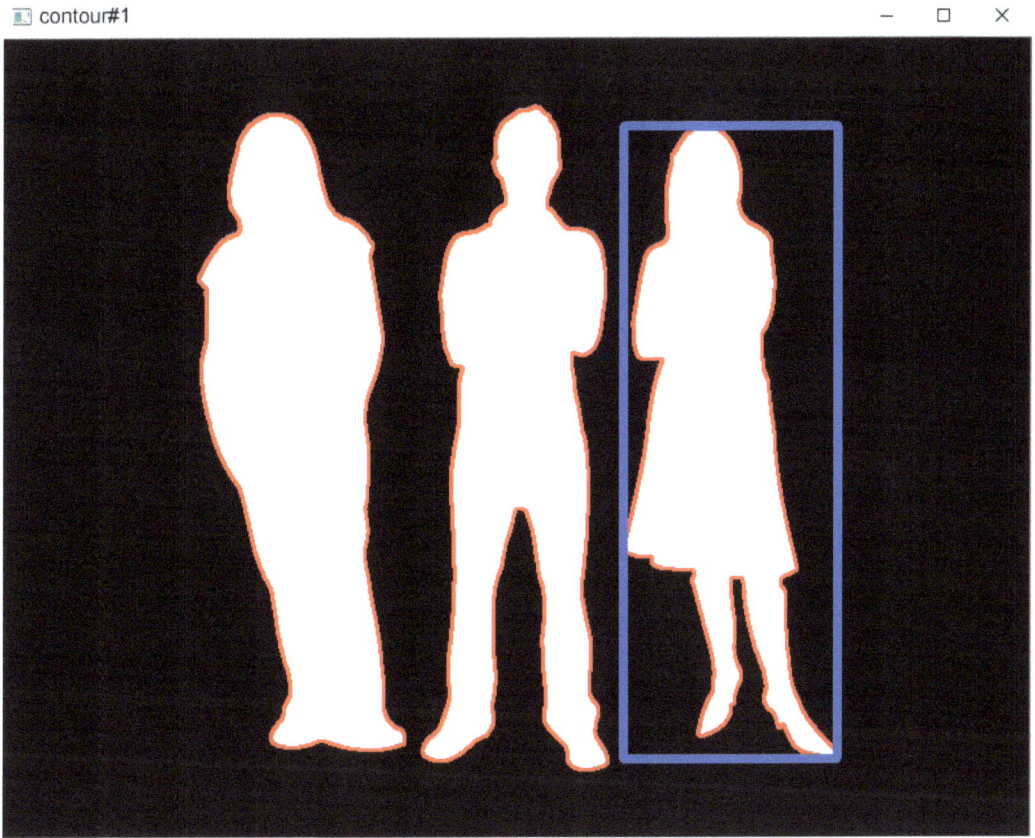

Figure 4.13: Contour number 1 enclosed in a blue bounding box

If you want to draw bounding boxes around all detected contours, then you should use a `for` loop. This `for` loop will iterate over all the contours and, for each detected contour, it will find **x**, **y**, **w**, and **h** and draw the rectangle over that contour.

AREA OF A CONTOUR

OpenCV also provides you with some handy commands to get the attributes of a contour (including the area, the **x** and **y** coordinates of its centroid, the perimeter, the moments, and so on). You can visit https://docs.opencv.org/trunk/dd/d49/tutorial_py_contour_features.html for more details. One of the most commonly used attributes is the area of a contour. To get the area of a contour, you can use the following command:

```
contour_area = cv2.contourArea(contour)
```

You can also retrieve the contour with the maximum area as follows:

```
max_area_cnt = max(contour_list, key = cv2.contourArea)
```

Similarly, you can fetch the contour with the minimum area as follows:

```
min_area_cnt = min(contours, key = cv2.contourArea)
```

DIFFERENCE BETWEEN CONTOUR DETECTION AND EDGE DETECTION

Looking at *Figure 4.5*, you may wonder that if these detections are of contours, then doesn't that make contour detection the same as edge detection? Not quite. At least not for images with subtle changes in intensity. According to the documentation relating to OpenCV, *a contour is the boundary of an object whereas an edge is essentially the portion of an image with a significant change in intensity*. For example, an object that has many different shades of color on its surface will have many distinct edges on its surface. But since it is one object, it will have only one boundary along its shape (contour).

For example, in the following image, have a close look at the surface of the banana:

Figure 4.14: Image of a banana

The intensity of color inside this banana varies significantly.

Contour detection will only give you the (closed) boundaries of the main object(s), whereas edge detection will give you all the detected regions where a change in intensity has taken place.

The results of contour detection are as follows:

Figure 4.15: Contour detection using the steps in the flowchart given in Figure 4.1

The results of performing edge detection on the sample image are as follows:

Figure 4.16: Edge detection

Edge detection for this sample banana image has marked all regions of the image where a change in intensity has occurred. Contour detection, on the other hand, gave us the outline of the shape according to the binary image generated. If you play around with different thresholds for binary conversion, you will get different binary images and, hence, different sets of contours.

You may observe that unlike edges, a contour is always a closed shape. In OpenCV, edge detection can directly be applied to an RGB image/grayscale image/binary image. Contour detection, on the other hand, is only for binary images.

HIERARCHY

If you remember, in the *Contours - Basic Detection and Plotting* section, we saw that one of the inputs to the **cv2.findContours** function was named **hierarchy**. Now, what exactly is this thing? Well, contours can have relationships with one another. One contour might lie inside another larger contour – it will be the child of this larger contour. Similarly, a contour might even have grandchildren and great-grandchildren as well. These are called nested contours.

Let's look at the following diagram:

Figure 4.17: Total contours

How many contours do you see? **1**, **2**, or **3**?

The answer is **3**. Remember what we talked about at the start of this chapter. *A contour is the boundary of a white object on a black background*. The preceding image has a hollow square and a filled circle. You might be sure of the fact that the filled circle is a single blob; however, you might get confused with the hollow square. A hollow square has two outlines: the outer border, and the inner border; making two white borders and, hence, two contours on the square. That makes a total of three contours in the entire image:

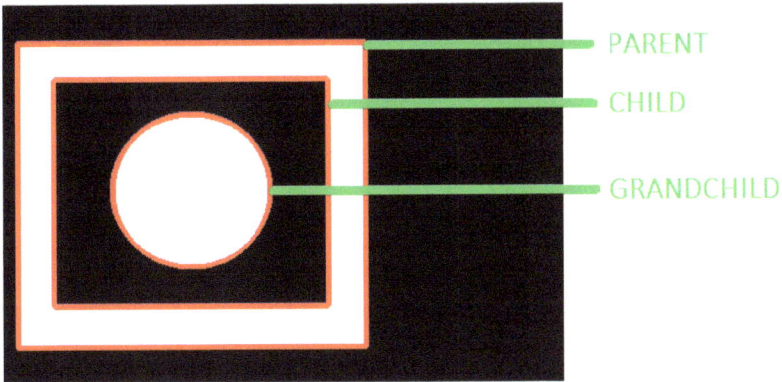

Figure 4.18: Contours with hierarchy

To access the hierarchy information, you need to set the `retrieval_mode` input of `cv2.findContours` according to your requirements. When you learned about the `cv2.findContours` command at the start of this chapter, you saw the different input options (`RETR_EXTERNAL`, `RETR_LIST`, `RETR_CCOMP`, and `RETR_TREE`) for setting the contour retrieval mode.

Now, let's look at its corresponding output variable: `hierarchy`. If you take a look at the value of this `hierarchy` variable for any sample image, you will notice that it is a matrix with several rows equal to the number of detected contours. Consider the following contour retrieval commands and hierarchy structure for our sample image:

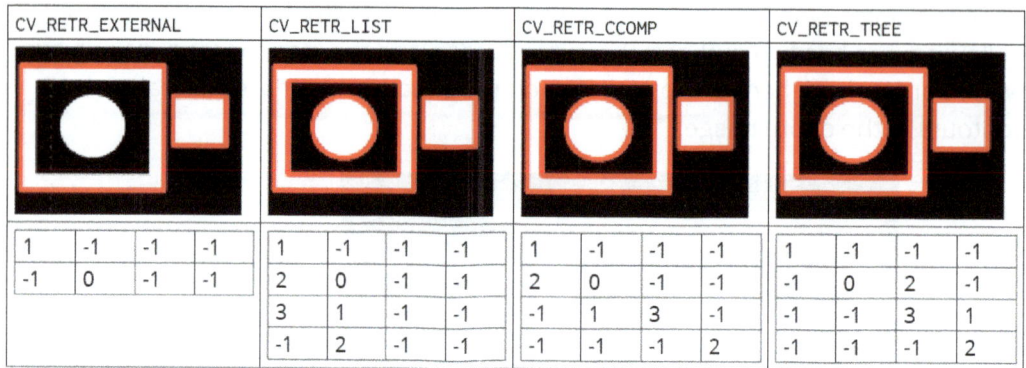

Figure 4.19: Contour retrieval commands and hierarchy structure for our sample image

Each row contains information about an individual contour. A row, as shown in the preceding figure, has four columns:

- **column 1** contains the numeric ID of its **next contour**, which is on an identical hierarchical level. If such a contour does not exist, then its value here is listed as **-1**.

- **column 2** contains the numeric ID of its **previous contour**, which is on an identical hierarchical level. If such a contour does not exist, then its value here is listed as **-1**.

- **column 3** contains the numeric ID of its **first child contour**, which is at the next hierarchical level. If such a contour does not exist, then its value here is listed as **-1**.

- **column 4** contains the numeric ID of its **parent contour**, which is at the next hierarchical level. If such a contour does not exist, then its value here is listed as **-1**.

Now, let's use these options on our sample image and see what happens. As a first step, let's read this sample image and save a copy.

> **NOTE**
>
> The sample image is available at https://packt.live/3etTonP.

Any changes made to the original image will not affect this copy. In the end, we can plot our results on this copy of the original image:

```
import cv2
image = cv2.imread('contour_hierarchy.png')
imagecopy= image.copy()
cv2.imshow('Original image', image)
cv2.waitKey(0)
cv2.destroyAllWindows()
```

The output is as follows:

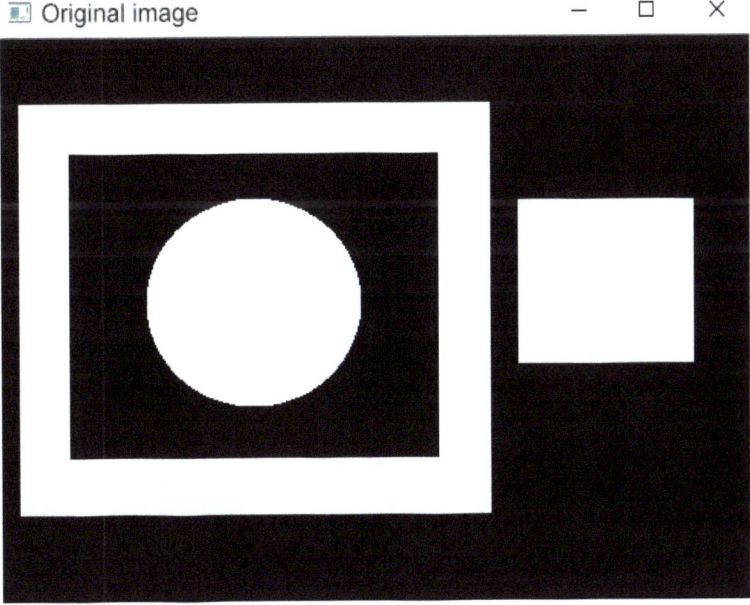

Figure 4.20: Sample image

208 | Working with contours

Such images are saved in a three-channel format. So, although this black and white image looks like a binary image, in reality, it is a three-channel BGR image. To apply contour detection to it, we would, of course, need to convert it to an actual single-channel binary image. We know that only grayscale images can be converted to binary.

We first need to convert the image to grayscale using the following code:

```
gray_image = cv2.cvtColor(image, cv2.COLOR_BGR2GRAY)
```

> **NOTE**
>
> In OpenCV, you can also read a BGR image directly as grayscale using the `im = cv2.imread(filename,cv2.IMREAD_GRAYSCALE)` command (note that the `im = cv2.imread(filename,0)` command will give identical results).

Now, we can convert it to binary:

```
ret, im_binary = cv2.threshold(gray_image, 00, 255, cv2.THRESH_BINARY)
```

The output is as follows:

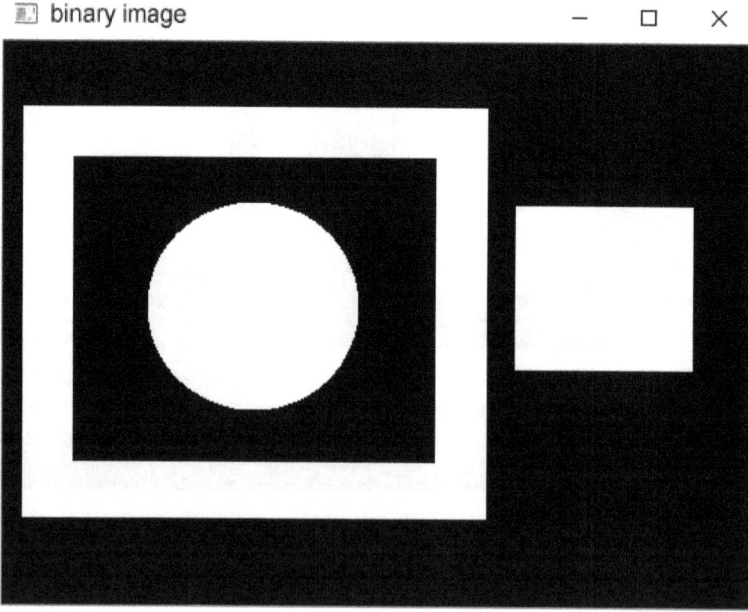

Figure 4.21: Binary image

Now, let's go back to *Figure 4.19*. You can see that for `cv2.RETR_EXTERNAL`, there are only two detected contours. All inner contours are ignored.

The first row of its hierarchy matrix corresponds to contour **0**. Its next contour is **1**. Its previous contour does not exist (**-1**). Its child and parent contours also do not apply (**-1** and **-1**) because only external contours are considered.

Similarly, the second row of its hierarchy matrix corresponds to contour **1**. Its next contour does not exist (**-1**) because there are no more contours in this image. Its previous contour was contour **0**. Its child and parent contours do not exist (**-1** and **-1** in the last two columns).

In *Figure 4.19*, for `cv2.RETR_LIST`, there are four detected contours. All inner contours are detected 'independently'. By 'independently' here, we mean that no parent-child relationship is formed (the last two columns are full of -1s). This means that all contours are considered to be at the same hierarchical level. There is a total of four contours in this image (since four rows are present). The first row has contour **0**. It has contour **1** after it and no contour before it. The second row has contour **1**. It has contour **2** after it and contour **0** before it. The third row has contour **2**. It has contour **3** after it and contour **1** before it. The fourth row has contour **3**. It has no contour after it (as there are no more contours left) and contour **2** before it.

For `cv2.RETR_CCOMP`, there are only two hierarchical levels. The outer boundaries of objects are on one level and the inner boundaries are on another level. Let's try to make sense of this hierarchy matrix here because it is not as simple as the first two. We know that a contour with only a single outer boundary (no hole) has no parent and no child. So, the last two columns for such a contour should be **-1**. The first two rows here have **-1** in their last two columns – from this, we can gather that these first two rows correspond to the filled square and the filled circle. We also know that the first two columns give the indices of the next and previous contours on the same hierarchical level. If there is no other contour on the same level, then these two values should be **-1**, right? Look at the last row. It must belong to the inner hole of the hollow square because there is no other contour on that level (that is, there is no other inner hole boundary). That leaves us with row **3** and the outer boundary contour of the hollow square. So, row **3** must be for this contour. Do you want to confirm this? Well, this contour has a child contour but no parent contour, so for this contour, there would be a child but no parent, and this is reflected in the last two columns of row **3**.

Finally, let's get to `cv2.RETR_TREE`. Look at the first row of its hierarchy matrix. Its last two columns have values of **-1** and **-1**, so it is a contour with no child and no parent. Looking at the figure, we can easily conclude that it is a contour of the filled square. On the second row, we have a contour that has a child but no parent. So, it is the outer boundary of the hollow square. On the third row, we have a contour that has a child and a parent. So, it must be the boundary of the hole of the hollow square. In the last column, we have a contour that has no child but one parent. Hence, it must have no contours inside it, but a contour outside it. This explanation, of course, points to the filled circle in the center.

EXERCISE 4.04: DETECTING A BOLT AND A NUT

Suppose you are designing a system where a bolt is to be inserted into a nut. Using the image provided, you need to figure out the exact location of the following:

- The bolt, so that the robot can pick it up.

- The inner hole of the nut, so that the robot knows the exact location where the bolt is to be inserted.

Given the following binary image of the bolt and nut, detect the bolt and the inner hole of the nut:

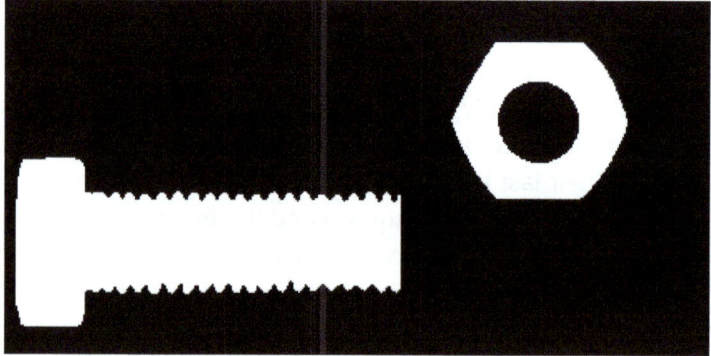

Figure 4.22: Bolt and nut

> **NOTE**
>
> The image can be downloaded from https://packt.live/2YQWlKJ.

The bolt and the inner hole of the nut should be marked in separate colors as follows:

Figure 4.23: Required result

1. Open a new file in Jupyter Notebook in the folder where the image file of the nut and bolt is present and save it as **Exercise4.04.ipynb**.

2. Import OpenCV because this will be required for contour detection:

   ```
   import cv2
   ```

3. Read the three-channel black and white image:

   ```
   image_3chan = cv2.imread('nut_bolt.png')
   ```

4. Later, we are going to convert it into a single-channel binary image, so save a copy of this 3-channel image now to be able to plot the contours on it at the end of the exercise:

   ```
   image_3chan_copy= image_3chan.copy()
   ```

5. Display the original image:

   ```
   cv2.imshow('Original image', image_3chan)
   cv2.waitKey(0)
   cv2.destroyAllWindows()
   ```

It will be displayed as follows:

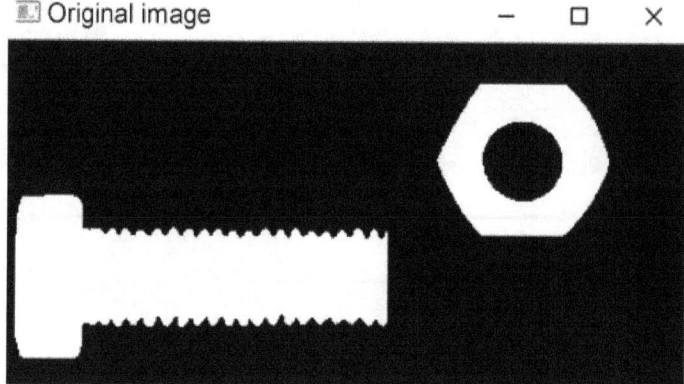

Figure 4.24: Original image of the nut and bolt

6. Convert it to grayscale:

```
gray_image = cv2.cvtColor(image_3chan, cv2.COLOR_BGR2GRAY)
```

7. Then, convert it to binary (using any suitable threshold) and display it as follows:

```
ret,binary_im = cv2.threshold(gray_image,250,255,cv2.THRESH_BINARY)
cv2.imshow('binary image', binary_im)
cv2.waitKey(0)
cv2.destroyAllWindows()
```

It will be displayed like this:

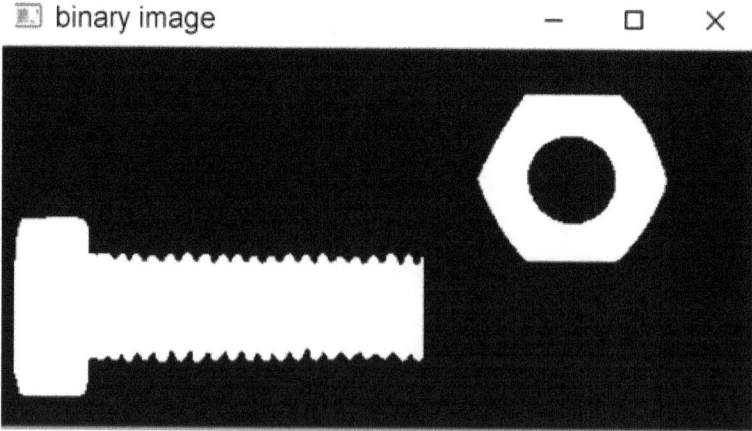

Figure 4.25: Binary image

In this binary image, we need to detect the outer boundary of the bolt and the inner boundary of the nut. Speaking in terms of hierarchy, the bolt is the only contour in this image that has no parent contour and no child contour. The inner hole of the nut, on the other hand, has a parent contour but no child contour. This is how we are going to differentiate between the two in the next steps.

8. Find all contours in this image:

```
contours_list,hierarchy = cv2.findContours(binary_im, \
                                           cv2.RETR_TREE, \
                                           cv2.CHAIN_APPROX_SIMPLE)
```

We used **RETR_TREE** mode to get the contours because we are going to need their parent-child relationships.

9. Let's print the **hierarchy** variable to see what it contains:

```
print('Hierarchy information of all contours:')
print (hierarchy)
```

The output is as follows:

```
Hierarchy information of all contours:
[[[ 1 -1 -1 -1]
  [-1  0  2 -1]
  [-1 -1 -1  1]]]
```

From here, we can see that this is a list containing one list inside it. This nested list is further made up of three lists. In the following diagram, you can see how to access the three individual lists:

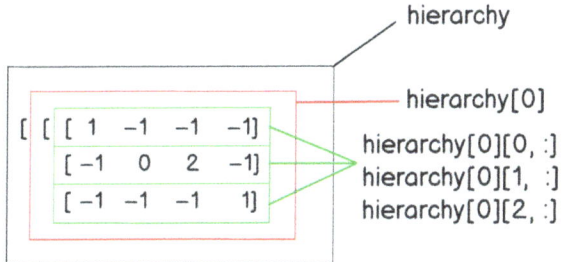

Figure 4.26: Accessing the hierarchy of individual contours in the hierarchy variable

10. In a `for` loop, we will now access the hierarchy information for each individual contour. For each contour, we will check the last two columns.

    If both are **-1**, then that means that there is no parent and no child of that contour. This means that it is the contour of the bolt in the image. We will plot it in green to identify it.

    If the third column is **-1** but the fourth column is not **-1**, then that means that it has no child but does have a parent. This means that it is the contour of the inner hole of the nut in the image. We will plot it in red to identify it:

```
for i in range(0, len(contours_list)):
    contour_info= hierarchy[0][i, :]
    print('Hierarchy information of current contour:')
    print(contour_info)
    # no parent, no child
    if contour_info[2]==-1 and contour_info[3]==-1:

        with_contours = cv2.drawContours(image_3chan_copy,\
                        contours_list,i,[0,255,0],thickness=3)

        print('Bolt contour is detected')
    if contour_info[2]==-1 and contour_info[3]!=-1:

        with_contours = cv2.drawContours(with_contours,\
                        contours_list,i,[0,0,255],thickness=3)

        print('Hole of nut is detected')
```

The output is as follows:

```
Hierarchy information of current contour:
[ 1 -1 -1 -1]
Bolt contour is detected
Hierarchy information of current contour:
[-1  0  2 -1]
Hierarchy information of current contour:
[-1 -1 -1  1]
Hole of nut is detected
```

11. Display the image with the marked contours:

```
cv2.imshow('Contours marked on RGB image', with_contours)
cv2.waitKey(0)
cv2.destroyAllWindows()
```

The preceding code will produce an output like this:

Figure 4.27: Bolt detected in green and the nut's hole in red

In this exercise, you implemented a practical example of accessing contours using their hierarchy information. This was meant to give you an idea of how the information stored in different contour retrieval modes can be useful in different scenarios.

> **NOTE**
>
> To access the source code for this specific section, please refer to https://packt.live/2CYL8iF.

In the next exercise, we will implement a scenario where you will detect the net of a basketball by selecting the largest contour in the image.

EXERCISE 4.05: DETECTING A BASKETBALL NET IN AN IMAGE

Your task is to draw a bounding box around the basket after detecting it. The approach we will follow here is that we will convert this image to binary using such a threshold that the entire white region of the basket is detected as a single object. This will undoubtedly be the contour with the largest area here. Then, we will mark this largest contour with a bounding box. Here is the `basketball.jpg` image:

Figure 4.28: Image of a basketball net

> **NOTE**
>
> The image can be downloaded from https://packt.live/3ihLjpt.

The output should be something like this:

Figure 4.29: Required output

Let's perform the following steps to complete this exercise:

1. Open Jupyter Notebook. Click on **File | New File**. A new file will open up in the editor. In the following steps, you will begin to write code in this file. Save the file by giving it any name you like.

   > **NOTE**
   >
   > The name must start with an alphabetical character and must not contain any spaces or special characters.

2. Jupyter Notebook requires us to import the necessary libraries and, for the current exercise, we will use OpenCV only. Import the OpenCV library using the following code:

   ```
   import cv2
   ```

3. Read the colored image using the following code:

   ```
   image = cv2.imread('basketball.jpg')
   ```

4. Make a copy of this image and save it to another variable using the following code:

   ```
   imageCopy= image.copy()
   ```

   This is just a safety precaution so that even if your original image is modified later in the code, you will have a copy saved in case you might need it later.

5. Display the image you have read using the following code:

   ```
   cv2.imshow('BGR image', image)
   cv2.waitKey(0)
   cv2.destroyAllWindows()
   ```

The output is as follows:

Figure 4.30: Input image

Press any key on the keyboard to close this image window and move forward.

6. Convert the image to grayscale and display it as follows:

```
gray_image = cv2.cvtColor(image,cv2.COLOR_BGR2GRAY)
cv2.imshow('gray', gray_image)
cv2.waitKey(0)
cv2.destroyAllWindows()
```

Running the preceding code will produce the following output:

Figure 4.31: Grayscale image

7. Convert this grayscale image to a binary image using a threshold such that the entire white boundary region of the basketball net is detected as a single blob:

```
ret,binary_im = cv2.threshold(gray_image,100, 255, \
                              cv2.THRESH_BINARY)
cv2.imshow('binary', binary_im)
cv2.waitKey(0)
cv2.destroyAllWindows()
```

Using trial and error, we found **100** to be the best threshold, in this case, to convert the image to binary. If you find that some other threshold works better, feel free to use that.

Running the preceding code will display the following:

Figure 4.32: Binary image

8. Detect all contours using the following code:

```
contours,hierarchy = cv2.findContours(binary_im,\
                    cv2.RETR_TREE,cv2.CHAIN_APPROX_SIMPLE)
```

9. Draw all the detected contours on the image and then display the image. Use the following code to plot all the contours in green:

```
contours_to_plot= -1
plotting_color= (0,255,0)
# if we want to fill the drawn contours with color
thickness= -1
with_contours = cv2.drawContours(image,contours, contours_to_plot, \
                                 plotting_color,thickness)
cv2.imshow('contours', with_contours)
cv2.waitKey(0)
cv2.destroyAllWindows()
```

A thickness of `-1` was given to fill the drawn contours with color for a better visual display.

This will give you the following image output:

Figure 4.33: Drawn contours

10. Next, we must plot bounding boxes around all contours. The code for this is given in the following snippet:

```
for cnt in contours:
    x,y,w,h = cv2.boundingRect(cnt)
    image = cv2.rectangle(image,(x,y),(x+w,y+h),\
                          (0,255,255),2)
```

In the preceding code, we looped over each contour one by one and computed the starting **x** and **y** coordinates along with the width and height of a bounding box that would fit over each contour. Then, we plotted the identified rectangular bounding box for each contour on the actual image. We used yellow to draw the bounding boxes. The BGR code for yellow is **(0, 255, 255)**. A thickness of **2** was used to draw the bounding boxes.

11. Now, display the image using the following code:

```
cv2.imshow('contours', image)
cv2.waitKey(0)
cv2.destroyAllWindows()
```

This will give you the following image output:

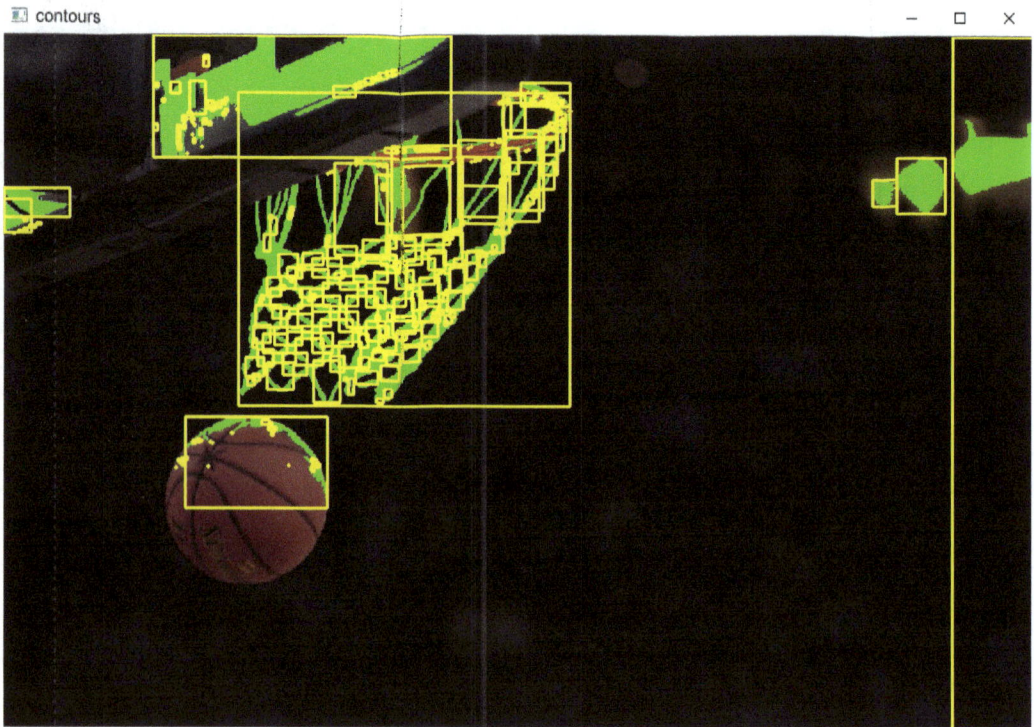

Figure 4.34: Detected contours bounding boxes around them

12. Find the contour with the largest area:

```
required_contour = max(contours, key = cv2.contourArea)
```

13. Find the starting **x** and **y** coordinates and the width and height of a rectangular bounding box that should enclose this largest contour:

```
x,y,w,h = cv2.boundingRect(required_contour)
```

14. Draw this bounding box on a copy of the original colored image that you had saved earlier:

```
img_copy2 = cv2.rectangle(imageCopy, (x,y),(x+w, y+h),\
                           (0,255,255),2)
```

15. Now, display this image with the bounding box drawn over it:

```
cv2.imshow('largest contour', img_copy2)
cv2.waitKey(0)
cv2.destroyAllWindows()
```

This will give you the following output image:

Figure 4.35: Final result

In this exercise, we learned how to apply contour detection to detect the net of a basketball hoop.

> **NOTE**
>
> To access the source code for this specific section, please refer to https://packt.live/31BlWJc.

In the next section, we are going to explore a method for finding the contour on an image that most closely matches a reference contour given in another image.

CONTOUR MATCHING

In this section, we will learn how to numerically find the difference between the shapes of two different contours. This difference is found based on Hu moments. Hu moments (also known as Hu moment invariants) of a contour are seven numbers that describe the shape of a contour.

> **NOTE**
>
> Visit https://docs.opencv.org/2.4/modules/imgproc/doc/structural_analysis_and_shape_descriptors.html for more details.

Hu moments can be computed using OpenCV in the following way:

Image moments can be computed with the following command:

```
img_moments= cv2.moments(image)
```

The **cv2.HuMoments** OpenCV function is given the following moments as input:

```
hu_moments= cv2.HuMoments(img_moments)
```

This array is then flattened to get the feature vector for the Hu moments:

```
hu_moments= hu_moments.flatten()
```

This feature vector has a row with seven columns (seven numeric values).

Hu moment vectors of some sample contour shapes are shown in the following table:

CONTOUR	Hu moment vector H						
	H[0]	H[1]	H[2]	H[3]	H[4]	H[5]	H[6]
U	$1.401e^{-03}$	$1.306e^{-08}$	$1.232e^{-10}$	$4.882e^{-11}$	$-3.720e^{-21}$	$5.302e^{-15}$	$-7.084e^{-22}$
H	$1.258e^{-03}$	$5.039e^{-09}$	$1.130e^{-16}$	$4.176e^{-13}$	$2.869e^{-27}$	$-2.964e^{-17}$	$0.000e^{+00}$
B	$1.453e^{-03}$	$2.918e^{-07}$	$4.502e^{-11}$	$1.201e^{-12}$	$8.828e^{-24}$	$-1.651e^{-17}$	$-4.462e^{-25}$
B	$6.744e^{-03}$	$2.940e^{-06}$	$2.968e^{-08}$	$6.638e^{-09}$	$-6.418e^{-17}$	$1.138e^{-11}$	$-6.754e^{-17}$
B	$1.573e^{-03}$	$5.884e^{-07}$	$4.206e^{-10}$	$1.400e^{-10}$	$2.985e^{-20}$	$1.019e^{-13}$	$-1.627e^{-20}$
B	$1.573e^{-03}$	$5.884e^{-07}$	$4.206e^{-10}$	$1.400e^{-10}$	$2.985e^{-20}$	$1.019e^{-13}$	$1.627e^{-20}$

Figure 4.36: Hu moment vectors of some sample contour shapes

For a Hu moment vector, consider the following:

- The first six values remain the same even if the image is transformed by reflection, translation, scaling, or rotation. What this means is that two contours of the same basic shape will have nearly equal values of these moments even if one of them is resized, rotated, flipped in any direction, or changes its position or location in the image (by way of an example, refer to the first six values in the last two rows of *Figure 4.36*).

- The seventh value changes sign (positive or negative) if the contour's image is flipped (for an example, refer to the last values in the last two rows of *Figure 4.36*).

Sounds interesting, doesn't it? Think of all the wonderful stuff you can do with it. **Optical Character Recognition** (**OCR**) is just the beginning.

The command to perform shape matching (the comparison of two shapes) using Hu moments is given in the following code. It will give you a numerical value describing how different the two shapes are from one another:

```
contour_difference = cv2.matchShapes (contour1, contour2, \
                                     compar_method, parameter)
```

The smaller the value of this output variable (`contour_difference`), the greater the similarity between the two compared contours. Concerning the aforementioned formula, note the following:

- `contour1` and `contour2` are the two individual contour objects you want to compare (alternatively, instead of giving contour objects, you can also give the cropped grayscale images of the two individual contours).

- `compar_method` is the method of comparison. It can either be an integer ranging from **1** to **3** or its corresponding string command, as outlined in the following table:

Comparison method	Numeric command	String command
CONTOURS_MATCH_I1	1	cv.CONTOURS_MATCH_I1
CONTOURS_MATCH_I2	2	cv.CONTOURS_MATCH_I2
CONTOURS_MATCH_I3	3	cv.CONTOURS_MATCH_I3

Figure 4.37: Three comparison methods of contours

You will find that **cv.CONTOURS_MATCH_I1** works best in most scenarios. However, when working on a project, do test all three of these to see which one works best for your data.

- `parameter` is a value related to the chosen comparison method, but you don't need to worry about it because it has become redundant in the latest versions of OpenCV. Do note, however, that you might get an error if you don't pass in a fourth input to the `cv2.matchShapes` function. You can pass any number in its place and it would work just fine. Usually, we pass **0**.

Let's compare the two contours as follows:

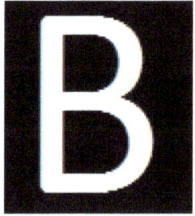

Figure 4.38: Contour 1 (binary_im1)

The second contour appears as follows:

Figure 4.39: Contour 2 (binary_im2)

If we want to find the numerical difference between them, then we can write the following command:

```
contour_difference = cv2.matchShapes (binary_im1, \
                                      binary_im2, 1, 0)
```

In the preceding code, **binary_im1** is the image of contour **1**, and **binary_im2** is the image of contour **2**.

If we execute this command, then the **contour_difference** output variable will tell us the difference in the Hu moment vectors of these two shapes using the **CONTOURS_MATCH_I1** distance. It comes out as 0.1137 for this example.

In this section, you learned how to match shapes that can be described by a single blob. What if you want to match a symbol that has two blobs instead of one? For example, a question mark:

Figure 4.40: Sample question mark image

Look at *Figure 4.40* of a question mark. It has two contours. If you want to find this symbol in an image, simple contour matching won't work. For that, you would use another similar technique called template matching. To read more about this, visit https://docs.opencv.org/2.4/doc/tutorials/imgproc/histograms/template_matching/template_matching.html.

EXERCISE 4.06: DETECTING FRUITS IN AN IMAGE

You are given the following image of a selection of fruits:

Figure 4.41: Image of fruits

Your task is to detect all fruits present in this image. Do this via contour detection and draw the shapes of the contours on the image like so, such that the fruit contours are drawn in red:

Figure 4.42: Detected fruits

> **NOTE**
>
> The image is available at https://packt.live/3iinnSF.

Perform the following steps:

1. Import the OpenCV library:

```
import cv2
```

2. Now, read the image and display it as follows:

```
image = cv2.imread('many fruits.png')
cv2.imshow('Original image', image)
cv2.waitKey(0)
cv2.destroyAllWindows()
```

The preceding code produces the following output:

Figure 4.43: Original image

3. Convert it to grayscale and display it as follows:

```
gray_image = cv2.cvtColor(image,cv2.COLOR_BGR2GRAY)
cv2.imshow('gray', gray_image)
cv2.waitKey(0)
cv2.destroyAllWindows()
```

The output is as follows:

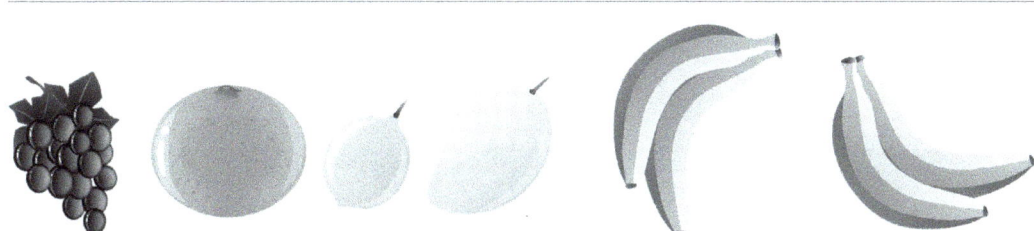

Figure 4.44: Grayscale image

4. Convert it to binary using a suitable threshold and display it. The threshold you select must give the outer boundaries of the fruits as single objects. It is okay if some holes remain in the middle of each fruit:

```
ret,binary_im = cv2.threshold(gray_image,245,\
                              255,cv2.THRESH_BINARY)
cv2.imshow('binary', binary_im)
cv2.waitKey(0)
cv2.destroyAllWindows()
```

The output is as follows:

Figure 4.45: Binary image

5. Since we require a white foreground on a black background to do contour detection in OpenCV, we will invert this image as follows:

```
binary_im= ~binary_im
cv2.imshow('inverted binary', binary_im)
cv2.waitKey(0)
cv2.destroyAllWindows()
```

The output is as follows:

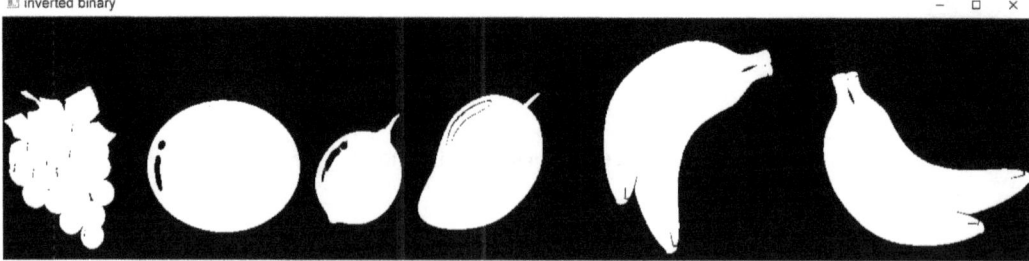

Figure 4.46: Inverted binary image

Note that there are some empty pixels inside the fruits, so we will do external contour detection next.

6. Find all the external contours as follows:

```
contours,hierarchy = cv2.findContours(binary_im,\
                    cv2.RETR_EXTERNAL,cv2.CHAIN_APPROX_SIMPLE)
```

7. Draw these contours in red on the original image and display it as follows:

```
with_contours = cv2.drawContours(image,contours,\
                    -1,(0,0,255),3)

cv2.imshow('contours marked on RGB image', with_contours)
cv2.waitKey(0)
cv2.destroyAllWindows()
```

The output is as follows:

Figure 4.47: Detected external contours

In this exercise, you practiced contour detection by implementing the steps mentioned in the flowchart of *Figure 4.1*.

> **NOTE**
>
> To access the source code for this specific section, please refer to https://packt.live/3ggehUR.

232 | Working with contours

EXERCISE 4.07: IDENTIFYING BANANAS FROM THE IMAGE OF FRUITS

This exercise is an extension of *Exercise 4.06, Detecting Fruits in an Image*. You are given the following reference image containing a banana:

Figure 4.48: Reference image of a banana

> **NOTE**
>
> This exercise is built on top of the previous exercise and should be executed in the same notebook. The image is available at https://packt.live/2NLBPog.

Your task is to identify all bananas present in the image of the previous exercise and mark them in blue, as follows:

Figure 4.49: Required output

Perform the following steps:

1. Read the reference image and display it using the following command:

   ```
   ref_image = cv2.imread('bananaref.png')
   cv2.imshow('Reference image', ref_image)
   cv2.waitKey(0)
   cv2.destroyAllWindows()
   ```

   The output is as follows:

   Figure 4.50: Reference image of a banana

2. Convert it to grayscale and display the result as follows:

   ```
   gray_image = cv2.cvtColor(ref_image,cv2.COLOR_BGR2GRAY)
   cv2.imshow('Grayscale image', gray_image)
   cv2.waitKey(0)
   cv2.destroyAllWindows()
   ```

The output is as follows:

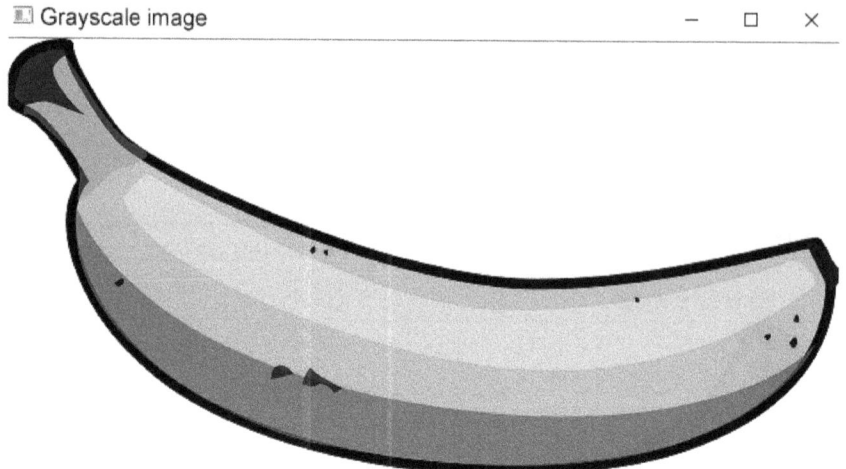

Figure 4.51: Grayscale version of the reference image

3. Using a suitable threshold, convert it to binary and display the result:

```
ret,binary_im = cv2.threshold(gray_image,245,255,\
                cv2.THRESH_BINARY)
cv2.imshow('Binary image', binary_im)
cv2.waitKey(0)
cv2.destroyAllWindows()
```

The output is as follows:

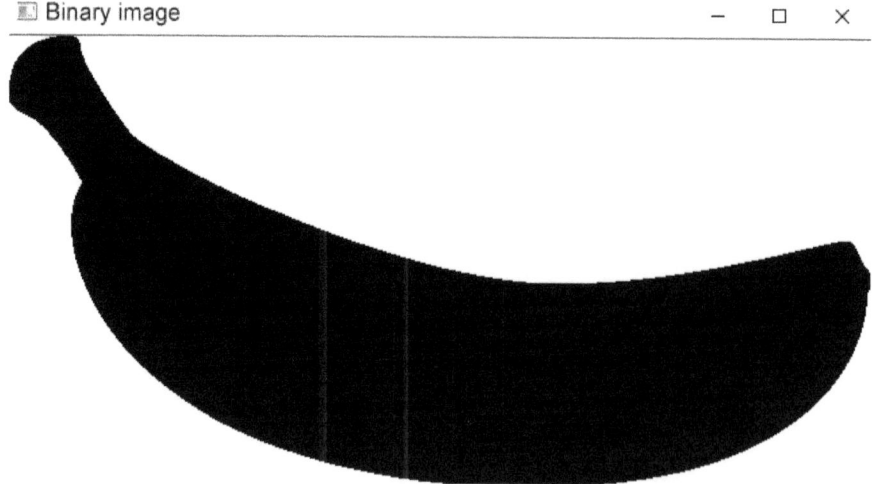

Figure 4.52: Binary version of the reference image

4. To make the object white and the background black (which is required for the contour detection command in OpenCV), we will invert the image:

```
binary_im= ~binary_im
cv2.imshow('inverted binary image', binary_im)
cv2.waitKey(0)
cv2.destroyAllWindows()
```

The output is as follows:

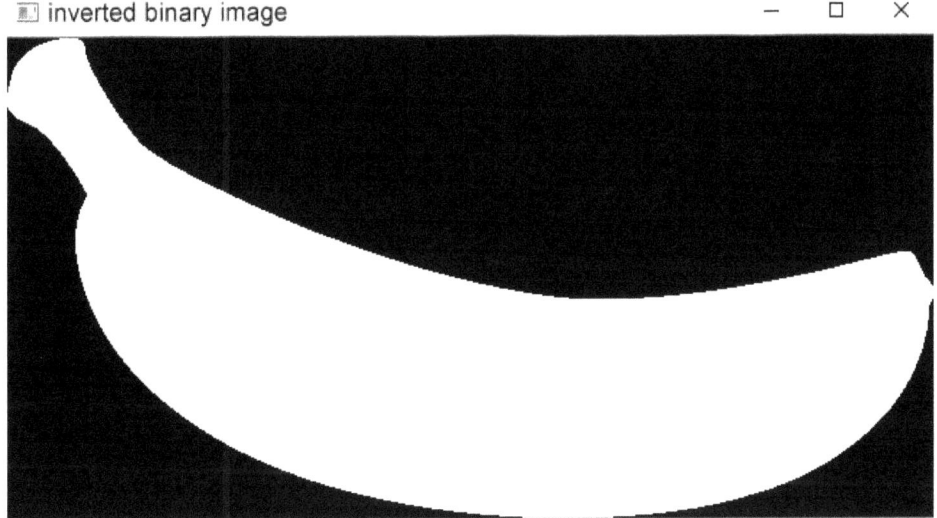

Figure 4.53: Inverted binary image

5. Find the external boundary of this shape and draw a red outline on it:

```
ref_contour_list,hierarchy = cv2.findContours(binary_im,\
                             cv2.RETR_EXTERNAL,\
                             cv2.CHAIN_APPROX_SIMPLE)

with_contours = cv2.drawContours(ref_image,\
            ref_contour_list,-1,(0,0,255),3)

cv2.imshow('contours marked on RGB image',\
        with_contours)
cv2.waitKey(0)
cv2.destroyAllWindows()
```

The output is as follows:

Figure 4.54: Detected contour in red

6. There should only be one contour in this image – the contour of the banana. To confirm it, you can check the number of contours in the Python list, **ref_contour_list**, using the **len(ref_contour_list)** command. You will find that there is only one contour present:

```
print('Total number of contours:')
print(len(ref_contour_list))
```

The output is as follows:

```
Total number of contours:
1
```

7. Put this contour at index **0** of the list because it is the banana. It is also the only contour in this list:

```
reference_contour = ref_contour_list[0]
```

8. Now, we have to compare each fruit contour we detected in the original image of the fruits with the reference contour of the banana. The **for** loop is the best way to go. Before starting the **for** loop, we will initialize an empty list by the name of **dist_list**. In each iteration of the **for** loop, we will append to it the numerical difference between the contour and the reference contour:

   ```
   dist_list= [ ]
   for cnt in contours:
       retval=cv2.matchShapes(cnt, reference_contour,1,0)
       dist_list.append(retval)
   ```

   The next task is to find the two contours at the smallest distances from the reference contour.

9. Make a copy of **dist_list** and store it in a separate variable:

   ```
   sorted_list= dist_list.copy()
   ```

10. Now, sort this list in ascending order:

    ```
    sorted_list.sort()
    ```

    Its first element now (**sorted_list[0]**) is the smallest distance and its second element (**sorted_list[1]**) is the second-largest distance – these two distances correspond to the two bananas present in the image.

11. In the original **dist_list** list, find the indices where the smallest and second smallest distances are present:

    ```
    # index of smallest distance
    ind1_dist= dist_list.index(sorted_list[0])
    # index of second smallest distance
    ind2_dist= dist_list.index(sorted_list[1])
    ```

12. Initialize a new empty list and append the contours at these two indices to it:

    ```
    banana_cnts= [ ]
    banana_cnts.append(contours[ind1_dist])
    banana_cnts.append(contours[ind2_dist])
    ```

13. Now, draw these two contours on the image in blue:

```
with_contours = cv2.drawContours(image, banana_cnts,\
                                 -1,(255,0,0),3)

cv2.imshow('contours marked on RGB image', \
           with_contours)

cv2.waitKey(0)
cv2.destroyAllWindows()
```

The output of the preceding code will be as follows:

Figure 4.55: Detected bananas in blue

In this exercise, we implemented a scenario where we had to detect two bunches of bananas present in an image. A reference image of the banana was provided to allow us to do the matching. There was only a single banana present in the reference image, whereas in the image of all the fruits, there were two bunches of the bananas present. Each bunch had two bananas in it. One bunch was facing from left to right, whereas the other was in an upright position, yet contour matching detected them both successfully. This is because the shape of a bunch of two bananas is quite similar to the shape of a single banana. This exercise demonstrates the power of the contour-matching technique to retrieve similar-looking objects from an image.

> **NOTE**
>
> To access the source code for this specific section, please refer to https://packt.live/38gFN1H.

EXERCISE 4.08: DETECTING AN UPRIGHT BANANA FROM THE IMAGE OF FRUITS

In this task, you will build on the work you did in the last exercise. In the previous exercise, you detected the two banana bunches in the image where multiple fruits were present. One bunch was in a horizontal position, and the other was in a vertical position. In this exercise, you will detect the banana bunch that is in an upright position as follows:

Figure 4.56: Required output

The task here is to draw a bounding box around that banana bunch whose height is greater than its width. Since this is an extension of the previous exercise, before proceeding with this, you must implement *steps 1* to *13* of the previous exercise. After that, you can proceed as follows:

1. Use a **for** loop to check each of these two contours one by one. If its height is greater than its width, plot a blue bounding box around it on the copy of the image you had saved earlier:

```
for cnt in banana_cnts:
    x,y,w,h = cv2.boundingRect(cnt)
    if h>w:
        cv2.rectangle(imagecopy,(x,y),(x+w,y+h),\
                    (255,0,0),2)
```

2. Display this image:

```
cv2.imshow('Upright banana marked on RGB image', imagecopy)
cv2.waitKey(0)
cv2.destroyAllWindows()
```

The output is as follows:

Figure 4.57: Banana contour with height greater than the width

In this exercise, we put all our newly acquired skills to the test. We practically implemented the detection of contours, accessed them by hierarchy, did contour matching, and filtered the contours using their widths and heights. This should give you an idea of the many situations where contour detection and matching can come in handy.

> **NOTE**
>
> To access the source code for this specific section, please refer to https://packt.live/2Vw9MgU.

ACTIVITY 4.01: IDENTIFYING A CHARACTER ON A MIRRORED DOCUMENT

When scanning documents, you know that you usually need to keep the paper steady, with the text in a straight horizontal orientation while the computer scans it. It would make for an interesting application to build a smart system that could read a document even if it is shown a mirror image or a lopsided image of it.

One of the basic tools you need for such an application is the ability to recognize a character even if it is inverted, flipped, mirrored, or slanted at an angle.

You are given two images. The first is of a rotated handwritten note that you captured through a mirror:

Figure 4.58: Handwritten note

> **NOTE**
>
> The image can be downloaded from https://packt.live/3ik9Lqc.

The second image is where you explicitly tell your program what the letter 'B' looks like. Your task is to write a program to identify where the letter 'B' occurs in the preceding image. You are given the following reference image for this:

Figure 4.59: Reference image for 'B'

> **NOTE**
>
> The image can be downloaded from https://packt.live/38hKDM1.

242 | Working with contours

Your task here is to create a program that will identify where the letter 'B' occurs in the handwritten note, given the reference image of 'B' provided.

Your program should generate the following output:

Figure 4.60: Required output

Perform the following steps to complete the activity:

1. Convert the reference image and the image of the handwritten note to binary form.

2. Find all the contours on both these images using the **cv2.RETR_EXTERNAL** method.

3. Compare each detected contour on the image of the handwritten note with the detected contour on the reference image to find the closest match.

4. On the image of the handwritten note, mark the contour most similar to the contour on the reference image.

> **NOTE**
>
> The solution for this activity can be found on page 491.

In this activity, we learned how to apply contour detection to recognize the letter 'B'. The same technique can be applied to detect other alphabetical characters, digits, and special characters too – anything that can be represented as a single blob. Note that the 'B' in the reference image is written in quite a stylish, complex format. This is to demonstrate to you the strength of this technique of contour matching.

If you want to explore it further, you can make images with a handwritten 'B' or 'B' typed in some other standard formats and then take those as reference images instead. Print the minimum distance of the match and see how it varies for the following reference images:

Figure 4.61: Other possible reference images you can try

SUMMARY

In this chapter, you learned that a contour is the outline of an object. This outline can either be on the outer border of the shape (an external contour) or it may be on a hole or other hollow surface inside the object (an inner contour). You learned how to detect contours of different objects and how to draw them on images using different colors. Also, you learned how to access different contours based on their area, width, and height.

Toward the end of the chapter, you gained hands-on experience of how to access the contours of a reference shape from an image containing multiple contours. This chapter can serve as a good foundation if you want to create smart systems, such as a license plate recognition system.

Up to *Chapter 3, Working with Histograms*, you learned about the basics of image processing using OpenCV. This chapter, on the other hand, was designed to give you a gentle push into the vast world of image processing and its practical applications in Computer Vision.

In the next chapter, you are going to learn how to detect a human face, the specific parts of a face, and even how to detect smiles in an image.

5

FACE PROCESSING IN IMAGE AND VIDEO

OVERVIEW

This chapter aims to show you how to detect faces in an image or a video frame using Haar Cascades, and then track the face when given an input video. Once a face has been detected, you will learn how to use techniques such as GrabCut for performing skin detection. We will also see how cascades can be extended to detect other parts of the face (for instance, the eyes, mouth, and nose). By the end of this chapter, you will be able to create applications for smile detection and Snapchat-like face filters.

INTRODUCTION

In the previous chapters, we learned how to use the OpenCV library to carry out basic image and video processing. We also had a look at contour detection in the previous chapter. Now, it's time to take it up a notch and focus on faces – one of the most interesting parts of the human body for computer vision engineers.

Face processing is a hot topic in artificial intelligence because a lot of information can be automatically extracted from faces using computer vision algorithms. The face plays an important role in visual communication because a great deal of nonverbal information, such as identity, intent, and emotion, can be extracted from human faces. This makes it important for you, as a computer vision enthusiast, to understand how to detect and track faces in images and videos and carry out various kinds of filtering operations, for instance, adding sunglasses filters, skin smoothing, and many more, to come up with interesting and significant results.

Whether you want to auto-focus on faces while clicking a selfie or while using portrait mode for pictures, frontal face detection is the foundational technique for building these applications. In this chapter, we will start by understanding the concepts of Haar Cascades and how to use them for frontal face detection. We will also see how we can extend the cascades for detecting features such as eyes, mouths, and noses. Finally, we will learn how to perform skin detection using techniques such as GrabCut and create interesting Snapchat face filters.

INTRODUCTION TO HAAR CASCADES

Before we jump into understanding what Haar Cascades are, let's try to understand how we would go about detecting faces. The **brute-force** method would be to have a window or a block that will slide over the input image, detecting whether there is a face present in that region or not. Detecting whether the block is a face or not can be done by comparing it with some sample face image. Let's see what the issue is with this approach.

The following is a picture of a group of people:

Figure 5.1: Picture of a group of people

Notice the variety of faces in this image. Now, let's look at the following self-explanatory representation of the brute-force method for face detection:

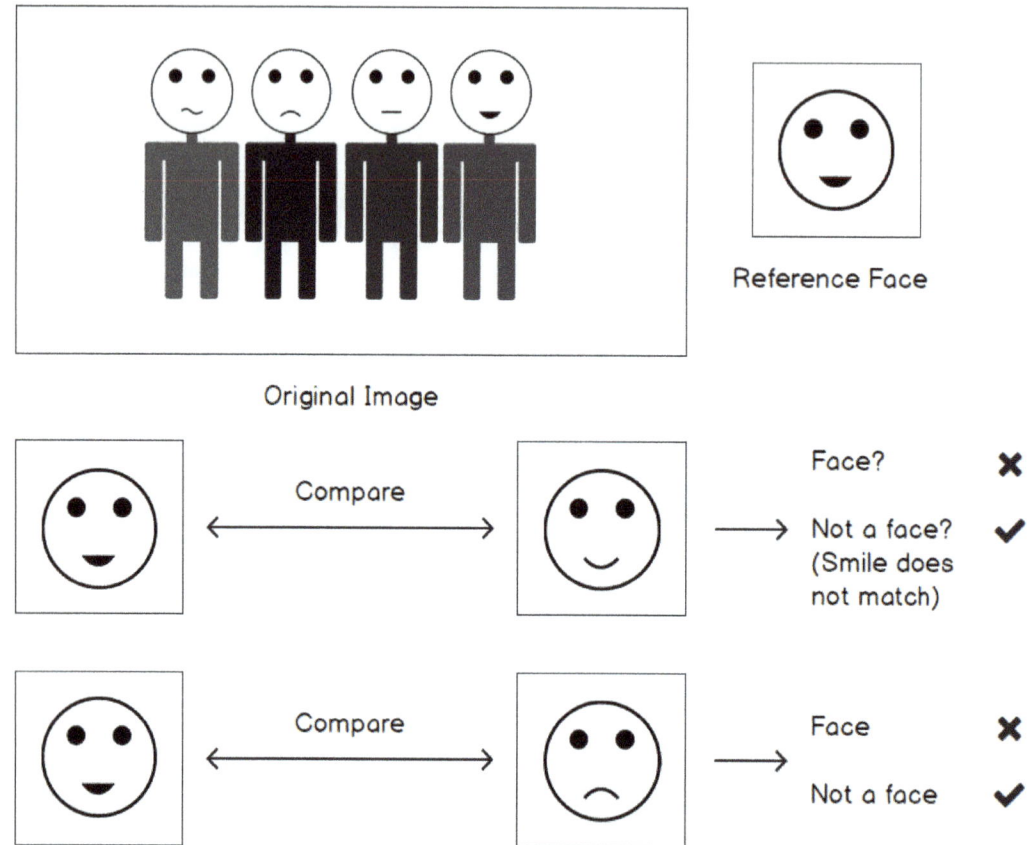

Figure 5.2: Brute-force method for face detection

Here, we can see how difficult it would be if we were to create one standard face that can be used as a reference face to check whether a region in an image has a face or not. For instance, think about the difference in skin tone, facial structure, gender, and other facial features that must be taken into consideration to distinguish faces between people. Another issue with this approach is encountered during the comparison of images. Unlike comparing numerical values, this is a very difficult process. You will have to subtract corresponding pixel values in both images and then see if the values are similar. Let's consider a scenario where both images are the same but have different brightness.

In this case, the pixel values will be very different. Let's consider another scenario where the brightness is the same in the images but the scale (size of the images) is different. A well-known way to deal with this issue is to resize both images to a standard size. Considering all these complexities, you can see that it's very difficult to come up with a general approach to achieve good accuracy by using the basic image processing steps that we learned about in the previous chapters.

To solve this issue, Paul Viola and Michael Jones came up with a machine learning-based approach to object detection in their paper titled *Rapid Object Detection Using a Boosted Cascade of Simple Features* (https://www.cs.cmu.edu/~efros/courses/LBMV07/Papers/viola-cvpr-01.pdf).

This paper proposed the use of machine learning for extracting relevant features from images. The relevancy here can be described by the role the features play in deciding whether an object is present in the image or not. The authors also came up with a technique called **Integral Image** for speeding up the computation related to feature extraction in images. Feature extraction was carried out by considering a large set of rectangular regions of different scales, which then move over the image. The trouble with this step, as you can imagine, is the number of features that will be generated by this process. At this step, **AdaBoost**, a well-known boosting technique, comes into the picture. This technique can reduce the number of features to a great extent and is able to yield only the relevant features out of the huge pool of possible features. Now, we can take this a step further by processing this in a video, which is basically a slideshow of images (termed as frames) within a time period. For instance, based on the quality/purpose of the video, it ranges from 24 frames per second (fps) to 960 fps. Haar Cascades can be used extensively in videos by applying them to each frame.

The basic process for generating such cascades, like any machine learning approach, is data-based. The data here consists of two kinds of images. One set of images doesn't have the object that we want to detect, while the second set of images has the object present in them. The more images you have, and the more variety of images you cover, the better your cascade will be. Let's think about faces here. If you are trying to build a frontal face cascade model, you need to make sure you have a fair share of images for all genders. You will also need to add images with different brightnesses (that is, images taken in the shade or a low-light area versus images taken in bright sunlight). For better accuracy, you would also need to have images showing faces with and without sunglasses, and so on. The aim is to have images that cover a lot of possible varieties.

Using Haar Cascades for Face Detection

Now that we have understood the basic theory behind Haar Cascades, let's understand how we can use Haar Cascades to detect faces in an image with the help of an example. Let's consider the following image as the input image:

Figure 5.3: Input image for the Haar Cascade example

> **NOTE**
>
> To get the corresponding frontal face detection model, you need to download the `haarcascade_frontalface_default.xml` trained model (obtained after being trained on a huge number of images with and without faces) from https://packt.live/3dO3RKx.

To import the necessary libraries, you will use the following code:

```
import cv2
import numpy as np
```

You have already seen in the previous chapters that the first statement is used for importing a Python wrapper of the OpenCV library. The second statement imports the popular NumPy module, which is used extensively for numerical computations because of its highly optimized implementations.

To load the Haar Cascade, you will use the following code:

```
haarCascadeFace = \
cv2.CascadeClassifier("haarcascade_frontalface_default.xml")
```

Here, we have used the **cv2.CascadeClassifier** function, which uses only one input argument – the filename of the XML Haar Cascade model. This returns a **CascadeClassifier** object.

The **CascadeClassifier** object has several methods implemented in it, but we will only focus on the **detectMultiScale** function. Let's have a look at the arguments it takes:

```
detectedObjects = \
cv2.CascadeClassifier.detectMultiScale(image, [scaleFactor, \
                                               minNeighbors, flags, \
                                               minSize, maxSize])
```

Here, the only mandatory is the image, which is the **grayscale** input image in which the faces are to be detected.

The optional arguments are as follows:

scaleFactor decides to what extent the image is going to be resized at every iteration. This helps in detecting faces of different sizes present in the input image. **minNeighbors** is another argument that decides the minimum number of neighbor's a rectangle should have to be considered for detection. **minSize** and **maxSize** specify the minimum and maximum size of the face to be detected. Any face with a size lying outside this range will be ignored.

252 | Face Processing in Image and Video

The output of the function is a list of rectangles (or bounding boxes) that contain the faces present in the image. We can use OpenCV's drawing functions – `cv2.rectangle` – to draw these rectangles on the image. The following diagram shows the pictorial representation of the steps that need to be followed:

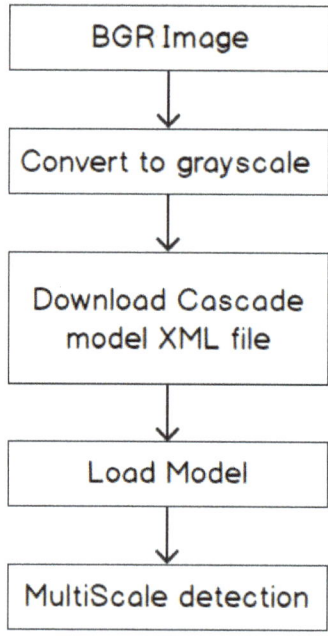

Figure 5.4: Using Haar Cascades for face detection

The following image shows the sample results for different parameter values that have been obtained by performing face detection on *Figure 5.3* using the frontal face cascade classifier:

Figure 5.5: The result obtained by using the frontal face cascade classifier

The preceding result is obtained by using the frontal face cascade classifier with `scaleFactor = 1.2` and `minNeighbors = 9` parameters

Once the parameters have been revised to **scaleFactor = 1.2** and **minNeighbors = 5**, you will notice how the quality of the results obtained degrades. Thus, the results highly depend on the parameter values. The result for the **scaleFactor = 1.2** and **minNeighbors=5** parameter values would be as follows:

Figure 5.6: The result obtained with the parameters revised to scaleFactor = 1.2 and minNeighbors = 5

Now that we have understood the required functions for performing face detection using Haar Cascades, let's discuss an important concept about bounding boxes before we have a look at an exercise. The bounding box is nothing but a list of four values – **x**, **y**, **w**, and **h**. **x** and **y** denote the coordinates of the top-left corner of the bounding box. **w** and **h** denote the width and height of the bounding box, respectively. You can see a diagrammatic representation of this for the **cv2.rectangle(img, p1, p2, color)** syntax in the following diagram:

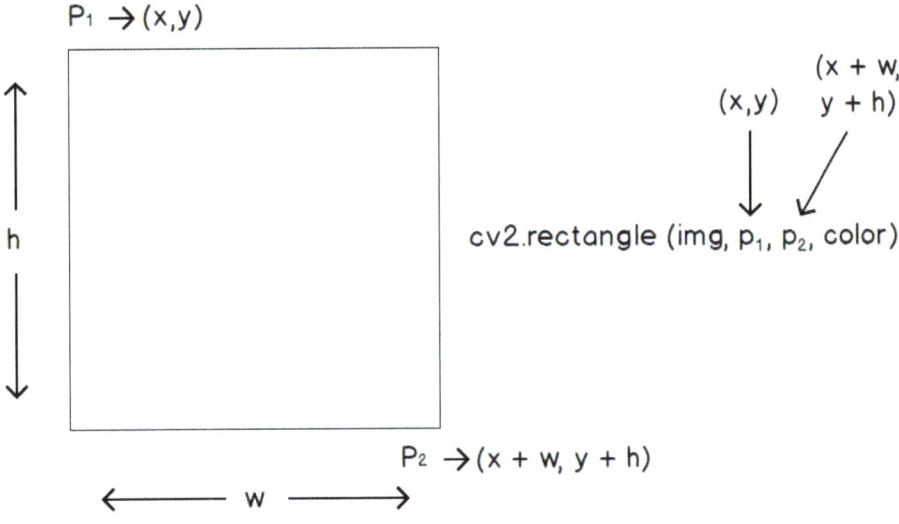

Figure 5.7: Plotting a bounding box using OpenCV's rectangle function

NOTE

Before we go ahead and start working on the exercises and activities in this chapter, please make sure you have completed all the installation steps and set up everything, as instructed in the *Preface*.

You can download the code files for the exercises and the activities in this chapter using the following link: https://packt.live/2WfR4dZ.

EXERCISE 5.01: FACE DETECTION USING HAAR CASCADES

Let's consider a scenario where you must find out the number of students present in a class. One approach is to count the students manually by visiting the class every hour, but let's try and automate this process using computer vision and see whether we can succeed.

In this exercise, we will implement face detection using Haar Cascades. Then, by counting the detected faces, we will see if we can get the number of students present in the class. The observations should focus on the **false positives** and **false negatives** discovered. In our scenario, false positive means that a face was not present in the bounding box that the function generated as an output, while false negative means that a face was present, but the cascade failed to detect it.

Since this process does not require a lot of computation, it can be easily carried out at the edge. By edge, we mean that the computation can be carried out at or near the input device, which in our case can be a very simple Raspberry Pi camera that will take a photo every hour, considering that a class in school typically lasts for an hour. Let's consider the following as the input image:

Figure 5.8: Input image

> **NOTE**
>
> The above image can be found at https://packt.live/2BVgEO5.

Here are the steps we will follow:

1. Open a Jupyter notebook, create a new file named **Exercise5.01.ipynb**, and type out your code in this file.

2. Load the required libraries:

```
import cv2
import numpy as np
```

3. Specify the path of the input image:

> **NOTE**
>
> Before proceeding, ensure that you can change the path to the image (highlighted) based on where the image is saved in your system.

```
inputImagePath = "../data/face.jpg"
```

4. Specify the path for the Haar Cascade XML file:

> **NOTE**
>
> Before proceeding, ensure that you can change the path to the file (highlighted) based on where the file is saved in your system.

```
haarCascadePath = "../data/haarcascade_frontalface_default.xml"
```

5. Load the image using the **cv2.imread** function:

```
inputImage = cv2.imread(inputImagePath)
```

6. Convert the image from BGR mode into grayscale mode using the **cv2.cvtColor** function:

```
grayInputImage = cv2.cvtColor(inputImage, cv2.COLOR_BGR2GRAY)
```

7. Load the Haar Cascade:

```
haarCascade = cv2.CascadeClassifier(haarCascadePath)
```

8. Perform multi-scale face detection using the **detectMultiScale** function:

   ```
   detectedFaces = haarCascade.detectMultiScale(grayInputImage, 1.2, 1)
   ```

   Here, **1.2** is the **scaleFactor** and **1** is the **minNeighbors** argument.

9. Iterate over all the detected faces and plot a bounding box using the **cv2.rectangle** function:

   ```
   for face in detectedFaces:
       # Each face is a rectangle representing
       # the bounding box around the detected face
       x, y, w, h = face
       cv2.rectangle(inputImage, (x, y), (x+w, y+h), (0, 0, 255), 3)
   ```

   Here, **detectedFaces** is the list of bounding boxes covering the faces present in the image. Thus, the **for** loop used here iterates over each detected face's bounding box. Then, we used the **cv2.rectangle** function to plot the bounding box. For this, we passed the image that we want to plot the rectangle on (**inputImage**), the top left corner of the rectangle **(x, y)**, the bottom right corner of the rectangle **(x+w, y+h)**, the color of the rectangle red = **(0, 0, 255)**, and the line thickness of the rectangle (**3**) as arguments.

10. Use the **cv2.imshow** function to display the image:

    ```
    cv2.imshow("Faces Detected", inputImage)
    cv2.waitKey(0)
    cv2.destroyAllWindows()
    ```

    Here, the **cv2.imshow** function will display **inputImage** in a window with the provided name, **"Faces Detected"**. **cv2.waitKey** will wait for infinite time (since we have provided **0** as input) until a key is pressed. Finally, the **cv2.destroyAllWindows** function will close all the windows and the program will exit.

Running the preceding code will result in the following output:

Figure 5.9: Result obtained from face detection using Haar Cascade

You can also save the image using `cv2.imwrite` so that you can see the image later.

From our observations of the preceding resultant image, we can see that although we were able to detect most of the faces in the image, we still have two false positives and five false negatives.

> **NOTE**
>
> To access the source code for this specific section, please refer to https://packt.live/3dQwmHw.

260 | Face Processing in Image and Video

Before we proceed to the next section, think about solving the following challenges:

We used a scale factor of **1.2** and a minimum neighbors parameter of **3**. Try different values for these parameters and understand their effect on the obtained results. The observations are present in the bounding boxes that the function generated as an output. False negative means that a face was present but the cascade failed to detect it.

Notice that the following image has no false positives. All the bounding boxes that were detected have a face in them. However, the number of faces detected has also reduced significantly here:

Figure 5.10: Possible resultant image with less detection but no false positives

Compared to the results shown in the preceding image, this following image has detected all the faces in the image successfully, but the number of false positives has increased significantly:

Figure 5.11: Possible resultant image with more detection and more false positives

There is a very common challenge in detection- and tracking-related problems in computer vision called **occlusion**. Occlusion means that an object is partially or completely covered by another object in the image. Think of it like this – if you are trying to take a selfie but someone comes and stands in front of you, then your image will be **occluded** by the second person. Now, take two images with you and your friend so that in the first image, both of your faces should be clearly visible, and in the second image, your face should be partially covered by your friend's face. For both images, use the frontal face detection Haar Cascade that we used in this exercise and see if the cascade can detect the occluded face or not. Such tests are always performed for any detection-related models to understand how well suited they are for real-life scenarios.

To understand this better, let's have a look at an example. The following image shows a picture of a soldier and a dog:

Figure 5.12: Picture of a soldier and a dog

Notice how the dog has partially covered the soldier in this image. Now, we as humans are aware that there is a person behind the dog, but an object detection tool might fail if it has not been trained on similar images.

DETECTING PARTS OF THE FACE

In the previous section, we discussed face detection and its importance. We also focused on Haar Cascades and how they can be used for object detection problems. One thing that we mentioned but didn't cover in detail was that Haar Cascades can be extended to detect other objects, not just faces.

If you have another look at the methodology for training the cascade model, you will notice that the steps are very general and thus can be extended to any object detection problem. Unfortunately, this requires acquiring a huge number of images, dividing them into two categories based on the presence and absence of the object, and, finally, training the model, which is a computationally expensive process. It is important to note that we cannot use the same `haarcascade_frontalface_default.xml` file to detect other objects, along with detecting faces at the same time, considering that this model is only trained to detect faces.

This is where OpenCV comes into the picture. OpenCV's GitHub repository has an extensive variety of cascades for detecting objects such as cats, frontal faces, left eyes, right eyes, smiles, and so on. In this section, we will see how we can use those cascades to detect parts of the face, instead of the face itself.

First, let's consider a scenario where detecting parts of the face such as eyes can be important. Imagine that you are building a very basic piece of iris recognition-based biometric security software. The power of this idea can be understood by the fact that the iris of a human being is believed to be much more unique than fingerprints and thus, iris recognition serves as a better means of biometric security. Now, imagine that you have a camera in place at the main gate of your office. Provided that the camera has a high resolution, an eye cascade model will be able to create a bounding box around the eyes present in the video frame. The detected eye regions can then be further processed to carry out iris recognition.

Now, let's try to understand how we can use the various cascade models available online to carry out detection. The process will be very similar to what we have discussed before regarding face detection. Let's take a look:

1. To download the relevant model, we will visit OpenCV's GitHub repository at https://packt.live/2VBvtMG.

   > **NOTE**
   >
   > At the time of writing this chapter, models for detecting eyes, left eyes, right eyes, smiles, and so on were available in this GitHub repository.

2. Once the model has been downloaded, the rest of the process will remain the same. We will read the input image.

3. As before, the image will have to be converted into grayscale since the cascade model will work only for a grayscale image.

4. Next, we will load the cascade classifier model (XML file).

5. The only new step will be for multi-scale object detection.

The best part of using Haar Cascades is that you can use the same OpenCV functions every time. This allows us to create a function that will complete all of the preceding steps and return the list of bounding boxes.

Even though the steps that are to be followed remain the same, the importance of parameters increases extensively. After performing *Exercise 5.01, Face Detection Using Haar Cascades*, and the two challenges, you know that by changing the minimum number of **neighbors** and the **scale factor**, the performance of the cascade can be varied. Specifically, in cases where we are trying to detect smaller parts such as eyes, these factors become much more important. As a computer vision engineer, it will be your responsibility to make sure that proper experimentation has been carried out to come up with the parameters that give the best results. In the upcoming exercises, we will see how we can detect eyes in an image, and we will also carry out the required experimentation to get the best set of parameters.

EXERCISE 5.02: EYE DETECTION USING CASCADES

In this exercise, we will perform eye detection using the `haarcascade_eye.xml` cascade model.

> **NOTE**
>
> The image used in this exercise can be found at https://packt.live/3dO3RKx.

Perform the following steps to complete this exercise:

1. First, open a Jupyter notebook and create a new file called **Exercise5.02.ipynb**. Then, type out your code in this file.

2. Import the required libraries:

   ```
   import cv2
   import numpy as np
   ```

3. Load **matplotlib**, a common Python module for visualization:

   > **NOTE**
   >
   > For instructions on how to install the **matplotlib** module, please refer to the *Preface*.

   ```
   import matplotlib.pyplot as plt
   ```

4. Specify the path to the input image and the **haarcascade_eye.xml** file:

   > **NOTE**
   >
   > Before proceeding, ensure that you can change the path to the image and file (highlighted) based on where they are saved in the system.

   ```
   inputImagePath = "../data/eyes.jpeg"
   haarCascadePath = "../data/haarcascade_eye.xml"
   ```

5. Create a custom function to carry out all the steps that we covered in *Exercise 5.01, Face Detection Using Haar Cascades*:

   ```
   def detectionUsingCascades(imageFile, cascadeFile):
       """ This is a custom function which is responsible
       for carrying out object detection using cascade model.
       The function takes the cascade filename and the image
       filename as the input and returns the list of
       bounding boxes around the detected object instances."""
   ```

   In the next seven steps, we will create this function. This function will also be used in the upcoming exercises and activities.

6. Load the image using the **cv2.imread** function as *Step 1* of the **detectionUsingCascades** custom function:

   ```
   # Step 1 - Load the image
   image = cv2.imread(imageFile)
   ```

7. Convert the image into grayscale as *Step 2* of the **detectionUsingCascades** custom function:

   ```
   # Step 2 Convert the image from BGR to Grayscale
   gray = cv2.cvtColor(image, cv2.COLOR_BGR2GRAY)
   ```

An important point to note here is that you should not convert an image if it is already in grayscale.

8. Load the cascade classifier as *Step 3* of the **detectionUsingCascades** custom function:

    ```
    # Step 3 - Load the cascade
    haarCascade = cv2.CascadeClassifier(cascadeFile)
    ```

    As we saw previously, the **cv2.CascadeClassifier** function takes in the path to the XML file as input. This path is provided to our custom function, **detectionUsingCascades**, as an input argument.

9. Perform multi-scale detection as *Step 4* of the **detectionUsingCascades** custom function:

    ```
    # Step 4 - Perform multi-scale detection
    detectedObjects = haarCascade.detectMultiScale(gray, 1.2, 2)
    ```

    Here, it's important to note that **detectedObjects** is nothing but a list of bounding boxes. Also, note that we have hardcoded the values of the parameters – **minNeighbors** and **scaleFactor**. You can either replace these values with better values that you might have obtained from the additional challenges after *Exercise 5.01*, *Face Detection Using Haar Cascades*, or you can manually tune these values to obtain the best results. A more finished application would display a set of results for different parameter values and will let the user choose one of the multiple results.

10. Draw the bounding boxes using the **cv2.rectangle** function as *Step 5* of the **detectionUsingCascades** custom function:

    ```
    # Step 5 - Draw bounding boxes
    for bbox in detectedObjects:
        # Each bbox is a rectangle representing
        # the bounding box around the detected object
        x, y, w, h = bbox
        cv2.rectangle(image, (x, y), (x+w, y+h), \
                      (0, 0, 255), 3)
    ```

11. Display the image with bounding boxes drawn over the detected objects as *Step 6* of the **detectionUsingCascades** custom function:

    ```
    # Step 6 - Display the output
    cv2.imshow("Object Detection", image)
    cv2.waitKey(0)
    cv2.destroyAllWindows()
    ```

12. Return the bounding boxes as *Step 7* of the **detectionUsingCascades** custom function:

    ```
    # Step 7 - Return the bounding boxes
    return detectedObjects
    ```

> **NOTE**
>
> Depending on your application, the returned bounding boxes can be used for further processing as well.
>
> Now, download the **haarcascade_eye.xml** file from https://packt.live/3dO3RKx and use the custom function to carry out eye detection on the same input image shown in *Figure 5.04*:

> **NOTE**
>
> Before proceeding, ensure that you can change the path to the image and file (highlighted) based on where they are saved in the system.

```
eyeDetection = detectionUsingCascades("../data/eyes.jpeg", \
                "../data/haarcascade_eye.xml")
```

Running the preceding code results in the following output:

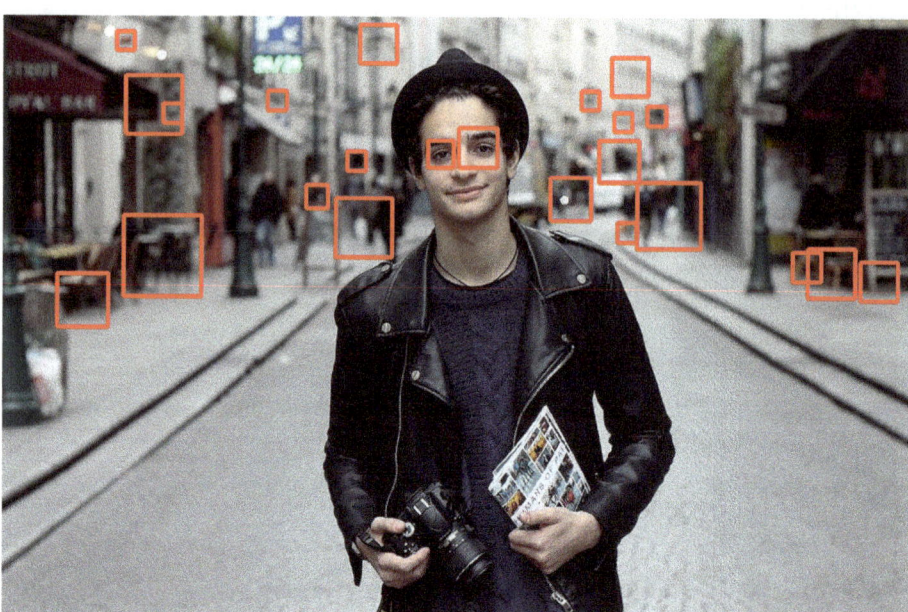

Figure 5.13: Result obtained using eye detection

As you can see, though the eyes have been detected, there are many false positives in the preceding result. We will carry out the necessary experimentation to improve the results in *Activity 5.01, Eye Detection Using Multiple Cascades*.

> **NOTE**
>
> To access the source code for this specific section, please refer to https://packt.live/2VBrXlj.

You might be wondering why we imported the `matplotlib` library but never used it. The reason for this is that we want to leave it up to you as a challenge. The finished application that we discussed in *Step 9* will use the `matplotlib` library to create subplots that will have the results for different values of the parameters in a multi-scale detection function. For more details on how to create subplots using `matplotlib`, you can refer to the documentation at https://matplotlib.org/3.1.1/api/_as_gen/matplotlib.pyplot.subplot.html. Here are a few of the results after varying the `minNeighbors` parameter:

Introduction to Haar Cascades | 269

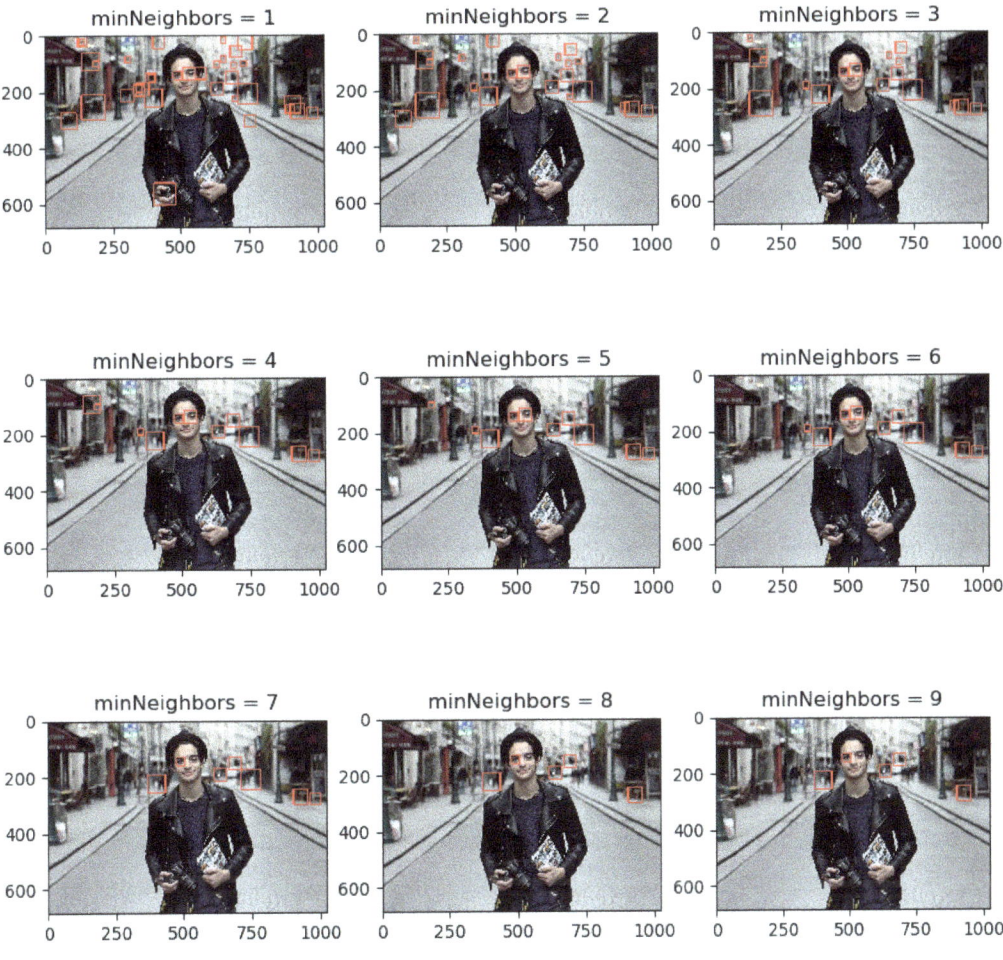

Figure 5.14: Results after the minNeighbors parameter has been varied

Notice how the results change as the **minNeighbors** parameter is varied. This experimentation can be carried out by using subplots offered by the **matplotlib** module.

At this point, let's ponder upon the downsides of cascades. We discussed how cascades are very well suited for operations performed at the edge. Coupled with devices such as Nvidia Jetson, these cascade models can give results with good accuracy and at a very fast speed (that is, for videos, they will have a high FPS).

In such a case, why are industries and researchers moving toward deep learning-based object detection solutions? Deep learning models take a much longer time to train, require more amount of data, and consume a lot of memory for training and sometimes during inference as well. Inference can be thought of as using the model that you got to obtain the results of operations such as object detection. One of the reasons for this is the higher accuracy that is offered by the deep learning models.

> **NOTE**
>
> As an interesting case study, you can go over this emotion recognition model (https://github.com/oarriaga/face_classification) based on Haar Classifiers, which can give real-time results and does not use deep learning.

CLUBBING CASCADES FOR MULTIPLE OBJECT DETECTION

So far, we've discussed how we can use cascade classifiers to perform object detection. We also discussed the importance of tuning the parameters to obtain better results. We also saw that one cascade model can be used to detect only one object.

If we consider the same iris recognition-based biometric verification case study, we can easily understand that a simple eye detection model is perhaps not the best approach. The reason behind this is that the eye is such a small object to detect that the model will need to check a lot of small regions to confirm if there is an eye present. We saw the same in the results shown in *Figure 5.14*. A lot of eyes were detected outside the face region. On the other hand, if you consider a face detection model, it will be larger than an eye and thus, we will be able to detect the object (face, in this case) much sooner and with higher accuracy. A better approach in this scenario would be to use a combined model. The steps for the same have been detailed here:

1. To detect faces in the image, we will use the face detection model.
2. To detect the eye region, we will crop out each face one by one and for each of those cropped faces, we will use the eye detection model.
3. To carry out iris recognition, for every detected eye, we can carry out the necessary steps to crop and process the iris region accordingly.

The advantage of using such an approach is that the eye detection model will need to search for an eye in a comparatively smaller region. This was possible only because we were able to use the logic that an eye will be present only in a face. Think of other applications where we can use a similar approach.

In the following activity, we will be resizing the image, which can be done using the **cv2.resize** function. The function is very easy to use and takes two arguments – the image you want to resize and the new size you want. So, if we have an image called "**img**" and we want it to resize to **100×100** pixels, we can use **newImg = cv2.resize(img, (100,100))**.

ACTIVITY 5.01: EYE DETECTION USING MULTIPLE CASCADES

We discussed the potential advantage we could get if we used multiple cascades to solve the eye detection problem, instead of directly using the eye cascade. In this activity, you will implement the multiple cascade approach and carry out a comparison between the results obtained using the multiple cascade approach and single cascade approach.

Perform the following steps to complete this activity:

1. Create a new Jupyter notebook file called **Activity5.01.ipynb**.
2. Import the necessary Python libraries/modules.
3. Download the frontal face cascade and the eye cascade from https://packt.live/3dO3RKx.
4. Read the input image and convert it into grayscale.
5. Load the frontal face and eye cascade models.
6. Use the frontal face cascade model to detect the faces present in the image. Recall that the multi-scale detection will give a list of bounding boxes around the faces as an output.
7. Now, iterate over each face that was detected and use the bounding box coordinates to crop the face. Recall that a bounding box is a list of four values – **x**, **y**, **w**, and **h**. Here, **(x, y)** is the coordinate of the top-left corner of the bounding box, **w** is the width, and **h** is the height of the bounding box.

8. For each cropped face, use the eye cascade model to detect the eyes. You can crop the face using simple NumPy array slicing. For example, if a box has the values `[x, y, w, h]`, it can be cropped by using `image[y:y+h, x:x+w]`:

Figure 5.15: The cropped face obtained using image slicing

9. Finally, display the input image with bounding boxes drawn around eyes and faces.

10. Now, use the eye cascade model directly to detect eyes in the input image. (This was carried out in *Exercise 5.02, Eye Detection Using Cascades*.)

11. For both cases, perform the necessary experimentation to tune the parameters – `scaleFactor` and `minNeighbors`. While you use a manual approach to tune the parameters by running the entire program again and again by changing the parameters, a much better approach would be to iterate over a range of values for both the parameters. As we discussed previously, `scaleFactor` can be kept as `1.2` as an initial value. Typically, `minNeighbors` parameter values can be varied from `1` to `9`. As an optional exercise, you can try mentioning the number of false positives and false negatives obtained for each experiment that's conducted. Such quantitative figures make your results more convincing.

12. Once you have obtained the best possible results using both approaches, try to find any differences in the results obtained using both approaches.

13. As an additional exercise to strengthen your understanding of this topic, you can try scaling up the cropped faces (use bilinear interpolation while using the `cv2.resize` function) to increase the size of the eyes and see if it improves the performance.

The following image shows the expected result of this activity:

Figure 5.16: Result obtained by using multiple cascades

Notice how the results improved significantly compared to the results shown in *Figure 5.13*.

> **NOTE**
>
> The solution for this activity can be found on page 497.

One of the problems that deep learning is used to solve is the emotion recognition problem. The basic idea is that, given an image of a face, you have to detect whether the person is happy, sad, angry, or neutral. Think about where this can be used; for example, to build a chatbot that is based on the emotion of the user to interact with them and try improving their mood. Now, of course, it won't be possible for us to solve this problem entirely using cascades because cascades are trained for detection-related problems and not recognition. But we can definitely get a head-start using the cascades. Let's see how. One of the key points that differentiate the various kinds of emotions is a **smile**. If a person is smiling, we can safely assume that the person is happy. Similarly, the absence of a smile can mean that the person is either angry, sad, or neutral. Notice how we have managed to simplify the **recognition** problem into a **detection** problem – specifically, a **smile detection** problem.

By the end of this activity, you have understood the details of cascades and their implementation for object detection problems. You have also learned how to use multiple cascades instead of just one to improve the performance. You have also carried out various experiments to understand the effect of parameters on the quality of the results obtained. Finally, you will have played around with the scale of the image using OpenCV's resize function to see if the performance improves if the image's size is increased.

ACTIVITY 5.02: SMILE DETECTION USING HAAR CASCADES

In the previous sections, we discussed using cascades and clubbing them to perform eye detection, face detection, and other similar object detection tasks. Now, in this activity, we will consider another object detection problem statement – smile detection.

Luckily for us, there is already a cascade model available in the OpenCV GitHub repository that we can use for our task.

> **NOTE**
>
> The image and files used in the activity can be found at https://packt.live/3dO3RKx.

We will break down this activity into two parts. The first will be detecting smiles directly using a smile detection cascade, while the second will be clubbing the frontal face cascade and the smile detection cascade.

Perform the following steps to complete this activity:

1. Create a new Jupyter notebook file called **Activity5.02.ipynb**.
2. Import the required libraries.
3. Next, read the input image and convert it into grayscale (if required).
4. Download the **haarcascade_smile.xml** file from https://packt.live/3dO3RKx.
5. Create a variable specifying the path to the smile detection cascade XML file. (You will need this variable to load the cascade.)
6. Now, load the cascade using the **cv2.CascadeClassifier** function. (You will have to provide the path to the XML file as the input argument. This will be the variable you created in the previous step.)

7. Next, carry out multi-scale detection, as seen in *Exercise 5.02, Eye Detection Using Cascades*.

8. Create a bounding box around the detected smile and display the final image.

This completes the first part of the activity. Now, use the two cascade classifiers by clubbing them. Perform the following steps to do so:

1. Continue in the same file, that is, `Activity5.02.ipynb`.

2. Create a variable specifying the path of the frontal face cascade classifier model.

3. Load the classifier by providing the path obtained in the previous step as an argument to the `cv2.CascadeClassifier` function.

4. Carry out multi-scale detection to detect the faces in the image.

5. Iterate over each face and crop it. Use the steps provided in *Activity 5.01, Eye Detection Using Multiple Cascades*, as a reference.

6. For each face, use the smile detection cascade classifier to detect a smile.

7. Create a bounding box over the detected smile.

   This way, you should have a bounding box over each face and for each face, a bounding box over the detected smile, as seen here:

Figure 5.17: Result obtained by using the frontal face and smile cascade classifiers

> **NOTE**
>
> The solution for this activity can be found on page 500.

This completes the entire activity. What do you think? Can you use this method to recognize whether a person is happy or not? The biggest drawback of the technique we discussed in this activity is that it assumes that a smile is enough to tell us whether a person is happy or not. There are other parts of the face such as facial muscles that come into the picture in regard to recognizing the emotion of a face. That's why deep learning models surpass simple machine learning models in these cases as they can take into account a large number of features to predict the output.

We can also replace the input image with a video stream from a webcam or a recorded video file using OpenCV functions such as `cv2.VideoCapture(0)` and `cv2.VideoCapture('recorded_video_file_name')`, respectively. Once you take inputs from the video stream, you will notice the performance of the technique as you make different faces. We will complete an interesting activity using a webcam at the end of this chapter.

In later chapters, we will discuss facial recognition using deep learning techniques where we will revisit this problem statement of emotion recognition and compare the performance of the cascade-based technique and deep learning-based technique.

In the next section, we will discuss a very important technique called GrabCut and how we can use it for skin detection problems.

GRABCUT TECHNIQUE

Before we go into the details of this amazing technique, we need to understand the term **Image Segmentation**, which will form the very basis of **GrabCut**. In layman's terms, an image segmentation task comprises dividing the pixels into different categories based on the class they belong to. Here, the classes we are looking for are just the background and foreground. This means that we want to segment the image by deciding what region of the image is part of the background and what region of the image is part of the foreground.

NOTE

In deep learning, image segmentation is a very big topic and is comprised of segmenting an image based on its classes, sometimes based on semantics, and so on. For now, we can consider GrabCut a simple background removal technique, though technically it can also be considered as an image segmentation technique with only two primary classes in mind – background and foreground. It's also important to note here that GrabCut uses not two, but four labels that we will discuss later in this section. But the basic idea will stay the same – remove the background from the image.

For example, consider the following image:

Figure 5.18: Person in a garden doing some exercise

If we focus on the foreground and background, we can say that the background is the garden or park in which the person is exercising, and the foreground consists of the person in the picture. Taking this into account, if we go ahead and perform image segmentation, we will end up with the following image:

Figure 5.19: Result after performing image segmentation

Notice how only the foreground is present in the image. The entire background has been removed and replaced with white pixels (or white background).

Hopefully, by now, some ideas have started popping up in your mind regarding where you can use this technique. For Adobe Photoshop users, a similar technique is used quite extensively in photo editing. The idea of green screen used in movies is also based on a similar concept.

Now, let's go into the details of the GrabCut technique. GrabCut was a technique for **interactive foreground extraction** (a fancy phrase meaning that a user can interactively extract foreground) proposed by Carsten Rother, Vladimir Kolmogorov, and Andrew Blake in their paper *"GrabCut" – Interactive Foreground Extraction using Iterated Graph Cuts* (http://pages.cs.wisc.edu/~dyer/cs534-fall11/papers/grabcut-rother.pdf). The algorithm used color, contrast, and user input to come up with excellent results.

The power of the algorithm can be understood by the number of applications built on top of it. While it might seem a manual procedure at first glance, the fact is that the manual part comes into the picture only to provide finishing to the results or to improve them further. Also, refer to the paper *Automatic Skin Lesion Segmentation Using GrabCut in HSV Color Space* by Fakrul Islam Tushar (https://arxiv.org/ftp/arxiv/papers/1810/1810.00871.pdf). This work is a very good example that shows the true power of GrabCut. The smart use of the HSV color space (refer to *Chapter 2, Common Operations When Working with Images*, for details regarding color spaces) instead of the usual BGR color space along with GrabCut's iterative procedure managed to segment (or separate) the skin lesion. As the paper mentions, this is the first step toward a digital and automatic diagnosis of skin cancer.

In the coming sections, we will use GrabCut to segment a person from an image for skin detection, which will then be used in the final activity of this chapter to create various face-based filters.

> **NOTE**
>
> Saving an image is an important step in computer vision and can be carried out using the `cv2.imwrite` function. The function takes two arguments – the name of the file and the image we want to save.

To carry out GrabCut, we will use the `cv2.grabCut()` function. The function syntax looks as follows:

```
outputmask, bgModel, fgModel = cv2.grabCut(image, mask, \
            rect, bgModel, fgModel, iterCount, mode)
```

Let's understand each of these arguments:

image is the input image we want to carry out GrabCut on. **iterCount** is the number of iterations of GrabCut you want to run. **bgModel** and **fgModel** are just two temporary arrays used by the algorithm and are simply two arrays of size **(1, 65)**.

There are four possible modes, but we will only focus on the two important ones, that is, **mask** and **rect**. You can read about the rest of them at https://docs.opencv.org/3.4/d7/d1b/group__imgproc__misc.html#gaf8b5832ba85e59fc7a98a2afd034e558. Let's take a look at them:

cv2.GC_INIT_WITH_RECT mode means that you are initializing your GrabCut algorithm using a rectangular mask specified by the **rect** argument. **cv2.GC_INIT_WITH_MASK** mode means that you are initializing your GrabCut algorithm with a general mask (a binary image, basically) specified by the **mask** input argument. **mask** has the same shape as the image provided as input.

The function returns three values. Out of the three values, the only important one for us is **outputmask**, which is a grayscale image of the same size as the image provided as input but has only the following four possible pixel values:

Pixel value 0 or **cv2.GC_BGD** means that the pixel surely belongs to the background. **Pixel value 1** or **cv2.GC_FGD** means that the pixel surely belongs to the foreground. **Pixel value 2** or **cv2.GC_PR_BGD** means that the pixel probably belongs to the background. **Pixel value 3** or **cv2.GC_PR_FGD** means that the pixel probably belongs to the foreground.

One more important function that we will be using extensively is OpenCV's **cv2.selectROI** function to select a rectangular region of interest when we are using the **cv2.GC_INIT_WITH_RECT** mode. The usage of the function is very simple. You can directly use your mouse to select the top-left corner and then drag your cursor to select the rectangle. Once done, you can press the *Spacebar* key to get the ROI.

Similarly, we use a **Sketcher** class, which has been provided in the upcoming exercises, to modify the mask interactively. We can download the class directly from OpenCV samples: https://github.com/opencv/opencv/blob/master/samples/python/common.py. You don't need to understand the details of this class, except that it lets us sketch an image using our mouse. The **Sketcher** class can be called as follows:

```
sketch = Sketcher('image', [img, mask], \
                  lambda : ((255,0,0), 255))
```

Here, **'image'** is the name of the display window that will pop up, and **img** and **mask** are two images that will be displayed – one where you will be sketching (**img**) and the other where the result is displayed automatically (**mask**). **(255,0,0)** means that when we sketch on **img**, we will be drawing in blue.

The **255** value means that whatever we draw on **img** will be displayed on **mask** in white. You can guess here that **img** is a BGR image, whereas **mask** is a grayscale image.

Once you have obtained the output mask, you can multiply it with your input image, which will give you only the foreground and remove the background from the image.

We will go into details of this topic with the help of a couple of exercises and then finally take up skin detection as an activity.

EXERCISE 5.03: HUMAN BODY SEGMENTATION USING GRABCUT WITH RECTANGULAR MASK

In this exercise, we will carry out foreground extraction to segment a person from the input image. We will use the **cv2.GC_INIT_WITH_MASK** mode in this problem with *Figure 5.20* as the input image.

> **NOTE**
>
> The image can be found at https://packt.live/2VzJ1bv.

Perform the following steps to complete this exercise:

1. Open your Jupyter notebook, create a new file called **Exercise5.03.ipynb**, and write your code in this file.

2. Import the required libraries:

```
import cv2
import numpy as np
```

3. Read the input image:

> **NOTE**
>
> Before proceeding, ensure that you can change the path to the image (highlighted) based on where the image is saved in your system.

```
# Read image
img = cv2.imread("../data/person.jpg")
```

4. Create a copy of the image:

```
imgCopy = img.copy()
```

5. Create a mask of the same size (width and height) as the original image and initialize **mask** with zeros using NumPy's **np.zeros** function. Also, specify the data type as an unsigned 8-bit integer, as shown in the following code:

```
# Create a mask
mask = np.zeros(img.shape[:2], np.uint8)
```

6. Create two temporary arrays:

```
# Temporary arrays
bgdModel = np.zeros((1,65),np.float64)
fgdModel = np.zeros((1,65),np.float64)
```

7. Use OpenCV's **cv2.selectROI** function to select the region of interest:

```
# Select ROI
rect = cv2.selectROI(img)
```

8. Draw the rectangle over the ROI and save the image:

```
# Draw rectangle
x,y,w,h = rect
cv2.rectangle(imgCopy, (x, y), (x+w, y+h), \
              (0, 0, 255), 3)
cv2.imwrite("roi.png",imgCopy)
```

Running the preceding code will result in the following output:

Figure 5.20: Region of interest selected using the cv2.SelectROI function

9. Next, we will perform GrabCut using **cv2.grabCut**:

```
# Perform grabcut
cv2.grabCut(img,mask,rect,bgdModel,fgdModel,5,cv2.GC_INIT_WITH_RECT)
```

10. Include all the confirmed and probable background pixels in the background by setting these pixels as **0**:

```
mask2 = np.where((mask==2)|(mask==0),0,1).astype('uint8')
```

11. Display both **mask** and **mask2** using the following code:

```
mask2 = np.where((mask==2)|(mask==0),0,1).astype('uint8')
cv2.imshow("Mask",mask*80)
cv2.imshow("Mask2",mask2*255)
cv2.imwrite("mask.png",mask*80)
cv2.imwrite("mask2.png",mask2*255)
cv2.waitKey(0)
```

The following shows the mask that was obtained after using the GrabCut operation:

Figure 5.21: Mask obtained after using the GrabCut operation

The brighter regions correspond to the confirmed foreground pixels, while the grey regions correspond to the probably background pixels. Black pixels are guaranteed to belong to the background.

The following image shows the mask that was obtained after combining the confirmed background pixels and probable background pixels:

Figure 5.22: Mask obtained after combining the confirmed background pixels and probable background pixels

The white region corresponds to the foreground. Unfortunately, the mask does not provide us with much information.

12. Multiply the mask with the image to obtain only the foreground using the following code:

```
img = img*mask2[:,:,np.newaxis]
```

We are introducing a new axis to simply match the shapes.

13. Display and save the final image using the following code:

```
cv2.imwrite("grabcut-result.png",img)
cv2.imshow("Image",img)
cv2.waitKey(0)
cv2.destroyAllWindows()
```

Running the preceding code displays the following output:

Figure 5.23: Foreground detected using GrabCut

Notice that though the result is of good quality, the shoes of the person are detected as the background.

> **NOTE**
>
> To access the source code for this specific section, please refer to https://packt.live/31AjOBE.

In *Figure 5.23*, we noticed how the GrabCut technique provided good results, but it still needed a finer touch. At this stage, there are two ways to improve the result:

- You can try changing the `iterCount` parameter in the `cv2.grabCut` function to see if the quality of the result improves:

Figure 5.24: Result obtained after running GrabCut for only one iteration

Notice that the shoes are still in the background and there is some additional grass that has been detected as part of the foreground.

- You can try selecting different ROIs to see if there is any difference in the result obtained.

But as you will see after experimenting, using these suggestions, there is no significant improvement in the result obtained using GrabCut. One option that can be used at this step is to modify the mask manually and provide it again as input to the GrabCut process for better results. We will see how to do that in the next exercise.

EXERCISE 5.04: HUMAN BODY SEGMENTATION USING MASK AND ROI

In the previous exercise, we saw how GrabCut was very easy to use but that the results needed some improvement. In this exercise, we will start by using a rectangular ROI mask to perform segmentation with GrabCut and then add some finishing touches to the mask to obtain a better result. We will be using *Figure 5.3* as the input image.

> **NOTE**
>
> The image can be found at https://packt.live/2BT1dFZ.

Perform the following steps to complete this exercise:

1. Open your Jupyter notebook, create a new file called **Exercise5.04.ipynb**, and start writing your code in this file.

2. Import the required libraries:

```
import cv2
import numpy as np
```

3. Read the input image:

> **NOTE**
>
> Before proceeding, ensure that you can change the path to the image (highlighted) based on where the image is saved in your system.

```
# Read image
img = cv2.imread("../data/grabcut.jpg")
```

4. Create a copy of the input image:

```
imgCopy = img.copy()
```

5. Create a mask of the same size as the input image, initialize it with zeros, and use unsigned 8-bit integers, as shown in the following code:

```
# Create a mask
mask = np.zeros(img.shape[:2], np.uint8)
```

6. Create two temporary arrays:

```
# Temporary arrays
bgdModel = np.zeros((1,65),np.float64)
fgdModel = np.zeros((1,65),np.float64)
```

7. Select a rectangular ROI to perform a crude GrabCut operation:

```
# Select ROI
rect = cv2.selectROI(img)
```

Running the preceding code will display the following output:

Figure 5.25: The GUI for selecting ROI using the cv2.selectROI function

8. Draw the rectangle on the copy of the input image and save it:

```
# Draw rectangle
x,y,w,h = rect
cv2.rectangle(imgCopy, (x, y), (x+w, y+h), \
              (0, 0, 255), 3)
cv2.imwrite("roi.png",imgCopy)
```

The following image will be saved as **roi.png**:

Figure 5.26: Final rectangular ROI selected for GrabCut

9. Next, perform the GrabCut operation, as we saw in the previous exercise:

```
# Perform grabcut
cv2.grabCut(img,mask,rect,bgdModel,fgdModel,5,cv2.GC_INIT_WITH_RECT)
```

10. Use both confirmed background and probable background pixels in the output mask as background pixels:

```
mask2 = np.where((mask==2)|(mask==0),0,1).astype('uint8')
```

11. Display and save both masks:

```
cv2.imshow("Mask",mask*80)
cv2.imshow("Mask2",mask2*255)
cv2.imwrite("mask.png",mask*80)
cv2.imwrite("mask2.png",mask2*255)
cv2.waitKey(0)
cv2.destroyAllWindows()
```

Note that we are multiplying **mask** by **80** since the pixel values in **mask** range only from **0** to **3** (inclusive) and thus won't be visible directly. Similarly, we are multiplying **mask2** by **255**.

The following image shows the mask obtained after performing GrabCut:

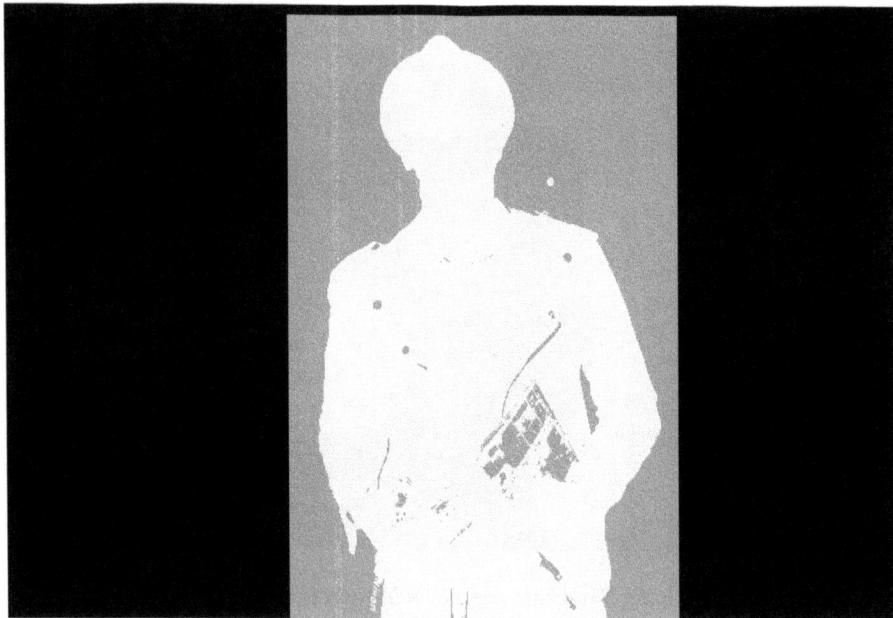

Figure 5.27: Mask obtained after performing the GrabCut operation

The brighter regions correspond to the confirmed foreground pixels, while the grey regions correspond to the probable background pixels. Black pixels are guaranteed to belong to the background.

The following image shows the mask that was obtained after clubbing the confirmed background and probable background pixels into the background pixels category:

Figure 5.28: Mask is obtained after clubbing the confirmed background

The preceding mask is obtained after clubbing the confirmed background and probable background pixels into the background pixels category.

Notice the white spot outside of the human body region and several black regions inside the foreground region. These patches will now need to be fixed.

12. Multiply the image with **mask2** to obtain the foreground region:

```
img = img*mask2[:,:,np.newaxis]
```

13. At this stage, create a copy of the mask and the image:

```
img_mask = img.copy()
mask2 = mask2*255
mask_copy = mask2.copy()
```

We are multiplying **mask2** by **255** because of the same reason mentioned in *Step 11*.

14. Use the **Sketcher** class provided by OpenCV for mouse handling. Using this class, modify the mask (that is, the foreground and background) with the help of your cursor:

```
# OpenCV Utility Class for Mouse Handling
class Sketcher:
    def __init__(self, windowname, dests, colors_func):
        self.prev_pt = None
        self.windowname = windowname
        self.dests = dests
        self.colors_func = colors_func
        self.dirty = False
        self.show()
        cv2.setMouseCallback(self.windowname, self.on_mouse)
```

We will be able to see the change in the result dynamically using this process. The preceding function, that is, __init__, initializes the **Sketcher** class object.

15. Use the **show** function to display the windows:

```
def show(self):
    cv2.imshow(self.windowname, self.dests[0])
    cv2.imshow(self.windowname + ": mask", self.dests[1])
```

16. Use the **on_mouse** function for mouse handling:

```
# onMouse function for Mouse Handling
def on_mouse(self, event, x, y, flags, param):
    pt = (x, y)
    if event == cv2.EVENT_LBUTTONDOWN:
        self.prev_pt = pt
    elif event == cv2.EVENT_LBUTTONUP:
        self.prev_pt = None
    if self.prev_pt and flags & cv2.EVENT_FLAG_LBUTTON:
        for dst, color in zip(self.dests, self.colors_func()):
            cv2.line(dst, self.prev_pt, pt, color, 5)
        self.dirty = True
        self.prev_pt = pt
        self.show()
```

17. Create a sketch using the **Sketcher** class, as shown here:

```
# Create sketch using OpenCV Utility Class: Sketcher
sketch = Sketcher('image', [img_mask, mask2], lambda : ((255,0,0),
255))
```

The parameters for the lambda function mean that any blue pixel (**255,0,0**) drawn in the image will be displayed as a white pixel (**255**) in the mask.

18. Next, create an infinite **while** loop since we want to modify the mask, and thus the results, for as long as we want. Use the **cv2.waitKey()** function to keep a record of the key that's been pressed:

```
while True:
    ch = cv2.waitKey()
    # Quit
    if ch == 27:
        print("exiting...")
        cv2.imwrite("img_mask_grabcut.png",img_mask)
        cv2.imwrite("mask_grabcut.png",mask2)
        break
    # Reset
    elif ch == ord('r'):
        print("resetting...")
        img_mask = img.copy()
        mask2 = mask_copy.copy()
        sketch = Sketcher('image', [img_mask, mask2], \
                  lambda : ((255,0,0), 255))
        sketch.show()
```

If the *Esc* key is pressed, we will break out of the **while** loop after saving both the mask and the image and if the *R* key is pressed, we will reset the mask and the image.

19. Use blue **(255,0,0)** to select the foreground and red (**0,0,255**) to select the background:

```
# Change to background
elif ch == ord('b'):
    print("drawing background...")
    sketch = Sketcher('image', [img_mask, mask2], \
                    lambda : ((0,0,255), 0))
    sketch.show()

# Change to foreground
elif ch == ord('f'):
    print("drawing foreground...")
    sketch = Sketcher('image', [img_mask, mask2], \
                    lambda : ((255,0,0), 255))
    sketch.show()
```

A background pixel in **mask** will be marked as black (**0**). By pressing *F* and *B*, you can switch between the foreground and background selections, respectively.

The result that's obtained by using the preceding code can be seen here:

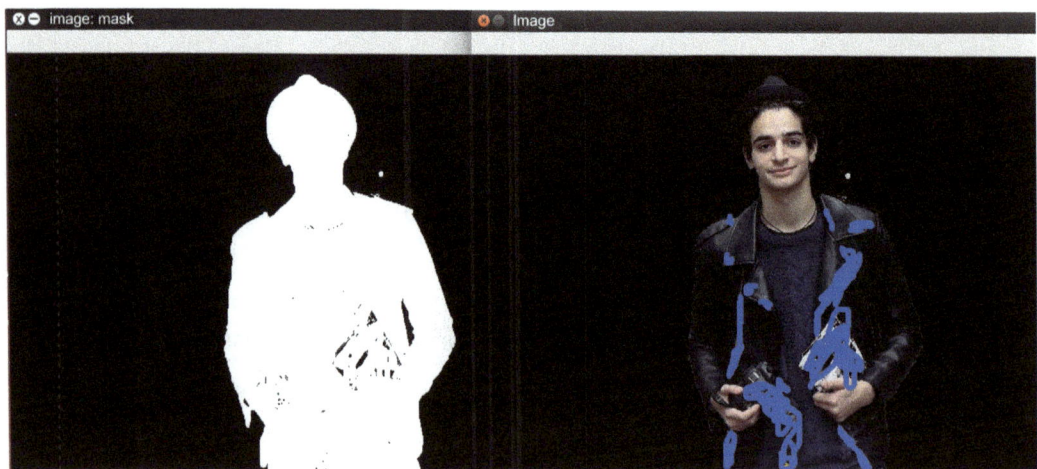

Figure 5.29: The foreground region correction marked in blue

The foreground region correction is marked in blue as this covers the black patches that were present in the mask in the foreground region shown in *Figure 5.27*. The regions marked in blue will now become a part of the foreground in the mask, which means they will now become white in the mask, as can be seen in the preceding image on the left.

The following image shows the background correction marked in red:

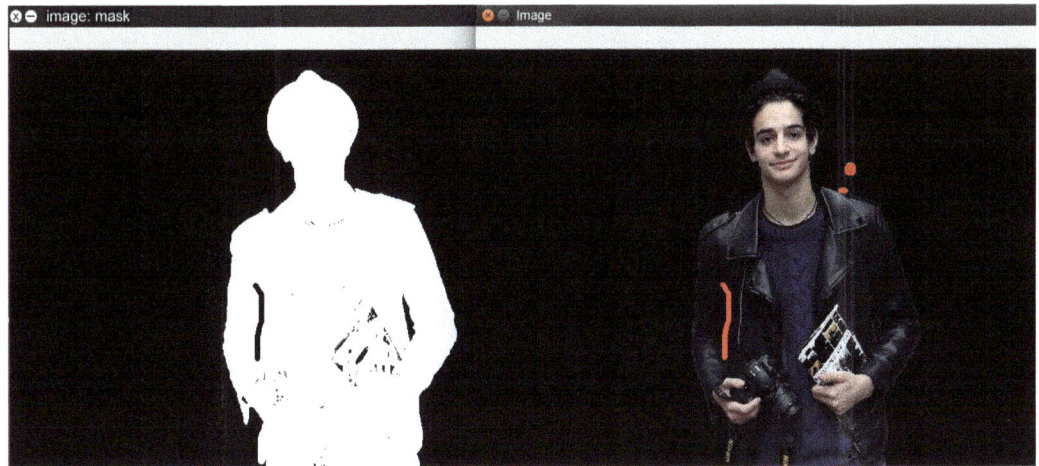

Figure 5.30: The background correction marked in red

Similar to the foreground correction, the background correction is marked in red. The marked region comprises the areas that truly belonged to the background but was labeled as foreground by GrabCut. These regions will be labeled as black in the mask. The same can be seen in the image on the left-hand side.

20. Carry out the GrabCut operation using the revised mask, in case any other key is pressed:

```
else:
    print("performing grabcut...")
    mask2 = mask2//255
    cv2.grabCut(img,mask2,None,
    bgdModel,fgdModel,5,cv2.GC_INIT_WITH_MASK)
    mask2 = np.where((mask2==2)|(mask2==0),0,1)\
            .astype('uint8')
    img_mask = img*mask2[:,:,np.newaxis]
    mask2 = mask2*255
    print("switching bank to foreground...")
    sketch = Sketcher('image', [img_mask, mask2], \
            lambda : ((255,0,0), 255))
    sketch.show()
```

Note that we are now using the **cv2.GC_INIT_WITH_MASK** mode since we are providing the mask instead of the ROI rectangle.

21. Close all the open windows and exit, once you are out of the **while** loop:

```
cv2.destroyAllWindows()
```

The following image shows the final mask that was obtained after foreground and background correction:

Figure 5.31: The final mask that was obtained after foreground and background correction

22. Compare the following final corrected image with the mask displayed in *Figure 5.27* to understand the corrections made:

Figure 5.32: The final image obtained after the corrections

This was a long exercise. Let's quickly summarize what we did in this exercise. First, we performed a crude GrabCut using a rectangular ROI. The mask was then corrected with the help of the **Sketcher** class. The corrections were broken down into two parts – foreground correction and background correction. The mask that was obtained was then used again by GrabCut to obtain the final foreground image.

> **NOTE**
>
> To access the source code for this specific section, please refer to https://packt.live/2NMv6dS.

ACTIVITY 5.03: SKIN SEGMENTATION USING GRABCUT

You must have watched those *James Bond*-type movies where a villain would put on a mask of the hero and commit a crime. Well, so far, we have learned how to detect faces and noses, but how do we crop out skin regions? In this activity, you will implement skin segmentation using the GrabCut technique. We will use the same code that we developed in the previous exercise and modify it slightly to create a frontal face cascade classifier to detect faces. You can also use the previous exercise as a reference to modify the mask using the mouse.

Let's learn how to carry out this activity:

1. Open your Jupyter notebook and create a new file called `Activity5.03.ipynb`.

2. Import the OpenCV and NumPy modules.

3. Add the code for the `Sketcher` class first as we will be using it in this activity to revise the mask obtained by GrabCut.

4. Read the input image, that is, `"grabcut.jpg"`, using the `cv2.imread` function. This is the same image that we used in the previous exercise.

    > **NOTE**
    >
    > The image can be found at https://packt.live/2BT1dFZ.

5. Convert the BGR image into grayscale mode using the `cv2.cvtColor` function.

6. Now, load the Haar Cascade for frontal face detection using the `cv2.CascadeClassifier` function.

7. Use the cascade classifier to detect a face in the image using the `detectMultiScale` function. Select the parameter values so that only the correct face is detected.

8. Use the bounding box obtained in the preceding step to crop the face from the image.

9. Now, resize the image and increase its dimensions to 3x. For this, you can use the `cv2.resize` function. We will now be working on this resized image.

10. Use the `cv2.grabCut` function to obtain the mask. The sample result can be seen here:

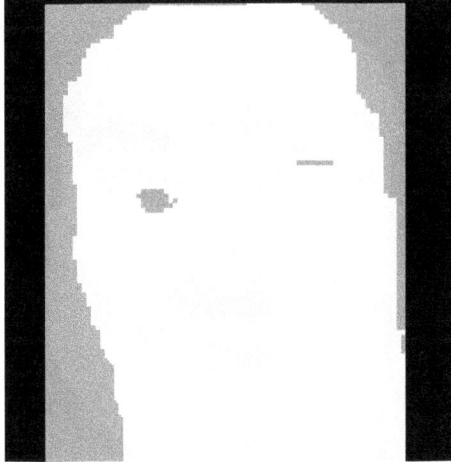

Figure 5.33: The mask obtained using the GrabCut function

11. Now, club the confirmed background and probable background pixels into confirmed background pixels. The following image shows the sample result:

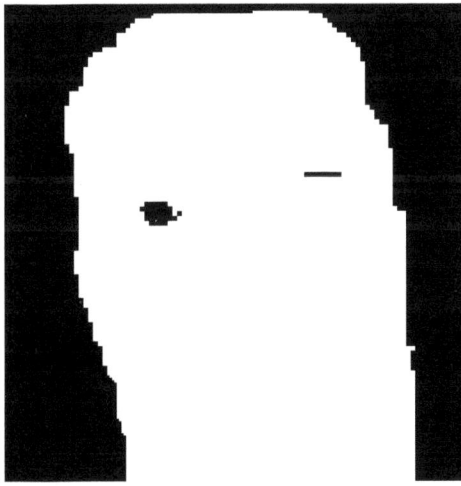

Figure 5.34: The mask obtained after clubbing the background and probable background pixels

12. Next, multiply the new mask with the image to obtain the foreground.

13. Now, use the steps described in the previous exercise to obtain a revised mask and finally the skin region:

Figure 5.35: Mask obtained after foreground and background correction

The segmented skin region will look as follows:

Figure 5.36: Segmented skin region

> **NOTE**
>
> The solution for this activity can be found on page 503.

To summarize, in this activity, we used frontal face cascade classifiers along with the GrabCut technique to perform skin segmentation.

Now, it's time to move on to the final activity of this chapter. In this activity, we will build an emoji filter, similar to Snapchat filters, using the techniques we studied in this chapter.

ACTIVITY 5.04: EMOJI FILTER

Whether it is Instagram, Facebook, or Snapchat, filters have always been in demand. In this activity, we will build an emoji filter using the techniques we have discussed in this chapter. With the help of the previously discussed **VideoCapture** function, use the input from your webcam.

Here are the steps you need to follow:

1. Open your Jupyter notebook and create a new file called **Activity5.04.ipynb**.

2. Import the OpenCV and NumPy modules.

3. Write the function responsible for applying the emoji filter. To start, copy the **detectionUsingCascades** function that we wrote in *Exercise 5.02, Eye Detection Using Cascades*. Modify this function slightly for this case study activity.

4. Change the function name to **emojiFilter** and add a new argument to it called **emojiFile**, which is the path to the emoji image file.

5. Inside the **emojiFilter** function, in *Step 1* (similar to how we did for the custom function earlier), add the code to read the emoji image using the **cv2.imread** function. Make sure you pass **-1** as the second argument since we also want to read the alpha channel in the image, which is responsible for transparency in images. We will also be providing the image directly as the first argument instead of the image file name, so remove the **cv2.imread** function call for reading the image.

6. There is no modification required in *Steps 2* and *3*.

7. Add an if-else condition to check if any object is even detected using the Haar Cascade in *Step 4*. If no object is detected, return **None**.

8. In *Step 5*, resize our emoji so that it matches the size of the detected face. The size of the detected face is nothing but **(w, h)**. For resizing, use the **cv2.resize** function, as we did previously.

9. Once the emoji has been resized, we will need to replace the face with the emoji. This is where the alpha channel comes into the picture. Our image has 100% transparency for the background and 0% transparency for the actual emoji region. The 100% transparency region will have a value of 0 (black region) in the alpha channel. You can use the following code as a reference:

```
(image[y:y+h, x:x+w])[np.where(emoji_resized[:,:,3]!=0)] = \
(emoji_resized[:,:,:3])[np.where(emoji_resized[:,:,3]!=0)]
```

The output is as follows:

Figure 5.37: An emoji that we are going to use in our emoji filter

> **NOTE**
>
> The image can be found at https://packt.live/3evTz1L.

The alpha channel of the emoji will look as follows:

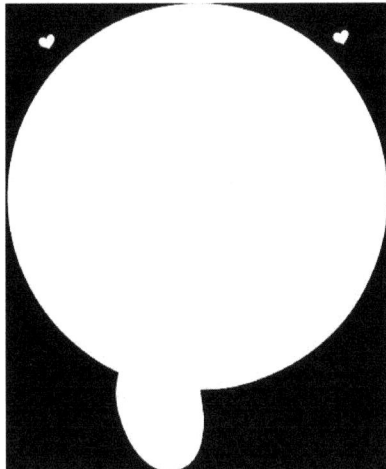

Figure 5.38: Alpha channel of the emoji we are using

10. We will exploit the fact mentioned in *Step 9* to overwrite only regions of the face corresponding to the non-transparent regions of the face with those non-transparent pixels.

11. Also, revise the function to return the final image instead of detected objects.

12. Now, since we are going to take input from a webcam, we will have to write some code for starting the webcam, capturing the frames, and then displaying the output.

13. Create a **VideoCapture** object using **cv2.VideoCapture**. Pass the only required argument as **0** since we want to take input from the webcam. If you wanted to provide input from a video file, you could have provided the filename. Name the **VideoCapture** object **cap**.

14. Now, create a **while** loop and in the loop, use **cap.read()** to capture a frame from the webcam. This returns two values – a return value signifying whether the frame capture was successful or not, and the captured frame.

15. If the return value from **cap.read()** was **True**, meaning that the frame was read successfully, pass it to the **emojiFilter** function and display the output using the **cv2.imshow** and **cv2.waitKey** functions. Provide a non-zero value as an argument to the **cv2.waitKey** function, since we don't want to wait infinitely long for the next frame to be displayed. Use a value of **10** or **20**:

Figure 5.39: Sample output of the emoji filter on a single frame

> **NOTE**
>
> The output shown in the preceding image has been generated using a pre-recorded video and not the webcam. We would recommend that you use your webcam to generate interesting results and only treat the preceding image as a reference.

Similar outputs will be generated for each frame of the video.

> **NOTE**
>
> The solution for this activity can be found on page 509.

That's it. You have prepared your own emoji filter. Of course, it's not a very good-looking filter, but it's a start. You can now make this much more interesting by switching between different emojis randomly or based on the presence of a smile. Switching between happy and sad emojis based on smile detection can help you build your own emotion detector emoji filter. Similarly, you can add other transparent images to the frames to give some nice makeup to your emoji. Once you have gathered and practiced the basic idea, the sky is the limit for you.

SUMMARY

In this chapter, we started off by seeing why the face is considered an important source that's used to identify different emotions, features, and so on. We also discussed why we cannot have a simple face-matching algorithm because of the large variety of faces available to us. We then approached the topic of Haar Cascades and built several applications using one or multiple cascade classifiers. We also talked in detail about the importance of experimentation when it comes to obtaining a set of arguments that will give us the best results using cascade classifiers. After that, we discussed the GrabCut technique and built a human body segmentation application using a rectangular ROI mask. After that, we modified the code to allow the user to correct the mask to obtain better results. We then covered how to carry out skin segmentation using GrabCut and finally ended the chapter with the emoji filter, which can be run on an input video taken from a webcam or a video file.

From the next chapter onward, we will start looking at the advanced topics of computer vision. The first topic we will be covering is object tracking, which will allow the user to track the motion of a selected object in a video. We will also discuss the various trackers available in OpenCV and discuss the differences between their performance.

6

OBJECT TRACKING

OVERVIEW

This chapter aims to show you how to track an object of your choice in an input video stream using various object trackers shipped along with the OpenCV library. We will start by understanding the basic idea behind object tracking and its applications in real life. Then, we will move on and look at the various filters present in OpenCV, understanding their algorithms, as well as their advantages and disadvantages. Finally, we will also have a look at another popular library – Dlib – and learn about the object trackers provided by it.

By the end of this chapter, you will be able to create your own tracking application where you will use cascade classifiers to track a human throughout a video using trackers such as Meanshift, CAMShift, and more. You will also carry out a comparison between trackers (Dlib-based object trackers and OpenCV object tracking API) in this chapter's activity.

INTRODUCTION

In the previous chapter, we discussed Haar Cascades and how to use them for object detection problems. We also used a variety of cascade classifiers for problems such as eye detection, people counters, and others. In this chapter, we will take this one step further and learn how to track the objects that have been detected. This tracking can be done for a pre-recorded video or a live video stream.

Object tracking is quite easy to understand. You have an object that either you have selected manually or was selected by some other algorithm, and now you want to track the object. The beauty of object tracking comes into the picture when we understand the difficulty of carrying out this very simple task in different scenarios. To understand this better, think of a scenario where you are with your friend in a busy market. Among a huge crowd, you need to detect and recognize your friend the entire time, and also need to keep track of where he or she is so that you don't end up losing them. Now, try to think of the steps that your brain is carrying out to solve this problem. First, your brain needs to coordinate with your eyes to detect humans among a large variety of objects. Then, you need to recognize your friend among many other humans, and then you need to track them continuously. There will be times when your friend will be away from you, in which case he or she will appear smaller. Similarly, sometimes, your friend will be obscured by other objects, so you must track them again once they are completely or partially visible. Now that you have understood the idea behind object tracking, you can think of scenarios where object trackers will come in handy. Whether you want to track a suspect in a video stream or track an endangered animal in a national park, such trackers can make your job easy. However, it's important to understand which tracker is suited for a specific task – how do they perform when the object you're tracking moves out of the frame and then comes back? How do they behave when the object is obscured? How much processing power do they take? How fast are they? All these questions form the base of the comparison of trackers.

In this chapter, we will start by going over the various trackers offered by the OpenCV library. We will investigate the details of each of these trackers, understanding how they work and what their advantages and disadvantages are. The chapter will consist of exercises detailing the use of trackers. We will end this chapter with an activity where we will use the cascade classifiers that we discussed in the previous chapter, along with the object trackers that we will discuss in this chapter, to create a complete object tracking application that can take input from a video stream or webcam.

NAÏVE TRACKER

A quick disclaimer before we start this section: there is no such thing as a naïve tracker; it's just a term that we are using to describe a very basic object tracker.

Object tracking is a very popular task, so there is a huge range of deep learning-based solutions that can give very good results. Later in this chapter, we will go over those object trackers, but for now, let's start with a very basic object tracker that is based on pure and simple image processing.

The idea that we are going to target here is that if there is a clear distinction between the color of the object that you want to track and the background, then ideally, you can use basic image processing tasks such as thresholding and color spaces to track the object. Of course, this will only work when there is only one instance of the object that you want to track. This is because two similarly colored instances will both end up appearing in the tracking – which we certainly don't want.

Let's look at a frame capture from a video. We'll be performing object tracking on this image in *Exercise 6.01, Object Tracking Using Basic Image Processing*:

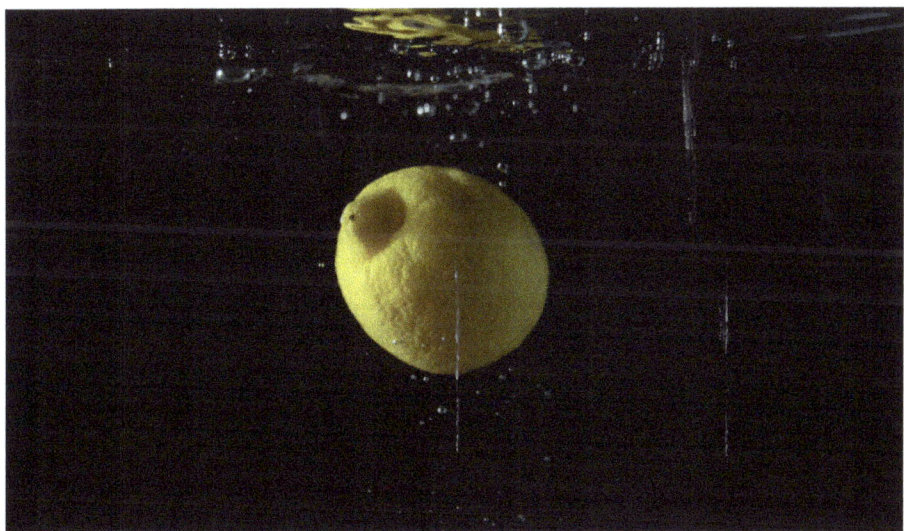

Figure 6.1: A frame capture showing a lemon falling in the water

> **NOTE**
>
> The clear distinction between the color of the lemon and the background makes it easy for us to track this object.

310 | Object Tracking

If you look at the preceding figure, you will notice that the lemon has a distinct yellow color compared to the black background and white bubbles. This clear difference in the color of the object and the background helps us in using simple thresholding techniques that can then be used to detect the lemon.

One more thing to note here is that while we are calling it object tracking, it's more like object detection being carried out on every frame of the video. But since there is a moving object here (in a video), we are discussing it in this chapter and not in the chapter regarding object detection.

Now, let's try to understand the concept behind this naïve tracker. The preceding image is an RGB image, which means that it has three channels – red, green, and blue. A very low value of all the three channels results in colors close to black, whereas values close to 255 for the three channels result in colors close to white. The yellow color of the lemon will lie somewhere in this range. Since we need to consider the effect of lighting, we cannot use just one RGB color value to denote the color of the lemon. So, we will start by finding out a range of pixel values that show only the lemon, and the rest of the background is converted into black. This process is nothing but thresholding. Once we have figured out that range, we can apply it to every frame of the video, separate the lemon in the video, and remove everything else. We are looking for an image that looks as follows:

Figure 6.2: The ideal result

In the preceding figure, we have separated the object (lemon) and replaced the entire background with black. Whether we can achieve this result using simple thresholding is a different question, though.

Now, let's get some hands-on experience with this technique by completing the following exercise.

EXERCISE 6.01: OBJECT TRACKING USING BASIC IMAGE PROCESSING

In this exercise, we will be using a frame from a video of a lemon being dropped into water to test object tracking (considering the lemon as the object to be tracked) based on the naïve tracker algorithm/technique. We will be using the basic color space conversion that we used in *Chapter 2, Common Operations When Working with Images*.

> **NOTE**
>
> The video can be downloaded from https://packt.live/2YPKhck.

Perform the following steps:

1. Open your Jupyter Notebook and create a new file called **Exercise6.01.ipynb**. We will be writing our code in this file.

2. Import the OpenCV and NumPy modules:

    ```
    import cv2
    import numpy as np
    ```

3. Create a **VideoCapture** object for the input video called **lemon.mp4**:

    > **NOTE**
    >
    > Before proceeding, ensure that you can change the path to the video (highlighted) based on where the video is saved in your system.

    ```
    # Create a VideoCapture Object
    cap = cv2.VideoCapture("../data/lemon.mp4")
    ```

4. Check whether the video was opened successfully or not:

```
if cap.isOpened() == False:
    print("Error opening video")
```

5. Create three windows. These will be used to display our input frame, our output frame, and the mask. Note that we are using the **cv2.WINDOW_NORMAL** flag so that we can resize the window as required. We have done this because our input video is in HD and thus the full-sized frame won't fit on our screen completely:

```
cv2.namedWindow("Input Video", cv2.WINDOW_NORMAL)
cv2.namedWindow("Output Video", cv2.WINDOW_NORMAL)
cv2.namedWindow("Hue", cv2.WINDOW_NORMAL)
```

6. Create an infinite **while** loop for reading frames, processing them, and displaying the output frames:

```
while True:
```

7. Capture a frame from the video using **cap.read()**:

```
    # capture frame
    ret, frame = cap.read()
```

8. Check whether the video has ended or not. This is equivalent to checking whether the frame was read successfully or not since, in a video that has finished, you can't read a new frame:

```
    if ret == False:
        break
```

9. Now comes the main part. Convert our frame that was in the BGR color space into the HSV color space. Here, H stands for hue, S stands for saturation, and V stands for value. We are doing this so that we can easily use the hue channel for thresholding:

```
    # Convert frame to HSV
    hsv = cv2.cvtColor(frame, cv2.COLOR_BGR2HSV)
```

10. Next, obtain the hue channel and apply thresholding. Set all the pixels that have a pixel value of more than **40** to **0** and all those that have a pixel value of less than or equal to **40** to **1**:

```
# Obtain Hue channel
h = hsv[:,:,0]
# Apply thresholding
hCopy = h.copy()
h[hCopy>40] = 0
h[hCopy<=40] = 1
```

The output is as follows:

Figure 6.3: Hue channel of a frame of the video

See how the lemon is a darker color, whereas the background is comparatively lighter. That's why we are setting pixels with values more than **40** (background) to **0** and those with values less than or equal to **40** (lemon) to **1**.

> **NOTE**
> We arrive at the value **40** by trial and error.

314 | Object Tracking

11. Now that we have obtained the binary mask we were looking for, we will use it to keep only the lemon region in the video and replace everything else with black. Then, we will display the new mask (**h**), input the video frame, and output the frame subsequently:

    ```
    # Display frame
    cv2.imshow("Input Video", frame)
    cv2.imshow("Output Video", frame*h[:,:,np.newaxis])
    cv2.imshow("Hue", h*255)
    ```

12. We can quit the process in the middle of it being run if we press the *Esc* key:

    ```
    k = cv2.waitKey(10)
    if k == 27:
        break
    ```

13. Outside of the **while** loop, release the **VideoCapture** object and close all the open display windows:

    ```
    cap.release()
    cv2.destroyAllWindows()
    ```

    The following frames are obtained after running the program:

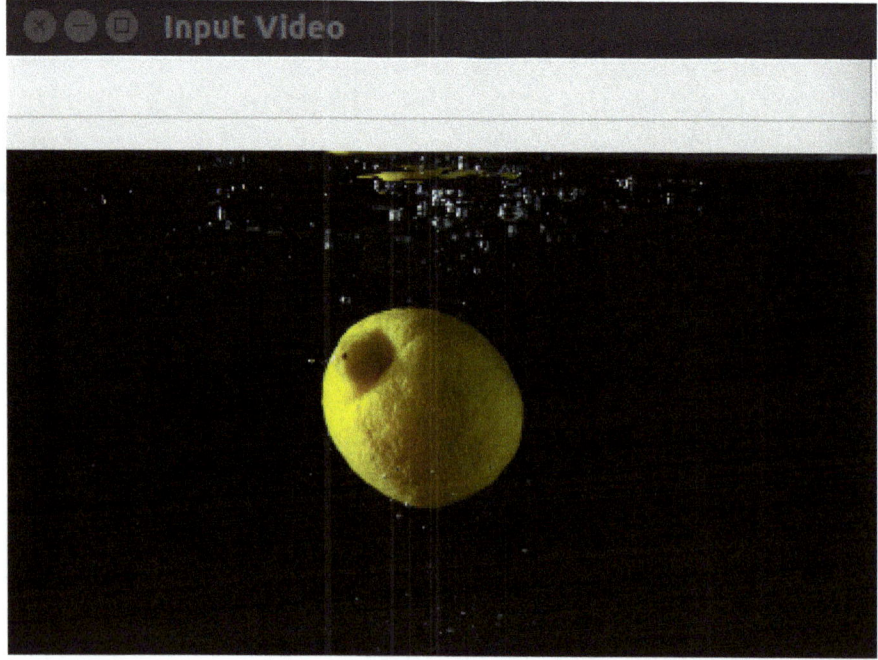

Figure 6.4: Input video frame

The following mask is obtained upon thresholding the hue channel, as performed in *Step 10*:

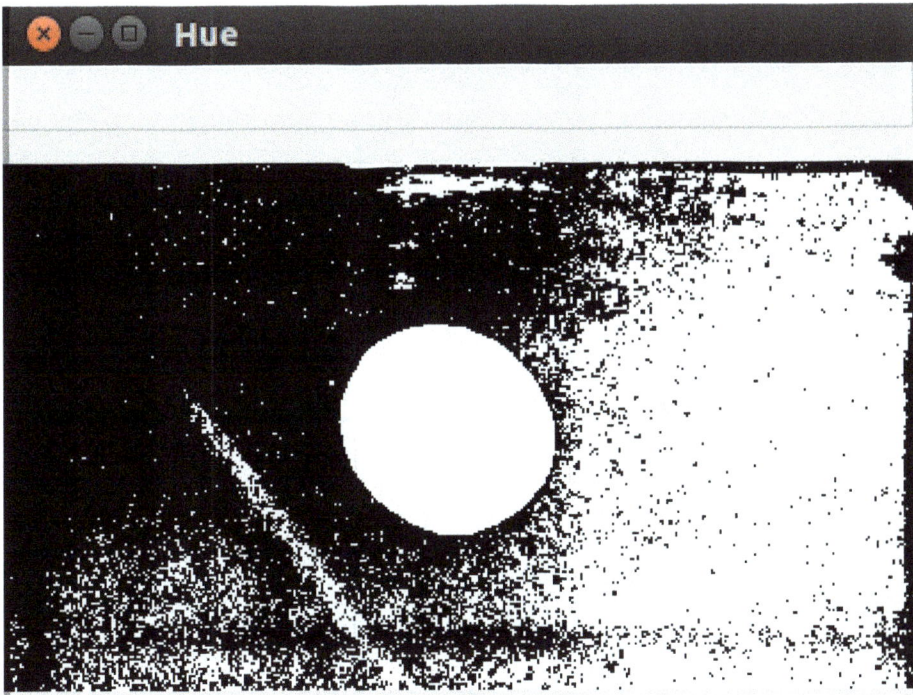

Figure 6.5: Mask obtained after thresholding the hue channel

316 | Object Tracking

You will obtain the following output frame after multiplying the mask by the input video frame, as performed in *Step 11*:

Figure 6.6: Output frame obtained after multiplying the mask by the input video frame

> **NOTE**
>
> If you look at the preceding figure, you will find that the majority of the bubbles have now been removed and that only the yellow-colored region is left.
>
> To access the source code for this specific section, please refer to https://packt.live/2Anr88s.

If you compare *Figure 6.4* and *Figure 6.6* carefully, you will notice that while the bubbles have been removed in the output frame, a reflection of the lemon has remained. This is because the reflection is also yellow. This kind of issue raises a major drawback with the basic image processing-based technique we discussed earlier. While such techniques are very fast and have very low computational power requirements, they cannot be used for real-life problems since coming up with one thresholding technique and one threshold value for all videos is impossible. That's why deep learning-based techniques are heavily used for object tracking problems as they don't struggle with the aforementioned issue.

NON-DEEP LEARNING-BASED OBJECT TRACKERS

In the previous section, we had a look at a very basic filter that used basic thresholding to detect objects. Tracking, in that case, was nothing but detecting an object in every frame. As we saw, this is not really object tracking, even though it might look like that, and will also fail in about every real-life scenario. In this section, we will have a look at some of the most commonly used non-deep learning-based object trackers that use motion modeling and some other techniques to track objects. We won't go into the code for these trackers here, but you can refer to the OpenCV documentation and tutorials for more information.

KALMAN FILTER – PREDICT AND UPDATE

Let's start with a tracker that also implements noise filtering and thus is referred to as a "filter." The Kalman filter was developed by Rudolf E. Kalman and Richard S. Bucy in the early 1960s. This work is more like a mechanical engineering concept rather than a core computer vision solution.

There are two main steps that are taken by the Kalman filter:

- Predict
- Update

Let's understand these steps one by one. First, let's get our terminology straight. We are going to extensively use the term **system** here. Now, a system, in this case, can be a ship, a person, or basically any object that you are interested in tracking. When we talk about the state of the system, we are referring to the state of motion of the system – the position and velocity of the object. Similarly, we will also talk about control inputs, which are responsible for changing the state of the system. This can be thought of as simple acceleration.

Now, recall the basic equation of the state of motion:

```
x= x + v * dt + 1/2 * a * dt²
```

Here, **x** is the current position of the object, **dt** is the time gap after which we want to find the position of the object again, **v** is the velocity of the object, and **a** is the acceleration.

If we know the velocity and acceleration of the object, we can **predict** the position of the object at any time step. This is the basic idea behind the **predict** step.

Unfortunately, in real-life scenarios, an object will change its position, velocity, and acceleration, or maybe even go out of the frame and then come back again. In such instances, it becomes impossible to predict the object's position. That's where the **update** step comes into the picture: whenever you encounter new information about the state of the object, you use it to initialize the **predict** step and then continue from there.

So, if we were to look at a basic flow diagram of this process, it would look as follows:

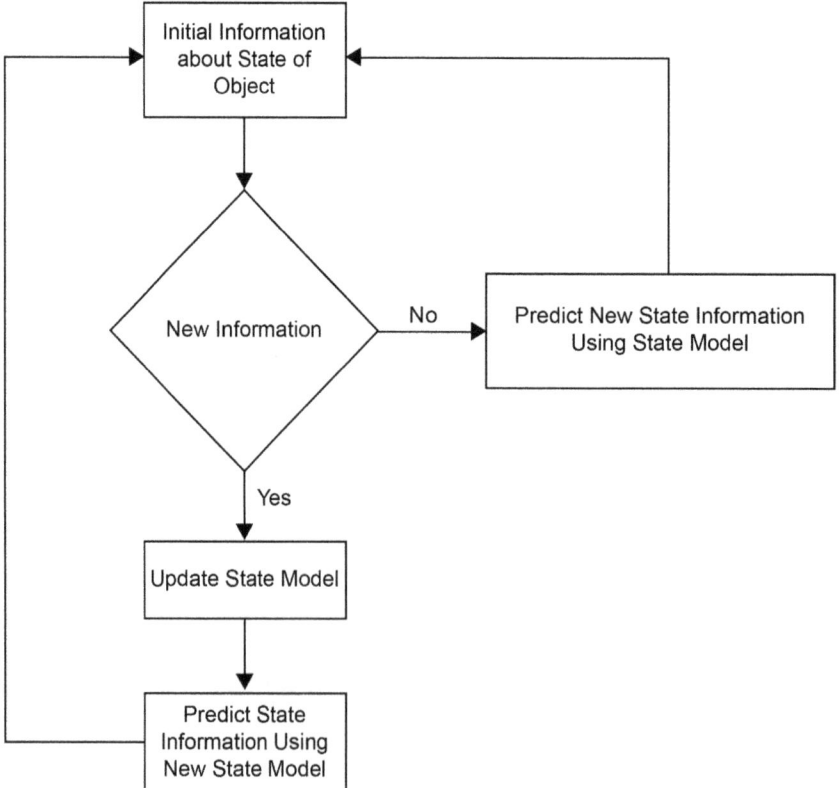

Figure 6.7: Kalman filter flowchart

Next, let's have a look at another interesting way of object tracking using a non-deep learning-based filter.

MEANSHIFT – DENSITY SEEKING FILTER

The meanshift filter is intuitive and easy to understand. It focuses on a very simple principle – we will go where the majority goes. To understand this, have a look at the following diagram. Here, the meanshift tracker keeps on following the object to be tracked (dotted circle) and can identify and track the object based on the density of the points:

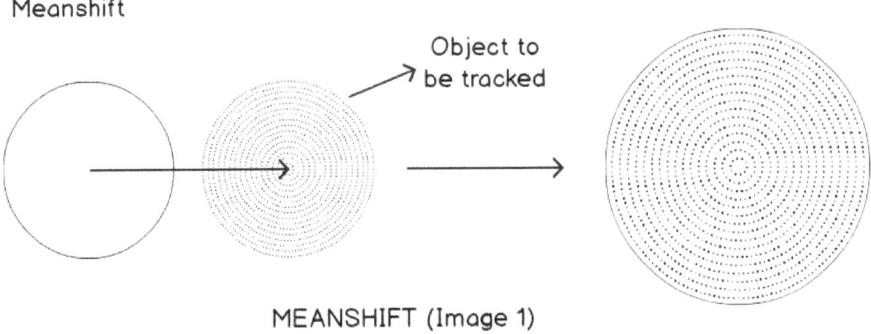

Figure 6.8: Meanshift's basic process

In the following figure, you can see that the meanshift filter is seeking the "mode" (even though it has the word "mean" in its name). This means that we are tracking where most of the points lie. This is exactly how it's able to track objects.

The object it wants to track looks like a highly dense scatterplot to the filter and when that dense scatterplot moves in a frame, the meanshift filter goes after it and thus ends up tracking the object:

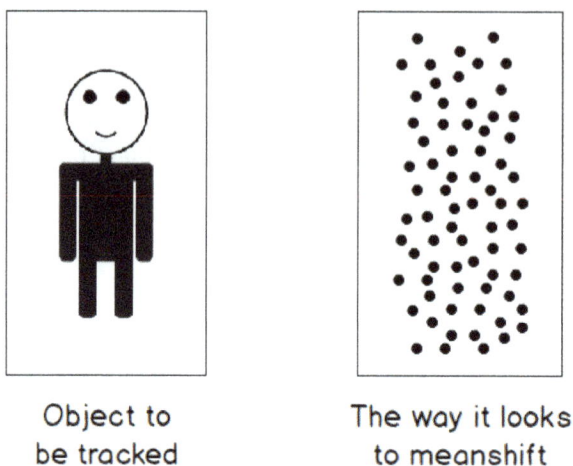

Object to   The way it looks
be tracked   to meanshift

Figure 6.9: The way an object to be tracked looks to meanshift

The advantage of this algorithm is that since it's just looking for the mode, it does not require a specific parameter to be tuned. This is easy to understand if you think back to Haar Cascades, where we had to specify **scaleFactor** and **minNeighbors** and the results would vary based on these parameters. Here, since we are tracking a mode, there are no parameters to vary.

So far, we've discussed meanshift purely in terms of scattered points, but practically, we are going to deal with objects present in images. That's why to go from a simple image of an object to a density function (mode), we need to carry out some pre-processing. One of the ways this can be done is as follows:

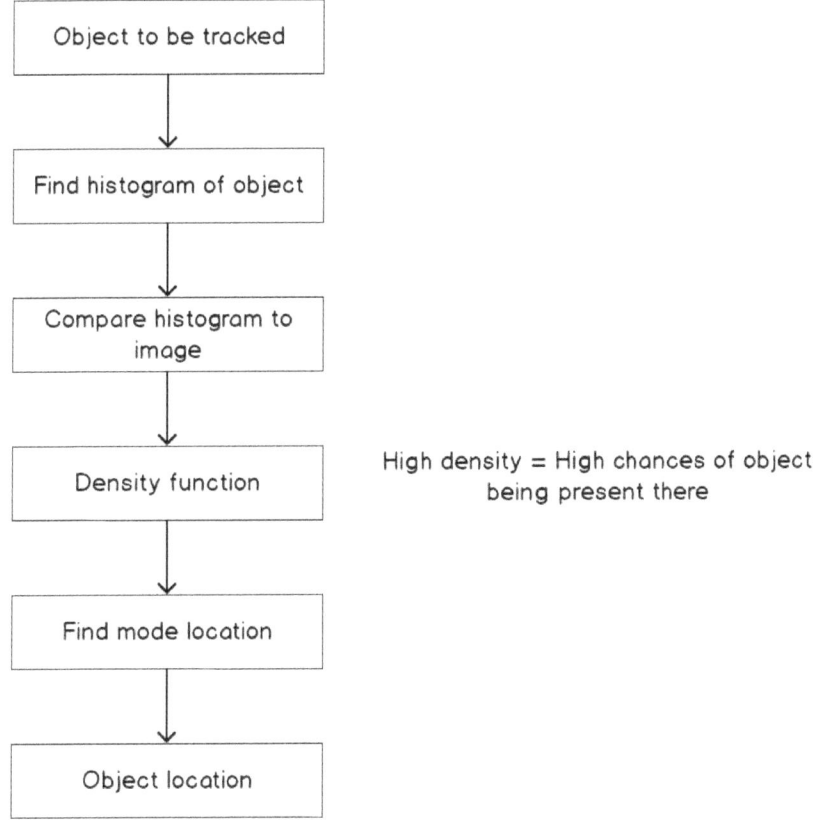

Figure 6.10: Converting an object into a density function

First, we obtain a histogram of the object, which we'll use to compare our object to a new image. The result we get from this is like a density function – a higher density will mean higher chances are that the object will be present in that location. So, if we want to find where an object is present, we just have to find the spot that has the highest density function value (this is the mode).

Finally, let's have a look at the modified version of the meanshift algorithm – continuously adaptive meanshift, also known as CAMshift.

CAMSHIFT – CONTINUOUSLY ADAPTIVE MEANSHIFT

CAMshift is a modification of meanshift that resolves the major downside of the parent algorithm – scale variation. Meanshift fails if the object's scale changes. To understand this with a real-life scenario, record yourself in a video using your mobile camera. Now, while recording, for some time, move toward the camera and then move away from the camera. In the recorded video, you will notice that as you move toward the camera, your image scale (or size) increases and that as you move away from your camera, the scale decreases. Meanshift does not give good results for this simply because it looks for the mode in a fixed-size window. This is where CAMshift comes into the picture. CAMshift uses the estimate of the mode to find out new window dimensions and then uses the new window to find the mode. This process continues until convergence is achieved, which means that the mode stops moving.

> **NOTE**
>
> One thing to note is that both meanshift and CAMshift depend a lot on convergence. They are iterative techniques and that's why, instead of finding out the object in one step, they come up with an intermediate answer that then changes further until the final location of the object is achieved. The primary difference between meanshift and CAMshift is that while meanshift keeps on going forward with these iterations while maintaining the same window size, CAMshift varies the window size, as per the intermediate result obtained, and thus can avoid any issues caused by scale variation. You can refer to the following figures to understand the difference.

Note how there is no window size variation in the following figure:

Figure 6.11: Meanshift output

Note the window size variation in the CAMshift output shown in the following figure:

Figure 6.12: CAMshift output

You might be wondering why there is no code present for meanshift, CAMshift, and the Kalman filter. The reason is that while these three techniques are quite well known and powerful, they lack the ease of use that comes with the deep learning-based models that we are going to cover next. There are implementations of these filters in OpenCV that we recommend you search for and try out yourself if you are interested in learning more.

THE OPENCV OBJECT TRACKING API

OpenCV provides support for a wide range of object trackers via its object tracking API. In this section, we will start by understanding the API, the important functions that we will be using in it, and the common trackers that we can use.

Let's start with the Tracking API. If you want to read about all the functions present in the Tracking API, you can refer to the documentation here: https://docs.opencv.org/4.2.0/d9/df8/group__tracking.html.

> **NOTE**
>
> The OpenCV object tracking API is part of `opencv-contrib`. If you are unable to run the code given in this section, this means you have not installed OpenCV properly. You can refer to the preface for the steps on how to install OpenCV.

Just by having a brief look at the documentation page, you will see that there is a huge number of trackers available in OpenCV. Some of the common ones are as follows:

- The `Boosting` tracker: `cv2.TrackerBoosting`
- The `CSRT` tracker: `cv2.TrackerCSRT`
- `GOTURN`: `cv2.TrackerGOTURN`
- `KCF (Kernelized Correlation Filter)`: `cv2.TrackerKCF`
- `Median Flow`: `cv2.TrackerMedianFlow`
- The `MIL` tracker: `cv2.TrackerMIL`
- The `MOSSE (Minimum Output Sum of Squared Error)` tracker: `cv2.TrackerMOSSE`
- The `TLD (Tracking, Learning, and Detection)` tracker: `cv2.TrackerTLD`

The interesting thing to note here is that all these trackers' functions start with **cv2.Tracker**. This is not really a coincidence. This is because all of these trackers have the same parent class – **cv::Tracker**. Because of good programming practice, all the trackers have been named in this format: **cv2.Tracker<TrackerName>**.

Now, before we move on and have a look at two of these trackers, let's have a look at the standard process of how to use them:

1. First, depending on our use case, we will decide what tracker we will use. This is the most important step and will decide the performance of your program. Typically, the choice of a tracker is made based upon the performance speed of the tracker, including occlusion, scale variation, and so on. We will talk about this in a bit more depth when we compare the trackers shortly.

   Once the tracker has been finalized, we can create the tracker object using the following function:

   ```
   tracker = cv2.Tracker<TRACKER_NAME>_create()
   ```

   If you are going to use the MIL tracker, for example, the preceding function will be **cv2.TrackerMIL_create()**.

2. Now, we need to tell our tracker what object we want to track. This information is provided by specifying the coordinates of the bounding box surrounding the object. Since it's difficult to specify the coordinates manually, the recommended way is to use the **cv2.SelectROI** function, which we used in the previous chapter to select the bounding box.

3. Next, we need to initialize our tracker. While initializing, the tracker needs two things – the bounding box and the frame in which the bounding box is specified. Typically, we use the first frame for this purpose. To initialize the tracker, use **tracker.init(image, bbox)**, where **image** is the frame, and **bbox** is the bounding box surrounding the object.

4. Finally, we can start iterating over each frame of the video that we want to track our object in and update the tracker using the **tracker.update(image)** function, where **image** is the current frame that we want to find the object in. The **update** function returns two values – a **success** flag and the bounding box. If the tracker is not able to find the object, the **success** flag will be **False**. If the object is found, then the bounding box tells us where the object is present in the image.

The preceding steps have been summarized in the following flowchart:

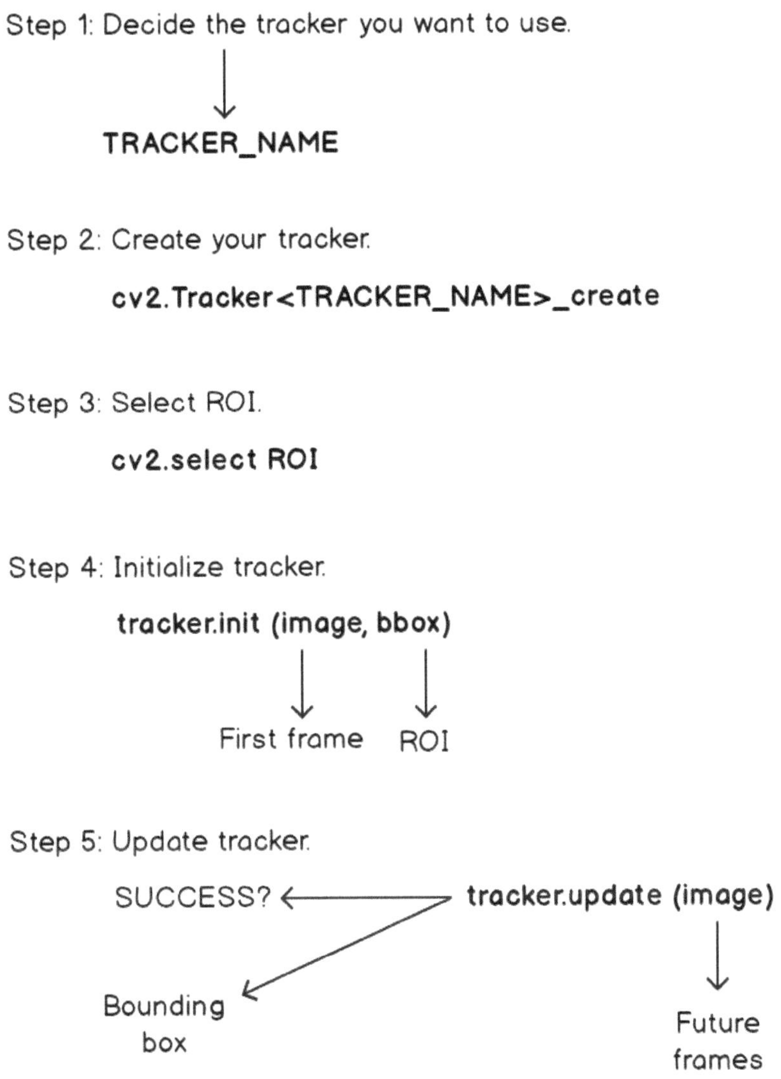

Step 1: Decide the tracker you want to use.

TRACKER_NAME

Step 2: Create your tracker.

cv2.Tracker<TRACKER_NAME>_create

Step 3: Select ROI.

cv2.select ROI

Step 4: Initialize tracker.

tracker.init (image, bbox)

First frame   ROI

Step 5: Update tracker.

SUCCESS?   tracker.update (image)

Bounding box   Future frames

Figure 6.13: Object tracking

OBJECT TRACKER SUMMARY

Let's summarize the pros and cons of the object trackers we mentioned previously. You can use this information to decide which tracker to choose, depending on your use case:

Tracker	Pros	Cons
Boosting	-	Average tracking performance. Note that this is a very old algorithm and thus is not recommended.
CSRT	High accuracy.	Low FPS (~20-30 FPS).
GOTURN	Comparatively faster than other trackers. Works fine, irrespective of changes in light and other deformations in the object.	Not suitable to use if occlusion is likely to happen. Requires downloading the `caffemodel` (which is pretty big) and `prototxt` files. Does not work with OpenCV versions > 3.4.1.
KCF	Good accuracy and speed (or FPS). Tracking performance is better than boosting. That's why it acts as a very good alternative to the Boosting tracker.	Does not perform well for occlusion.
Median Flow	Excellent tracker and can also identify when tracking has failed. This is really important because it avoids giving wrong bounding boxes.	Does not perform well for occlusion.
MIL	A good performance tracker. Works well when there is partial occlusion.	Does not work well with complete occlusion. Also, the results after partial occlusion are not good and the object is not tracked completely.
MOSSE	Can track the object after it has vanished from the frame and comes back again. Has a high performance (FPS). Works fine, irrespective of changes in light and other deformations in the object.	Lags behind GOTURN in terms of accuracy and performance. Also fails to deal with scale variation.
TLD	Best tracker to go for if high occlusion and scale variation are present in the video.	Since TLD plays it safe, it has a habit of detecting an object even when it's not present (false positive).

Figure 6.14: Table representing the pros and cons of object trackers

As stated in the preceding table, **GOTURN** is a deep learning-based object tracking model and thus it requires us to download the model files. Technically speaking, the model is a Caffe model, which is why we need to download two files – the `caffemodel` file (model weights) and the `prototxt` file (model architecture). You don't need to worry about what these files are and what their purpose is. The short version of the answer is that these are the trained model weights that we will be using.

> **NOTE**
>
> You can download the `prototxt` file from here: https://packt.live/2YP7bRp.

Unfortunately, the `caffemodel` file is pretty big (~350 MB). To get that file, we will have to carry out the following steps.

We need to download the smaller zip files from the following links:

File 1: https://packt.live/2NN51ex.

File 2: https://packt.live/2Zv8aoK.

File 3: https://packt.live/2BuN2qT.

File 4: https://packt.live/3imh54s.

Next, we need to concatenate these files to form one big ZIP file:

```
cat goturn.caffemodel.zip* > goturn.caffemodel.zip
```

Windows users can use **7-zip** to carry out the preceding step.

Next, unzip the ZIP file to get the `goturn.caffemodel` file. Both the `caffemodel` and `prototxt` files should be present in the folder where you are running your code.

> **NOTE**
>
> Unfortunately, **GOTURN Tracker does not work in OpenCV 4.2.0**. One option is to change to OpenCV 3.4.1 to make it work or to wait for a fix.

Now that we have seen the pros and cons of the various kinds of trackers, let's have a look at an example that shows how to use trackers in OpenCV.

EXERCISE 6.02: OBJECT TRACKING USING THE MEDIAN FLOW AND MIL TRACKERS

Now, let's proceed with the knowledge we gained in the previous section about OpenCV's object tracking API and track an object in a video using the Median Flow and MIL trackers.

> **NOTE**
>
> Try and make some observations regarding the performance of the trackers in terms of **Frames Per Second** (**FPS**) and the way they deal with occlusions and other similar distortions.

We will be using a video of a running man to test object tracking (considering the running man as the object to be tracked) based on OpenCV's object tracking API algorithm/technique. In this exercise, we will be using the Median Flow and MIL trackers to produce the results. After evaluating the performance of these trackers, we need to decide which one works better. Follow these steps to complete this exercise:

> **NOTE**
>
> The video is available at https://packt.live/31yP2ZV.

1. Create a new Jupyter Notebook and name it `Exercise6.02.ipynb`. We will be writing our code in this notebook.

2. Import the required modules:

```
# Import modules
import cv2
import numpy as np
```

3. Next, create a **VideoCapture** object using the **cv2.VideoCapture** function:

> **NOTE**
>
> Before proceeding, ensure that you can change the path to the video (highlighted) based on where the video is saved in your system.

```
# Create a VideoCapture Object
video = cv2.VideoCapture("../data/people.mp4")
```

4. Check whether the video was opened successfully using the following code:

```
# Check if video opened successfully
if video.isOpened() == False:
    print("Could not open video!")
```

5. Next, read the first frame and also check whether we were able to read it successfully:

```
# Read first frame
ret, frame = video.read()
# Check if frame is read successfully
if ret == False:
    print("Cannot read video")
```

6. Display the first frame that we read:

```
# Show the first frame
cv2.imshow("First Frame", frame)
cv2.waitKey(0)
cv2.imwrite("firstFrame.png", frame)
```

The output is as follows:

Figure 6.15: First frame of the video

7. Use the **cv2.selectROI** function to create a bounding box around the object. Print the bounding box to see the values:

```
# Specify the initial bounding box
bbox = cv2.selectROI(frame)
print(bbox)
```

The output is **(41, 10, 173, 350)**.

8. Close the window, as follows:

```
cv2.destroyAllWindows()
```

The image will look as follows:

Figure 6.16: Bounding box drawn around the person to track

9. Create our tracker. Use the **MedianFlow** tracker, but if you want to use the MIL tracker, you just have to replace **cv2.TrackerMedianFlow_create()** with **cv2.TrackerMIL_create()**:

```
# MedianFlow Tracker
tracker = cv2.TrackerMedianFlow_create()
```

10. Initialize our tracker using the frame and the bounding box:

```
# Initialize tracker
ret = tracker.init(frame,bbox)
```

11. Create a new display window. This is where we are going to display our tracking results:

```
# Create a new window where we will display
# the results
cv2.namedWindow("Tracker")
# Display the first frame
cv2.imshow("Tracker",frame)
```

12. Next, start iterating over the frames of the video:

```
while True:
    # Read next frame
    ret, frame = video.read()
    # Check if frame was read
    if ret == False:
        break
```

13. Update the tracker. This will give us a flag that tells us whether the object was found or not. There will be a bounding box around the object if it was found:

```
    # Update tracker
    found, bbox = tracker.update(frame)
```

14. If the object was found, display the bounding box around it:

```
    # If object found, draw bbox
    if found:
        # Top left corner
        topLeft = (int(bbox[0]), int(bbox[1]))
        # Bottom right corner
        bottomRight = (int(bbox[0]+bbox[2]), \
                       int(bbox[1]+bbox[3]))
        # Display bounding box
        cv2.rectangle(frame, topLeft, bottomRight, \
                      (0,0,255), 2)
```

15. If the object was not found, display a message on the screen telling the user that the object was not found:

```
    else:
        # Display status
        cv2.putText(frame, "Object not found", (20,70), \
                    cv2.FONT_HERSHEY_SIMPLEX, 0.75, (0,0,255), 2)
```

16. Finally, display the frame in the display window we created previously:

    ```
    # Display frame
    cv2.imshow("Tracker",frame)
    k = cv2.waitKey(5)
    if k == 27:
        break
    cv2.destroyAllWindows()
    ```

At this point, we recommend that you try out the different kinds of trackers that we looked at previously and see how the results change. After running the preceding code for the MIL tracker and Median Flow tracker, here's what we found.

As shown in the following figure, the trackers were able to detect the object, irrespective of the scale variation:

Figure 6.17: Object being tracked, irrespective of the scale variation

Problems arise when the object is hidden due to occlusion. While the Median Flow tracker displays that it was not able to find the object, the MIL tracker displays a false positive – a bounding box where the object is not present, as can be seen in the following figure:

Figure 6.18: False positive given by the MIL tracker after occlusion

As shown in the following figure, the Median Flow tracker displays a message stating that it was not able to detect the object after occlusion:

Figure 6.19: Median Flow tracker displaying an "Object not found" message

In this exercise, we saw how we can use OpenCV's object tracking API to track an object that had been selected using OpenCV's **selectROI** function. We also saw how easy it is to use the API: all we need to do is replace one tracker with another and the rest of the code will stay the same.

> **NOTE**
>
> To access the source code for this specific section, please refer to https://packt.live/2YPSa1D.

So far, we have been talking about OpenCV's object tracking API. Now, let's have a look at another popular computer vision library – **dlib**. First, we will install Dlib and then use it to carry out object tracking in the upcoming sections.

INSTALLING DLIB

Since we will be discussing object tracking using Dlib, let's see how we can install it. We will use the recommended approach of installing Dlib using the following `pip` command:

```
pip install dlib
```

You should be able to see the following on your Command Prompt or Terminal, depending on your operating system:

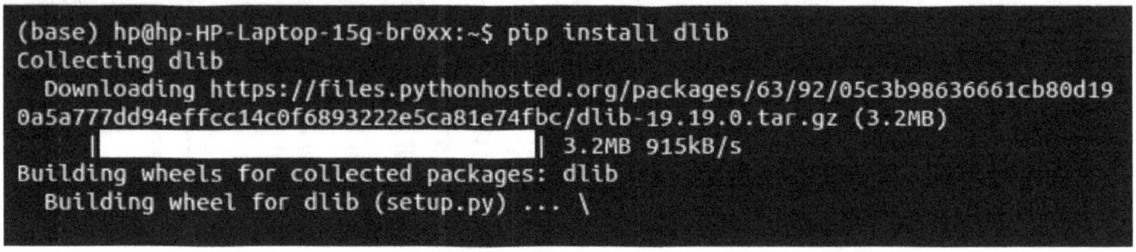

```
(base) hp@hp-HP-Laptop-15g-br0xx:~$ pip install dlib
Collecting dlib
  Downloading https://files.pythonhosted.org/packages/63/92/05c3b98636661cb80d19
0a5a777dd94effcc14c0f6893222e5ca81e74fbc/dlib-19.19.0.tar.gz (3.2MB)
     |                                | 3.2MB 915kB/s
Building wheels for collected packages: dlib
  Building wheel for dlib (setup.py) ... \
```

Figure 6.20: Output of the pip install dlib command

Once the installation is successful, you will see the following message on your screen:

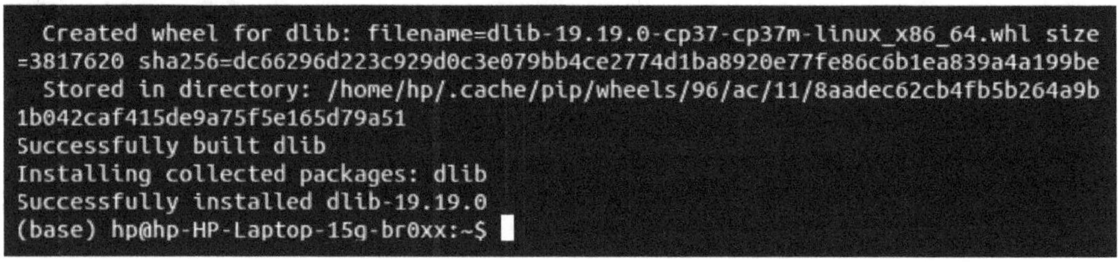

```
  Created wheel for dlib: filename=dlib-19.19.0-cp37-cp37m-linux_x86_64.whl size
=3817620 sha256=dc66296d223c929d0c3e079bb4ce2774d1ba8920e77fe86c6b1ea839a4a199be
  Stored in directory: /home/hp/.cache/pip/wheels/96/ac/11/8aadec62cb4fb5b264a9b
1b042caf415de9a75f5e165d79a51
Successfully built dlib
Installing collected packages: dlib
Successfully installed dlib-19.19.0
(base) hp@hp-HP-Laptop-15g-br0xx:~$
```

Figure 6.21: Output after Dlib installation is complete

> **NOTE**
>
> Note that `dlib` is a very heavy module and can take a lot of time and RAM during installation. So, if your system appears to be stuck during the installation process, there is no need to panic. Typically, installation can take anywhere between 5 to 20 minutes.

We can check whether the installation was successful or not by trying to import `dlib`, as follows:

1. Open your Command Prompt or Terminal and type **python**.

2. In the Python shell, enter **import dlib** to import the **dlib** module.

3. Finally, print the version of Dlib that has been installed using **print(dlib.__version__)**.

The output of the preceding commands is shown in the following screenshot:

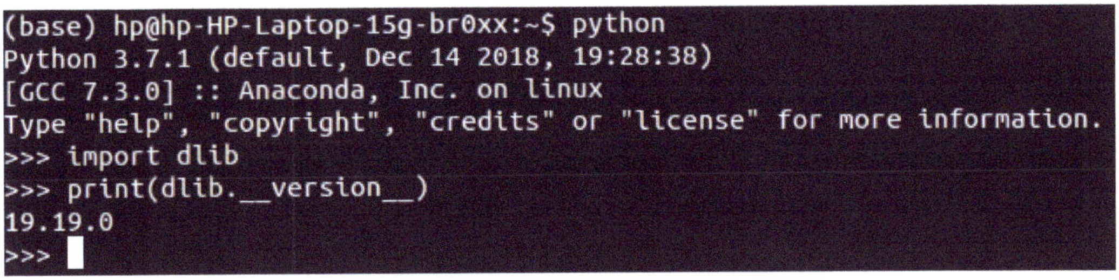

```
(base) hp@hp-HP-Laptop-15g-br0xx:~$ python
Python 3.7.1 (default, Dec 14 2018, 19:28:38)
[GCC 7.3.0] :: Anaconda, Inc. on linux
Type "help", "copyright", "credits" or "license" for more information.
>>> import dlib
>>> print(dlib.__version__)
19.19.0
>>>
```

Figure 6.22: Verifying the Dlib installation by printing the version of dlib that's installed

If you can see the preceding output, then you have successfully installed Dlib. Now, let's learn how to use Dlib for object tracking.

OBJECT TRACKING USING DLIB

Dlib's object tracking algorithm is built upon the **MOSSE** tracker. If you refer to the object tracker summary we provided previously, you will notice that MOSSE doesn't have the best performance when there is scale variation. To cater for this, Dlib's object tracking algorithm uses the technique described in the paper *Accurate Scale Estimation for Robust Visual Tracking*, which was published in 2014.

> **NOTE**
>
> **Citation**: Martin Danelljan, Gustav Häger, Fahad Shahbaz Khan, and Michael Felsberg. *Accurate Scale Estimation for Robust Visual Tracking*. Proceedings of the British Machine Vision Conference 2014, 2014.
> https://doi.org/10.5244/c.28.65.

This technique involves estimating the scale of the object after every update step. Because of this, we can now deal with scale variation and utilize the pros of the MOSSE tracker. The following flowchart demonstrates the object tracking process when using Dlib:

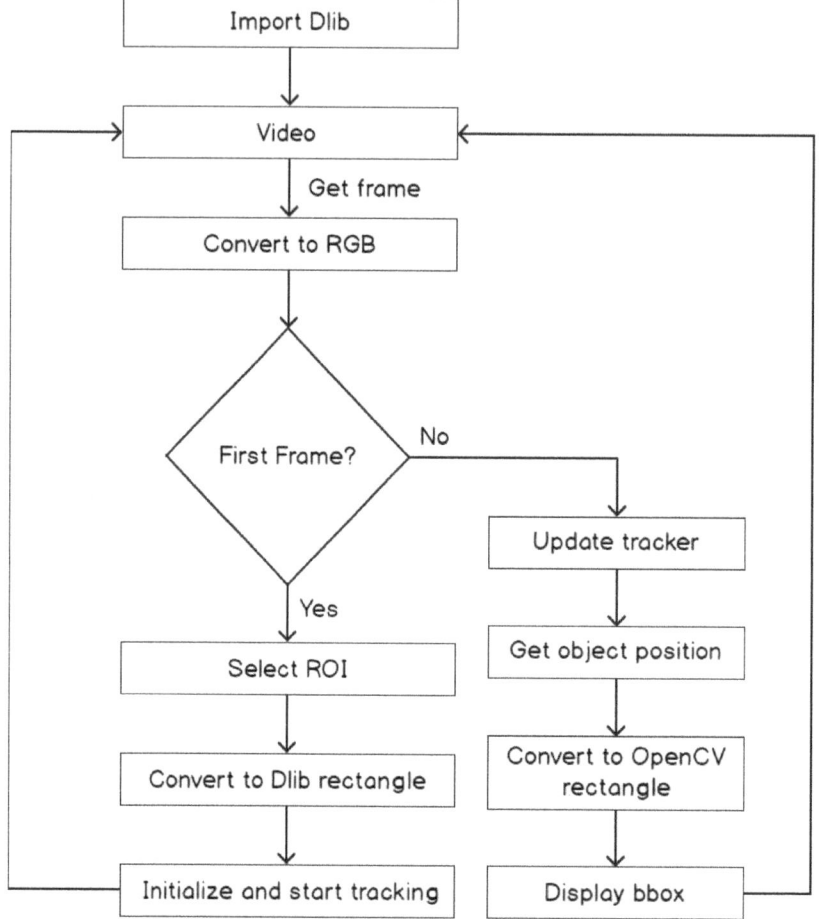

Figure 6.23: Object tracking using Dlib

Let's have a look at the process of using object tracking in Dlib. Notice how the first few steps remain the same as when using object tracking in OpenCV.

First, we'll need to import the **dlib** module into our code:

```
import dlib
```

Next, just like in OpenCV, we will have to create a bounding box around the object that we want to track, but only for the first time that we do this. For this, we can use the `cv2.selectROI` function.

Now comes a very important point. Dlib uses images in **RGB** format, whereas OpenCV uses images in **BGR** format. So, we need to convert images that have been loaded by OpenCV into RGB format using the following code:

```
rgb = cv2.cvtColor(bgr, cv2.COLOR_BGR2RGB)
```

Next, we will also need to convert the rectangle given by the `cv2.selectROI` function into Dlib's `rectangle` type, as follows:

```
topLeftX, topLeftY, w, h) = bbox
bottomRightX = topLeftX+w
bottomRightY = topLeftY+h
dlibRect = dlib.rectangle(topLeftX, topLeftY, \
           bottomRightX, bottomRightY)
```

Now, we can create our `dlib` tracker, as follows:

```
tracker = dlib.correlation_tracker()
```

Since we have the tracker ready, we can start the tracking process. The first time we do this, we will have to provide the bounding box around the object and the image:

```
tracker.start_track(rgb, dlibRect)
```

For the later frames, we can directly update the tracker and get the position of the object. Again, we'll need to make sure that we are using RGB frames here:

```
tracker.update(rgb)
objectPosition = tracker.get_position()
```

`objectPosition` is an instance of the Dlib rectangle. So, we need to convert it back into OpenCV's `rectangle` format so that we can create a bounding box using the `cv2.rectangle` command:

```
topLeftX = int(objectPosition.left())
topLeftY = int(objectPosition.top())
bottomRightX = int(objectPosition.right())
bottomRightY = int(objectPosition.bottom())
```

Finally, to draw the rectangle, we can use the `cv2.rectangle` command, as we did in the previous chapter:

```
cv2.rectangle(rgb, (topLeftX, topLeftY), (bottomRightX, bottomRightY), \
              (0,0,255), 2)
```

And that's it. That's all you need to do to carry out object tracking using Dlib. Now, let's strengthen our understanding of this process with the help of an example. You can also refer to *Figure 6.23* as a quick reference to the preceding process.

EXERCISE 6.03: OBJECT TRACKING USING DLIB

As a continuation of *Exercise 6.02, Object Tracking Using the Median Flow and MIL Trackers*, we will track the running man based on the Dlib technique/algorithm and use a modified MOSSE tracker to analyze the performance with respect to the same parameters that we calculated earlier. Also, we will evaluate the performance in comparison with the trackers used in *Exercise 6.02, Object Tracking Using the Median Flow and MIL Trackers*. As we saw previously, the video has lighting changes, partial and complete occlusion, and scale variation. This will help you compare the three trackers using a total of four aspects – performance in terms of FPS and their ability to deal with the three issues – lighting change, occlusion, and scale variation. We will leave this as an open exercise for you to try out yourself. See whether your observations match the observations we stated previously in this chapter. Follow these steps to complete this exercise:

> **NOTE**
>
> The video is available at https://packt.live/31yP2ZV.

Perform the following steps:

1. Import the required modules (OpenCV, NumPy, and Dlib):

   ```
   # Import modules
   import cv2
   import numpy as np
   import dlib
   ```

2. Create a **VideoCapture** object and see whether the video was loaded successfully:

   > **NOTE**
   >
   > Before proceeding, ensure that you can change the path to the video (highlighted) based on where the video is saved in your system.

   ```
   # Create a VideoCapture Object
   video = cv2.VideoCapture("../data/people.mp4")
   # Check if video opened successfully
   if video.isOpened() == False:
       print("Could not open video!")
   ```

3. Next, read the first frame and check whether it was read successfully:

   ```
   # Read first frame
   ret, frame = video.read()
   # Check if frame read successfully
   if ret == False:
       print("Cannot read video")
   ```

4. Now, display the first frame and then select the ROI:

   ```
   # Show the first frame
   cv2.imshow("First Frame", frame)
   cv2.waitKey(0)
   # Specify the initial bounding box
   bbox = cv2.selectROI(frame)
   cv2.destroyAllWindows()
   ```

5. Convert the frame into RGB since that's what Dlib uses:

   ```
   # Convert frame to RGB
   rgb = cv2.cvtColor(frame, cv2.COLOR_BGR2RGB)
   ```

6. Convert the rectangle into a Dlib rectangle:

   ```
   # Convert bbox to Dlib rectangle
   (topLeftX, topLeftY, w, h) = bbox
   bottomRightX = topLeftX + w
   bottomRightY = topLeftY + h
   dlibRect = dlib.rectangle(topLeftX, topLeftY, \
                             bottomRightX, bottomRightY)
   ```

7. Create the tracker and initialize it using the Dlib rectangle and the RGB frame:

   ```
   # Create tracker
   tracker = dlib.correlation_tracker()
   # Initialize tracker
   tracker.start_track(rgb, dlibRect)
   ```

8. Create a new display window. This is where we will be displaying our tracking output:

   ```
   # Create a new window where we will display the results
   cv2.namedWindow("Tracker")
   # Display the first frame
   cv2.imshow("Tracker",frame)
   ```

9. Start iterating over the frames of the video:

   ```
   while True:
       # Read next frame
       ret, frame = video.read()
       # Check if frame was read
       if ret == False:
           break
   ```

10. Convert the frame that we just read into RGB:

    ```
        # Convert frame to RGB
        rgb = cv2.cvtColor(frame,cv2.COLOR_BGR2RGB)
    ```

11. Update the tracker so that it can find the object:

    ```
        # Update tracker
        tracker.update(rgb)
    ```

12. Use the position given by the tracker to create a bounding box over the detected object:

```
objectPosition = tracker.get_position()
topLeftX = int(objectPosition.left())
topLeftY = int(objectPosition.top())
bottomRightX = int(objectPosition.right())
bottomRightY = int(objectPosition.bottom())
# Create bounding box
cv2.rectangle(frame, (topLeftX, topLeftY), (bottomRightX, \
                     bottomRightY), (0,0,255), 2)
```

13. Finally, display the frame on the display window we created previously:

```
# Display frame
cv2.imshow("Tracker",frame)
k = cv2.waitKey(5)
if k == 27:
    break
cv2.destroyAllWindows()
```

The output is as follows:

Figure 6.24: Dlib object tracking output

Notice how the Dlib object tracking algorithm was able to accurately track the object of interest, unlike the two algorithms we discussed in the previous exercise. This does not mean that this algorithm is superior compared to the other two. It only means that it's a more suitable choice for the specific case study we are talking about. Such experiments are hugely important in the computer vision domain.

> **NOTE**
>
> To access the source code for this specific section, please refer to https://packt.live/2ZvbVKS.

So, in this exercise, we saw how we can use an object tracker using Dlib. Now, to wrap up this chapter, let's consider a practical case study. This will be especially fun for people who either love playing football or watching it.

ACTIVITY 6.01: IMPLEMENTING AUTOFOCUS USING OBJECT TRACKING

In the previous chapter, we discussed face detection using Haar Cascades and also mentioned that we could use frontal face cascades to implement the autofocus feature in cameras. In this activity, we will implement an autofocus utility (we discussed it briefly in *Chapter 5, Face Processing in Image and Video*) that can be used in football to track the ball throughout the game. Now, while this might appear very strange at first, such a feature could come in handy. Right now, cameramen have to pan their cameras wherever the ball is and that becomes a cumbersome job. If we had an autofocus utility that could track the ball throughout the game, the camera could pan automatically to follow the ball. We will be using a video of a local football game. We want you to try out the two computer vision libraries we have covered so far – OpenCV and Dlib – for object tracking and see which specific algorithm gives you the best result in this case. This way, you will get a chance to carry out the experimentation that we discussed previously, after *Exercise 6.02, Object Tracking Using the Median Flow and MIL Trackers*. We've also included an additional challenge for you to try out on your own. Follow these steps to complete this activity:

1. Open your Jupyter Notebook and create a new file called `Activity6.01.ipynb`. We will be writing our code in this notebook. Also, use the `Exercise6.02.ipynb` and `Exercise6.03.ipynb` files as references for this activity.

2. First, import the necessary libraries – OpenCV, NumPy, and Dlib.

3. Next, create a **VideoCapture** object by reading the video file present at **../data/soccer2.mp4**.

   > **NOTE**
   >
   > The video can be downloaded from https://packt.live/2NHXwpp.

4. Check whether the video was loaded successfully.

5. Next, we will read the first frame and verify whether we were able to read the frame properly or not.

6. Now that we have read the frame, we can display it to see whether everything has worked fine so far.

7. Next, select the ROI (the bounding box around the football) using the **cv2.selectROI** function. In this case, we will be using OpenCV's object trackers for object tracking, but it is recommended to try both OpenCV and Dlib's object trackers.

8. Create an object tracker using the functions we used previously. Use proper experimentation to find out which object tracker gives the best performance.

9. Once the object tracker has been created, we can initialize it using the bounding box we created in *Step 7* and the first frame.

10. Next, create a display window to display the tracking output.

11. Now, you can use the same **while** loop that we used in **Exercise6.02.ipynb** to iterate over each frame, thus updating the tracker and displaying a bounding box around the object.

12. Once you are out of the **while** loop, close all the display windows.

The following figure shows the output that will be obtained for OpenCV's `selectROI` function, which is used to select the object (football) that we want to track:

Figure 6.25: ROI selection

The output of the object tracker will be as follows for one of the frames of the video. This will appear inside the `while` loop, which is where you will display the object's position that was obtained using the object tracker:

Figure 6.26: Output of the object tracker

Additional Challenge:

At this point, we want you to think about why such a simple object tracking program will fail in real life. We have got two kinds of trackers – ones that are very strict, which means that even if they have a slight doubt about the object, they will just say that the object was not found, and then we have got trackers such as the Dlib object tracker and the TLD object tracker, which give a lot of false positives. This calls for a hybrid tracker pipeline. The basic idea behind this is that we will use the strict object tracker to track the object and if the object is not detected, we will switch to the other object tracker to provide the location of the object. The advantage of such an approach is that for most frames, you will get a very accurate location of the object and when the strict object tracker fails, the other object tracker will still be able to give some information about the object's location. We recommend that you think about this approach more deeply and implement it on your own.

> **NOTE**
>
> The solution for this activity can be found on page 513.

In this activity, we saw how we can use different trackers based on the situation we are dealing with. The best part of using the OpenCV object tracking API is that it requires minimal code changes when shifting from one tracker to another. It is recommended that you try using other trackers supported by OpenCV on this case study and carry out a comparison between them.

SUMMARY

This was our first advanced computer vision topic where we used statistical and deep learning-based object tracking models to solve case studies involving object surveillance and other similar examples. We started with a very basic object tracker that was based on a very naive approach. We also saw how the tracker, while computationally very cheap, has very low performance and can't be used for most real-life scenarios. We used the HSV color space in *Exercise 6.01, Object Tracking Using Basic Image Processing*, to track a lemon across various frames on the input video. We then went ahead and discussed common non-deep learning-based trackers – the Kalman filter and the meanshift and CAMshift filters. Next, we discussed the OpenCV object tracking API, where we listed the eight commonly used object trackers and their pros and cons. We also had a look at how we can use those trackers and how **GOTURN** requires a slightly different process since we need to download the Caffe model and the `prototxt` file. We learned how to use these object trackers in *Exercise 6.02, Object Tracking Using the Median Flow and MIL Trackers*, where we tracked a jogger across various frames in a video. We also saw how we can shift between different object trackers with the help of OpenCV's object tracking API. We then went ahead and discussed Dlib, which is another well-known and commonly used computer vision library by Davis King. We saw how we can use object tracking using Dlib. Finally, we wrapped up this chapter by having a look at an activity involving tracking the ball in a football game. In the next chapter, we will talk about object detection and face recognition. The importance of this topic can be understood by the fact that, right now, we have to manually provide the bounding box, but with the help of an object detection pipeline, we can automate the process of finding an object in a frame and then track it in the subsequent frames.

7

OBJECT DETECTION AND FACE RECOGNITION

OVERVIEW

The objective of this chapter is to explain the concepts of object detection and facial recognition techniques and implement these techniques on input images or videos. The first part of this chapter addresses facial recognition techniques based on Eigenfaces, Fisherfaces, and **Local Binary Pattern** (**LBP**) methods. The second part focuses on object detection techniques, and it begins with a widely practiced method called MobileNet **Single Shot Detector** (**SSD**) and subsequently addresses the LBP and Haar Cascade algorithms. By the end of this chapter, you will be able to determine and implement the most suitable algorithms for object detection and facial recognition to solve practical problems.

INTRODUCTION

In the previous chapter, we learned how to implement various object tracking algorithms using the OpenCV library. Also, in *Chapter 5, Face Processing in Image and Video*, we discussed the concept of face detection. Now, it's time to discuss the next technique of face processing, called **facial recognition**. We will also discuss a few popular object detection techniques.

Facial recognition is a technique used to identify a person using their face in images or videos, including real-time image data. It is widely used by social media companies, companies manufacturing mobile devices, law enforcement agencies, and many others. One of the advantages of face recognition is that compared to other biometric processes, such as fingerprint or iris recognition, facial recognition is non-intrusive. It captures various facial features from an image and then analyzes and compares these features with facial features captured from other images to find the best match. In the first part of this chapter, we will discuss classical computer vision techniques used to implement facial recognition. It will provide a good foundation for you to start to explore more complex algorithms.

Object detection is a technique in which we aim to locate and identify all instances of objects (for example, people, vehicles, animals, and so on) present in a given image or video. It has many applications in the field of computer vision, such as pedestrian detection, vehicle detection, and ball tracking in sports.

Traditionally, object detection techniques were more reliant on feature engineering to find patterns in objects, and then these features would be used as an input to machine learning algorithms to locate objects. For example, suppose we needed to find a square-shaped box in an image. What used to happen is we would look for an object that has equal sides that are perpendicular to each other. Similarly, if another task was to find a face in an image, then we would look for features such as the distance between the eyes or the bridge of the nose (as with the *Viola-Jones* face detection algorithm). Since these approaches predominantly use rigidly engineered features, they often fail to generalize patterns of similar objects.

In 2012, the **AlexNet** architecture increased the accuracy of image classification tasks significantly in comparison to traditional methodologies, and this result led researchers to shift their focus to deep learning. With the advent of deep learning, the field of computer vision has seen great progress and improvement in the performance of object detection tasks. In the second part of this chapter, we will discuss traditional feature-based algorithms and deep learning algorithms (using OpenCV) for object detection.

We will start with face recognition, where we will discuss the **Eigenface**, **Fisherface**, and **LBP** methods for identifying faces. Then, we will discuss a deep learning algorithm called **MobileNet SSD** for object detection. Later on, we will discuss and implement the **Local Binary Patterns Histograms** (**LBPH**) method, a descriptor-based approach to detecting objects. In *Chapter 5*, *Face Processing in Image and Video*, we went through the concepts of **Haar Cascade**, so we will briefly revisit a few of the concepts and will conclude the discussion on object detection by applying the Haar Cascade method.

FACE RECOGNITION

Face recognition mainly involves the following stages:

1. **Pre-processing the images**: Generally, when we work with unstructured data such as images, we are working with raw data. In the pre-processing step, we aim to improve the quality of the raw data either by removing any irregularities, such as distortions, non-alignments, and noise, or by enhancing some features using geometric transformation, such as zooming in, zooming out, translation, and rotation.

2. **Face detection**: In this step, we aim to detect all the faces present in each image and store the faces with correct labels.

3. **Training (creating the face recognition model)**: In the training step, we take the images and their corresponding labels and feed them into a parametric algorithm that learns how to recognize a face. As a result, we get a trained parametric model that eventually helps us to recognize faces. The trained model contains some characteristic features corresponding to each unique label used during training.

4. **Face recognition**: This step uses the trained model to verify whether the faces in two images belong to the same person. In other words, we pass an unknown image to the trained model, whereupon it extracts some characteristic features from that image and matches them with the features of training images and outputs the closest match.

The following figure depicts the preceding points:

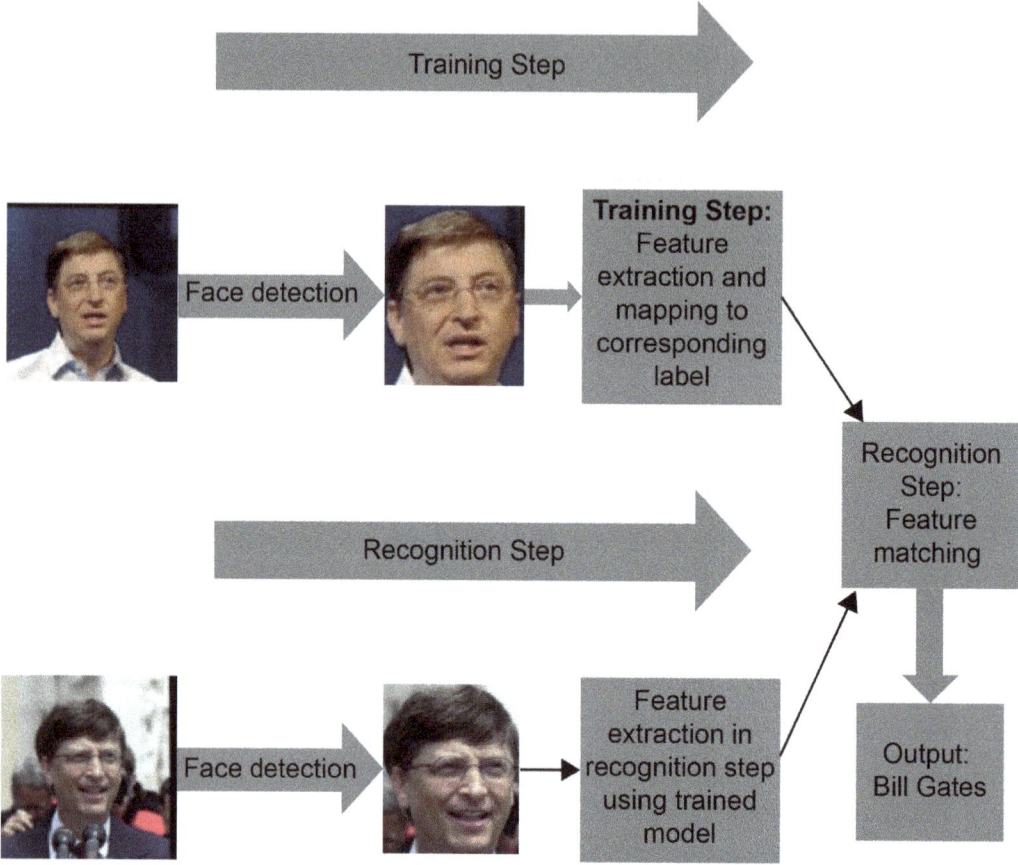

Figure 7.1: Face recognition steps

In the preceding figure, the upper part represents the training step and the lower part represents the recognition step. Note that the sizes of the images in the left-most part of the figure are nearly the same (the size should be the same for algorithms such as Eigenface and Fisherface) as that of the images that come under the pre-processing step.

> **NOTE**
>
> The images used in *Figure 7.1* are sourced from
> http://vis-www.cs.umass.edu/lfw/#download.

FACE RECOGNITION USING EIGENFACES

In this section, we will see how face recognition is achieved using Eigenfaces. Before diving into Eigenfaces, let's discuss an important concept in detail: dimensionality reduction and the **Principal Component Analysis (PCA)** technique.

PRINCIPAL COMPONENT ANALYSIS

Dimensionality reduction can be defined as a mapping of a higher-dimensional feature space (data points) to a lower-dimensional feature space, such that the maximum variance of the features remains intact. Principal component analysis, commonly known as PCA, is one of the more popular techniques used for dimensionality reduction. The objective of PCA is to identify the subspace in which the data can be approximated. In other words, the objective is to transform correlated features of large dimensions to the uncorrelated features of smaller dimensions.

Suppose we have a dataset of vehicles where each data point, x, has n different attributes (features); that is, $x = x_1, x_2, x_n$. The attributes of the dataset can include `speed in kph`, `speed in mph`, `engine capacity`, `size`, and `seating capacity`. If we observe the attributes, we can see that the speed attributes are linearly dependent (meaning we can get `speed in mph` by applying some linear transformation to `speed in kph`) and hence are redundant. So, eventually, the dimension of the input features is $n-1$, not n.

Let's take another example. Suppose we have another dataset of football players with many attributes, such as salary, club, age, and skills (sliding tackle, standing tackle, and dribbling). Again, our goal is to reduce the dimensionality of the features such that we keep the more informative features and discard the less informative ones. Intuitively, we can say that if a player is good at sliding tackles, they may be good at standing tackles. So, these two features are correlated with each other. So, again, we may need a new single feature that represents these two original features, especially in regression analysis, which is used to understand and examine the relationship (usually a linear relationship, but it can be a higher degree polynomial relationship as well) between two variables, that is, a dependent variable and an independent variable.

The question that arises is this: how can we automatically detect redundant features? PCA helps to remove redundant features by transforming the original features of dimension n to new features of dimension d such that $d<<n$. The new feature dimensions are known as principal components, which are orthogonal (have no correlation) to each other and ordered in such a way that the variance of data decreases as we move down.

We can also see that PCA helps to represent images efficiently by discarding less informative features, eventually reducing complexity and memory requirements.

EIGENFACES

Eigenfaces can be defined as an appearance-based method that extracts the variation in facial images and uses this information to compare images of individual faces (holistic, not part-based) in databases. The Eigenfaces algorithm relies on the underlying assumption that not all facial details or features are equally significant. For example, when we see faces, we mainly notice prominent and distinct facial details, such as the nose, eyes, cheeks, or lips. In other words, we observe details that are distinct due to significant differences between facial features, such as the difference between a nose and an eye. Mathematically, this can be interpreted as the variance between those facial features being high. So, when we come across multiple faces, these distinct and important features help us to recognize people.

The Eigenfaces algorithm works in this way for facial recognition tasks. Its face recognizer takes training images and extracts the characteristics and features from those images, and then applies PCA to extract the principal components with maximum variance. We can refer to these principal components as Eigenfaces, or the eigenvectors of the covariance matrix of the set of images.

In the recognition phase, an unknown image is projected to the space formed by the Eigenfaces. Then, we evaluate a Euclidean distance between the eigenvectors of unknown images and the eigenvectors of the Eigenfaces. If the value of the Euclidean distance is small enough, then we tag that image with the most appropriate class. On the other hand, if we find that the Euclidean distance is large, then we infer that the individual's face was not present during the training phase and hence we model the images to accommodate the new image.

> **NOTE**
>
> The terms *Eigenfaces* and *principal components* will be used interchangeably in this chapter.

Until now, we have discussed the concepts of the Eigenfaces method. Now we will build a face recognizer using this method. Fortunately, OpenCV provides a facial recognition module called **cv2.face.EigenFaceRecognizer_create()**. We will use this facial recognition module to implement this exercise. We will feed the images of the respective classes to this module during the training phase and will use the trained model to recognize a new face.

This module mainly takes the following two parametric inputs:

- **num_components**: The number of components for PCA. There is no rule regarding how many components, but we aim for an amount that will maintain maximum variance. It also depends on the input data, so we can experiment with a few amounts.

- **threshold**: The **threshold** value is used while recognizing a new image. If the distance between the new image and its neighbor (another image used during training) is greater than the **threshold** value, then the model returns **-1**. This means that the model has failed to recognize the new image.

The following functions are used by the **EigenFaceRecognizer** module to train and test a model, respectively:

- **eigenfaces_recognizer.train()**: **EigenFaceRecognizer** uses this function to train a model using the training images. It takes the previously mentioned parameters as input arguments.

- **eigenfaces_recognizer.predict()**: **EigenFaceRecognizer** uses this function to predict the label of a new image. It takes a test image as an input argument.

Let's see the steps involved in the Eigenfaces algorithm. *Figure 7.2* and *Figure 7.3* depict the training steps and test steps for the Eigenfaces algorithm using OpenCV, respectively:

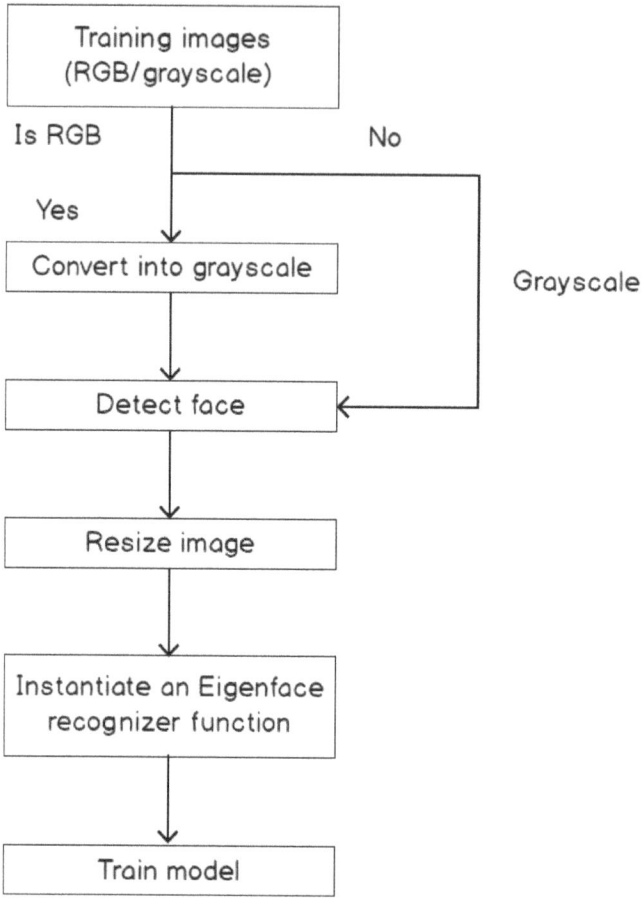

Figure 7.2: Training steps in the Eigenface algorithm

As shown in *Figure 7.2*, the training phase takes color or grayscale images as input. If the input is a color image, then the next step converts the input image into grayscale; otherwise, the input image is passed to the face detection function. As we discussed earlier, the Eigenface algorithm requires all images to be the same size; so, in the next step, all images are resized to the same dimensions. Then, the next step instantiates the recognizer model using the OpenCV module. After that, the model is trained using training images. The next figure shows the test steps in the Eigenfaces algorithm:

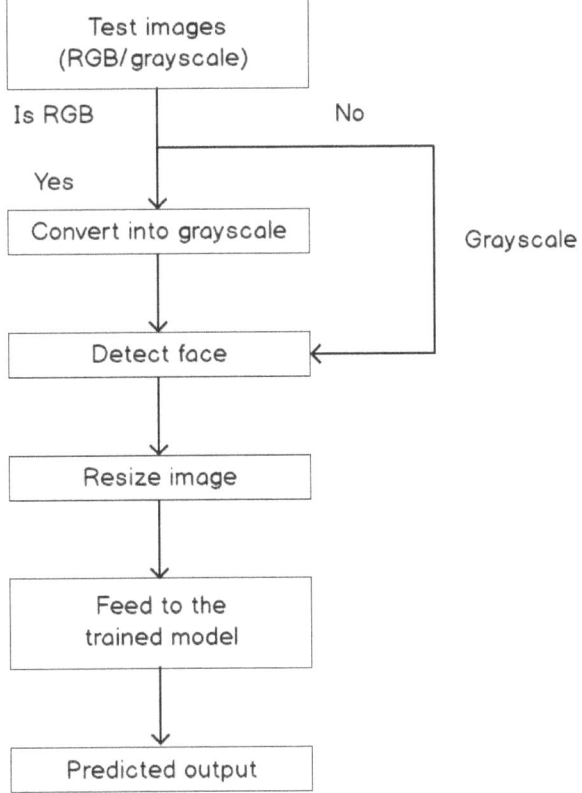

Figure 7.3: Steps during the test phase of the Eigenfaces algorithm

As shown in the preceding figure, the test phase first takes a new color or grayscale image as input. If the input is a color image, then the next step converts the input image into grayscale. Otherwise, the input image is passed to the face detection function. As we discussed earlier, the Eigenfaces algorithm requires all images to be the same size. Thus, in the next step, the test image is resized to the same size as the training images. Then, the next step feeds the resized image to the trained model (which was trained during the training steps), and finally outputs the category of the image.

> **NOTE**
>
> Before we go ahead and start working on the exercises and activities, please make sure that you have installed the **opencv** and **opencv-contrib** modules. If you are using Windows, it is recommended to use Anaconda. Please note that the code should be written in Python 3 or higher and the version of OpenCV should be 4 or higher.

EXERCISE 7.01: FACIAL RECOGNITION USING EIGENFACES

In this exercise, you will be performing facial recognition using the Eigenfaces method. We will be using a training dataset to train the facial recognition model and a testing dataset to test the performance of the model. We will also be using Haar Cascades for face detection, which we studied in *Chapter 5, Face Processing in Image and Video*.

> **NOTE**
>
> The images used in this exercise are sourced from http://vis-www.cs.umass.edu/lfw/#download. Note that in this exercise, the faces of a few famous personalities have been used, but you can implement the exercise with any facial dataset.

Follow these steps to complete the exercise:

1. Firstly, open an untitled Jupyter Notebook and name the file **Exercise7.01**.

2. Start by importing all the required libraries for this exercise using the following code:

```
import cv2
import os
import numpy as np
import matplotlib.pyplot as plt
%matplotlib inline
```

3. Next, specify the path of the dataset. Create a folder named **dataset** and, inside that, create two more folders: **training-data** and **test-data**. Put the training and test images into **training-data** and **test-data**, respectively.

> **NOTE**
>
> The **training-data** and **test-data** folders are available at https://packt.live/3dVDy5i.

The following code shows the path names and folder structure that needs to be maintained in your local system:

```
training_data_folder_path = 'dataset/training-data'
test_data_folder_path = 'dataset/test-data'
```

The structure of **training-data** looks as follows:

```
# directory of training data
dataset
        training-data
        |------0
        |       |--name_0001.jpg
        |       |--... ...
        |       |--name_0020.jpg
        |------1
        |       |--name_0001.jpg
        |       |--... ...
        |       |-- name_0020.jpg
        test-data
        # similar structure
```

The idea is simple. The higher the number of images used in training, the better the result we will get from the model.

4. Write the following code to check whether the images from the dataset are displayed properly:

```
random_image = cv2.imread('dataset/training-data'\
                          '/1/Alvaro_Uribe_0020.jpg')
fig = plt.figure()
ax1 = fig.add_axes((0.1, 0.2, 0.8, 0.7))
ax1.set_title('Image from category 1')
plt.imshow(cv2.cvtColor(random_image, cv2.COLOR_BGR2RGB))
plt.show()
```

The above code produces the following output:

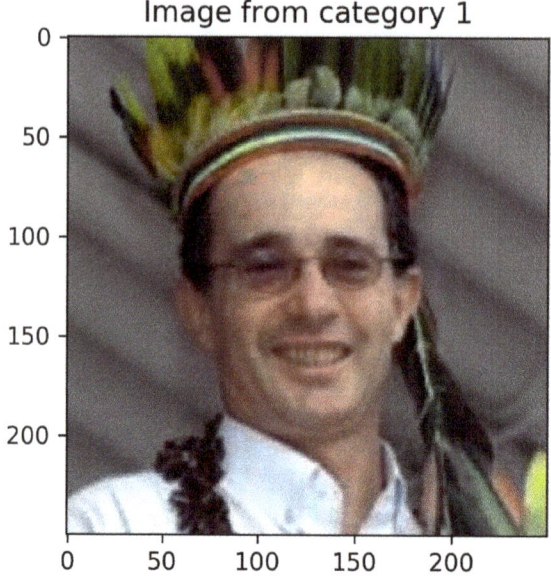

Figure 7.4: An image from the training dataset

The preceding code shows how to display an image from the dataset. Note that the folders titled **0**, **1**, **2** represent the image categories. The directory path should be in **'dataset/training-data/image_category/example_image.jpg'** format, where **example_image.jpg** represents the filename. In the preceding code, the **Alvaro_Uribe_0020.jpg** image is displayed. To display images from another directory, just change the folder and image names to ones that correspond to your system setup.

5. Now, specify the path of the OpenCV XML files that will be used to detect faces in each image. Create a folder named **opencv_xml_files** and save the **haarcascade_frontalface.xml** file inside that folder.

> **NOTE**
>
> The **haarcascade_frontalface.xml** file is available at https://packt.live/3gW5zeB.

The code is as follows:

```
haarcascade_frontalface = 'opencv_xml_files/\
                          'haarcascade_frontalface.xml'
```

6. Detect the face in the image using the **detect_face** function. This function first takes a color image as an input and converts it to grayscale. Then, it loads the XML cascade files that will detect faces in the image. Finally, it returns the faces and their coordinates in the input image:

```
def detect_face(input_img):
    image = cv2.cvtColor(input_img, cv2.COLOR_BGR2GRAY)
    face_cascade = \
    cv2.CascadeClassifier('opencv_xml_files'\
                          '/haarcascade_frontalface.xml')
    faces = face_cascade.detectMultiScale(image, scaleFactor=1.2, \
                                          minNeighbors=5)
    if (len(faces)==0):
        return -1, -1
    (x, y, w, h) = faces[0]
    return image[y:y+w, x:x+h], faces[0]
```

7. Define a function named **prepare_training_data**. First, initialize two empty lists to store the detected faces and their respective categories. Read the images from the directories, as shown in *step 3*. Then, for each image, try to detect the faces. If a face is detected in the image, then resize the face to **(121, 121)** and append it to the list:

```
def prepare_training_data(training_data_folder_path):
    detected_faces = []
    face_labels = []
    traning_image_dirs = os.listdir(training_data_folder_path)
    for dir_name in traning_image_dirs:
        label = int(dir_name)
        training_image_path = training_data_folder_path + "/" \
                              + dir_name
        training_images_names = os.listdir(training_image_path)
        for image_name in training_images_names:
            image_path = training_image_path + "/" + image_name
            image = cv2.imread(image_path)
            face, rect = detect_face(image)
            if face is not -1:
```

```
                resized_face = cv2.resize(face, (121,121), \
                    interpolation = cv2.INTER_AREA)
                detected_faces.append(resized_face)
                face_labels.append(label)
    return detected_faces, face_labels
```

> **NOTE**
>
> The Eigenfaces method expects all images to be the same size. To achieve that, we have fixed the dimensions of all images to **(121,121)**. We can change this value. We can set the minimum size of the image as fixed dimensions or we can also take an average of dimensions.

8. Use this code to call the function defined in the previous step:

```
detected_faces, face_labels = prepare_training_data(\
                    "dataset/training-data")
```

9. Print the number of training images and labels:

```
print("Total faces: ", len(detected_faces))
print("Total labels: ", len(face_labels))
```

The preceding code produces the following output:

```
Total faces: 105
Total labels: 105
```

Note that the number of training images and the number of labels should be equal.

10. OpenCV is equipped with face recognizer modules. So, use the Eigenfaces recognizer module from OpenCV. Currently, the default parameter values are selected. You can specify the parameter values to see the result:

```
eigenfaces_recognizer = cv2.face.EigenFaceRecognizer_create()
```

You now have prepared the training data and initialized the face recognizer.

11. Now, in this step, train the face recognizer model. Convert the labels into a NumPy array before passing it into the recognizer, as OpenCV expects labels to be a NumPy array:

```
eigenfaces_recognizer.train(detected_faces,np.array(face_labels))
```

12. Define the **draw_rectangle** and **draw_text** functions. This will help to draw a rectangular box around the detected faces with the predicted class or category:

```
def draw_rectangle(test_image, rect):
    (x, y, w, h) = rect
    cv2.rectangle(test_image, (x, y), (x+w, y+h), (0, 255, 0), 2)

def draw_text(test_image, label_text, x, y):
    cv2.putText(test_image, label_text, (x, y), \
                cv2.FONT_HERSHEY_PLAIN, 1.5, (0, 255, 0), 2)
```

13. Let's predict the category of a new image. Define a function named **predict** that takes a new image as input. Then, it passes the image to the face detection function to detect faces in the test image. The Eigenface method expects an image of a certain size. So, the function will resize the test image to match the size of the training images. Then, in the next step, it passes the resized image to the face recognizer to identify the category of the image. The **label_text** variable will return the category of the image. Use the **draw_rectangle** function to draw a green rectangular box around the face using the four coordinates obtained during face detection. Alongside the rectangular box, the next line uses the **draw_text** function to write the predicted category of the image:

```
def predict(test_image):
    detected_face, rect = detect_face(test_image)

    resized_test_image = cv2.resize(detected_face, (121,121), \
                          interpolation = cv2.INTER_AREA)

    label = eigenfaces_recognizer.predict(resized_test_image)
    label_text = tags[label[0]]

    draw_rectangle(test_image, rect)
    draw_text(test_image, label_text, rect[0], rect[1]-5)

    return test_image, label_text
```

14. Training images are made up of the following categories. These categories represent unique people to whom the faces that are detected belong. You can think of them as names corresponding to the faces:

```
tags = ['0', '1', '2', '3', '4']
```

15. So far, you have trained the recognizer model. Now, you will test the model with test images. In this step, you will first read a test image.

> **NOTE**
>
> The image is available at https://packt.live/2Vv3uyb.

The code is as follows:

```
test_image = cv2.imread('dataset/test-data/1/Alvaro_Uribe_0021.jpg')
```

16. Call the **predict** function to predict the category of the test image:

```
predicted_image, label = predict(test_image)
```

17. Now, let's display the result of the test image. *Step 16* returns the predicted image and the category label. Use **matplotlib** to display the images. Note that OpenCV generally represents RGB images in multi-dimensional arrays but in reverse order, that is, BGR. So, to display an image in the original order, use OpenCV's built-in **cv2.COLOR_BGR2RGB** function to convert BGR to RGB. Note that for each new test image, we must change the value of **i** in **tags[i]** to find out the actual class:

```
fig = plt.figure()
ax1 = fig.add_axes((0.1, 0.2, 0.8, 0.7))
ax1.set_title('actual class: ' + tags[1]+ ' | ' \
              + 'predicted class: ' + label)
plt.axis('off')
plt.imshow(cv2.cvtColor(predicted_image, cv2.COLOR_BGR2RGB))
plt.show()
```

The output is as follows:

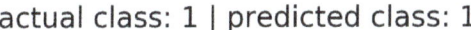

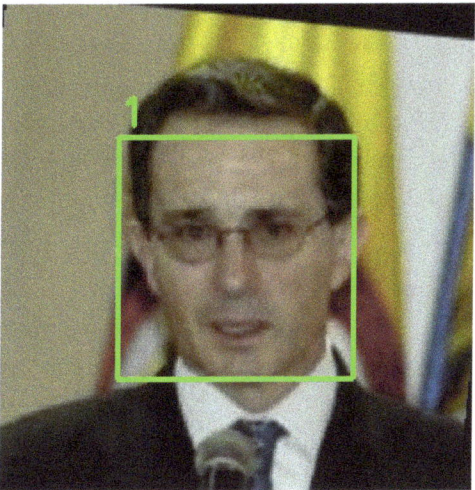

Figure 7.5: Recognition result from the test data using the Eigenface method

You have predicted the category of a new image from `class 1`. Similarly, you can predict the category of different test images from other classes also.

> **NOTE**
>
> The test images of other classes are available at https://packt.live/38IUQaj.
>
> To access the source code for this specific section, please refer to https://packt.live/2YN0ChZ.

In this exercise, you have learned how to implement a facial recognition model using the Eigenface method. You started off using Haar Cascade frontal face detection to detect the faces, and then you trained a facial recognition model using the training dataset. Finally, you used the testing dataset to find out about the performance of the model on an unseen dataset. You can also implement a similar model using a different dataset. You can extend the solution to retrieve similar images from the database.

But there are certain limitations to the Eigenfaces method. In the next section, we will discuss the limitations of this method, and then we will move on to look at the next method, called the Fisherface method.

LIMITATIONS OF THE EIGENFACE METHOD

We have discussed the underlying concepts of the Eigenface method and have implemented a face recognizer using Eigenfaces. But there are some limitations to this method. Before moving to the next method, let's discuss a few drawbacks of the Eigenface approach. We know that an Eigenface recognizer takes the facial features from all images at once and then it applies PCA to find the principal components (Eigenfaces). Intuitively, we can say that by considering all the features at once, we are not strictly capturing the distinct features of each image that would eventually help to distinguish between the faces of two different people.

Another drawback of this approach is that the effectiveness of PCA varies with illumination (lighting) levels. So, if we feed images with varying light levels to PCA, then the projected feature space will contain principal components that retain the maximum variation in lighting. It is possible that variation in the images due to the varying illumination levels may be greater than the variation due to the facial features of different people. Consequently, the recognition accuracy of an Eigenface recognizer will suffer.

For example, suppose we have images of people with varying illumination levels, as shown in *Figure 7.6*. *Figure 7.6(a)* shows the same face at varying illumination levels. *Figure 7.6(b)* and *Figure 7.6(c)* show two different people:

Figure 7.6: Faces of three different people

If we pass these images to an Eigenface face recognizer, then chances are high that PCA will give more importance to the faces shown in *Figure 7.6(a)* and may discard the features of faces shown in *Figure 7.6(b)* or *Figure 7.6(c)*. Hence, an Eigenfaces face recognizer may not contain features of all the people and eventually will fail to recognize a few faces.

Variation due to illumination levels can be reduced by removing the 3 most significant principal components. But we do not know whether the three significant principal components are due to variation in lighting. So, the removal of the three significant features may lead to information loss.

We have discussed how facial recognition using Eigenfaces works. We have also gone through a few drawbacks. Now, we need to look for ways to improve the limitations of the Eigenfaces-based face recognizer and to do so, we will investigate another facial recognition technique called the Fisherfaces facial recognition method in the next section. We will see whether this method can solve the limitations of the Eigenface method.

FISHERFACE

In the previous section, we talked about how one of the limitations of the Eigenface method is that it captures the facial features of all the training images at once. We can solve this limitation if we extract the facial features of all the unique identities separately. So, even if there is variation in lighting, the facial features of one person will not dominate the facial features of others. The Fisherface method does this for us.

In facial recognition techniques, a low-dimensional feature representation with an enhanced discriminator function is always preferred. The Fisherface method uses **Linear Discriminant Analysis** (**LDA**) for dimensionality reduction. It aims to maximize inter-class separation and minimize separation within classes.

For example, suppose we have 10 unique identities, each with 10 faces. The Eigenface method will create 100 Eigenfaces corresponding to each image, but the Fisherface method will create 10 Fisherfaces corresponding to each class, not to each image. Consequently, variation inside one class will not affect other classes.

So, if PCA can perform dimensionality reduction, then why use LDA? We know that PCA aims to maximize the total variance in data. Though it is a widely practiced technique for representing data, it does not consider categories of data and eventually loses certain information. Due to this limitation, PCA fails under certain conditions.

372 | Object Detection and Face Recognition

LDA is also a dimensionality reduction technique, but LDA is a supervised method. It uses a label set during the training steps to build a more reliable dimensionality reduction method. Suppose we have two classes. LDA will compute the mean, μ, and variance, σ, of each class and project data points from the current feature dimensions to new dimensions such that it maximizes the difference between μ_1 and μ_2 and minimizes the variances, σ_1 and σ_2, within the class. *Figure 7.7* shows how LDA dimensionality reduction works:

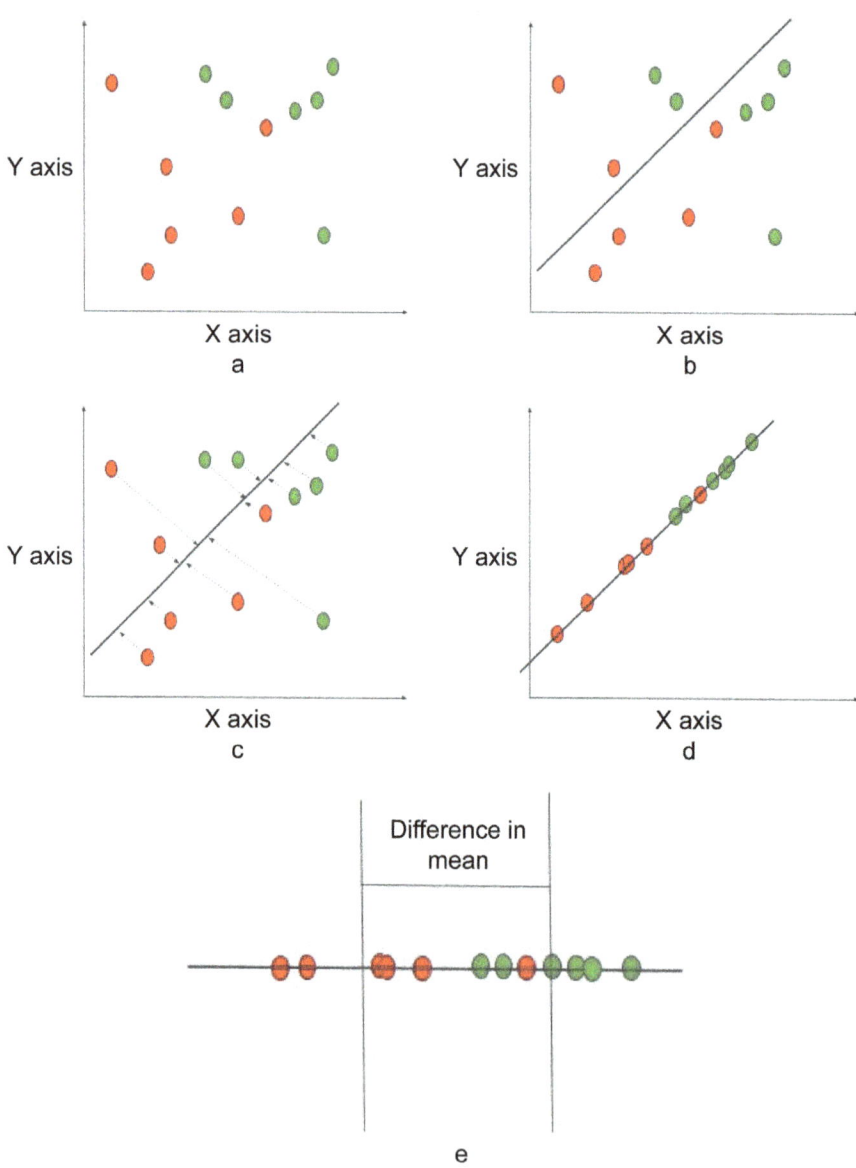

Figure 7.7: Dimensionality reduction in LDA

Let's look at an example to understand this. Suppose we have two-dimensional data as shown in *Figure 7.7(a)*. Now, the aim of LDA is to find a line that will linearly separate the two-dimensional data, as shown in *Figure 7.7(b)*. One way to do this could be projecting all the data points onto either one of the two axes, but this way, we will ignore information from the other axis. For example, if we project all the data points on the *x* axis, then we will lose information from the *y* axis. As discussed, LDA provides a better way. It projects the data points onto a new axis such that it maximizes the separation between the two categories, as shown in *Figure 7.7(c)* and *Figure 7.7(d)*. *Figure 7.7(e)* shows the projected data on the new dimension.

So far, we have discussed the concepts of the Fisherface method. Now we will build a face recognizer using this method. Fortunately, OpenCV provides a face recognizer module called **cv2.face.FisherFaceRecognizer_create()**. We will use the Fisherface face recognizer module to implement the next exercise. We will feed images and their respective categories to this module during the training phase, and then we will use the trained model to recognize a new face.

This module takes two parameters. **num_components** is the number of components for LDA. Generally, we keep the number of components close to the number of categories. The default value for the number of components is **c-1**, where **c** represents the number of categories. So, if we don't provide a number, it is automatically set to **c-1**. If we provide a value less than **0** or greater than **c-1**, it will be set to the default. The other parameter is **value.threshold**. Refer to the threshold part of the *Eigenfaces* section of this chapter.

The following functions are used by the **FisherFaceRecognizer** module to train and test a model respectively:

- **fisherfaces_recognizer.train()**: This function is used to train the model. It takes the previously mentioned parameters as input arguments.

- **fisherfaces_recognizer.predict()**: This function is used to predict the label of a new image. It takes an image as an input argument.

374 | Object Detection and Face Recognition

We have discussed the underlying concepts of LDA. Let's discuss the major steps of the Fisherface method in face recognition. *Figure 7.8* and *Figure 7.9* depict the training steps and test steps of the Fisherface algorithm using OpenCV, respectively:

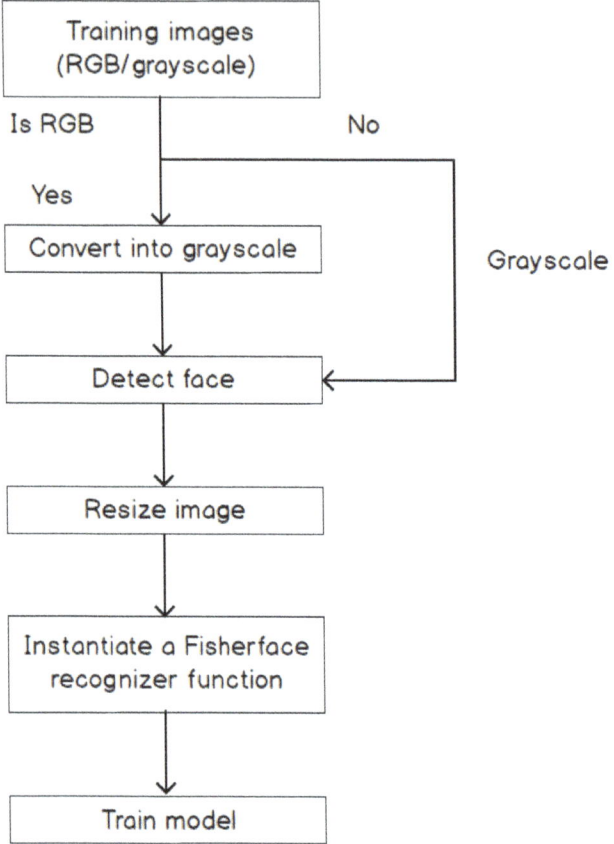

Figure 7.8: Training steps of the Fisherface algorithm

As shown in *Figure 7.8*, the training starts with taking a color or grayscale image as input. If the input is a color image, then the next step converts the input image into grayscale; otherwise, the input image is passed to the face detection function. As we discussed earlier, the Fisherface algorithm requires images to be of the same size; hence, in the next step, all images are resized to the same dimensions. Then, the next step instantiates the recognizer model using the OpenCV module. Finally, the model is trained using the training images:

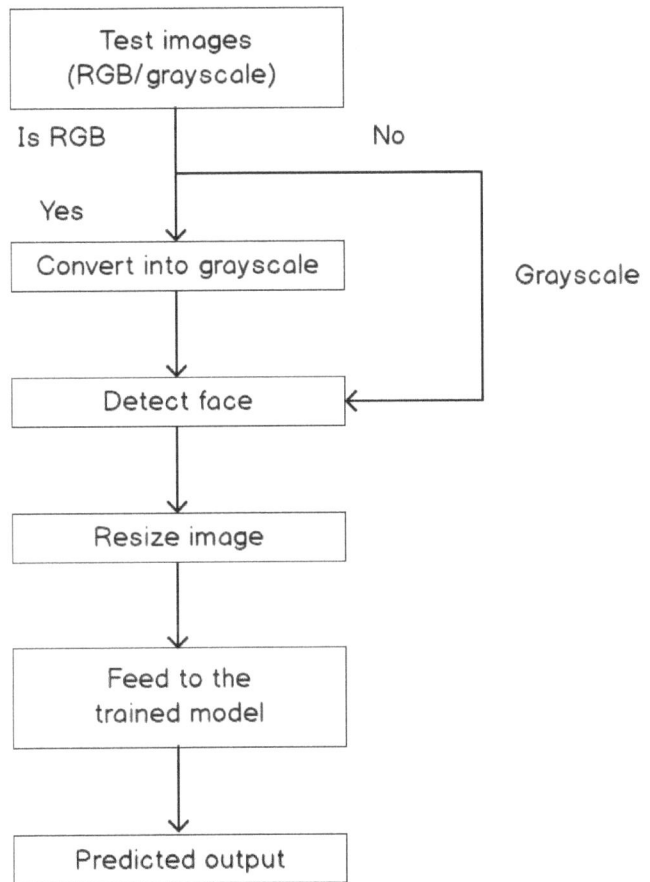

Figure 7.9: Steps during the test phase of the Fisherface algorithm

As shown in *Figure 7.9*, the testing starts with the model taking a new color or grayscale image as input. If the input is a color image, then the next step converts the input image into grayscale; otherwise, the input image is passed to the face detection function. As we discussed earlier, the Fisherface algorithm requires images to be of the same size; hence, in the next step, all images are resized to the same dimensions. Then, the next step feeds the resized image to the trained model (which was trained during the training steps), and, finally, it outputs the category of the input image.

EXERCISE 7.02: FACIAL RECOGNITION USING THE FISHERFACE METHOD

In this exercise, we will be performing facial recognition using the Fisherface method. The images used in this exercise are sourced from http://vis-www.cs.umass.edu/lfw/#download. Note that in this exercise, the faces of a few famous personalities have been used, but you can implement the exercise with any facial dataset. For this exercise, the basic pipeline will stay the same as in *Exercise 7.01, Facial Recognition Using Eigenfaces*. We will start by using Haar Cascades for frontal face detection and then the Fisherface method for face recognition:

1. Firstly, open an untitled Jupyter Notebook and name the file **Exercise7.02**.

2. Import all the required libraries for this exercise:

```
import cv2
import os
import numpy as np
import matplotlib.pyplot as plt
%matplotlib inline
```

3. Next, specify the path of the dataset. Create a folder named **dataset** and, inside that, create two more folders, **training-data** and **test-data**. Put the training and test images in **training-data** and **testing-data**, respectively.

   > **NOTE**
   >
   > The **training-data** and **test-data** folders are available at https://packt.live/2WibmU4.

The code is as follows:

```
training_data_folder_path = 'dataset/training-data'
test_data_folder_path = 'dataset/test-data'
```

4. Display a training sample using **matplotlib**. To check whether the images from the **dataset** are being displayed correctly, use the following code:

```
random_image = cv2.imread('dataset/training-\
            data/2/George_Robertson_0020.jpg')
fig = plt.figure()
ax1 = fig.add_axes((0.1, 0.2, 0.8, 0.7))
ax1.set_title('Image from category 2')
# change category name #accordingly
plt.imshow(cv2.cvtColor(random_image, cv2.COLOR_BGR2RGB))
plt.show()
```

The output is as follows:

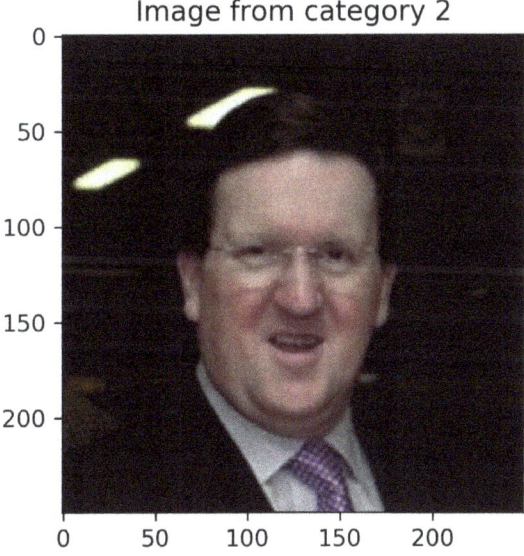

Figure 7.10: An image from the training dataset

The preceding code shows how to display an image from the dataset. Note that the folders titled **0,1,2** represent the image categories. The directory path should be in **'dataset/training-data/image_category/example_image.jpg'** format, where **example_image.jpg** represents the filename. In the preceding code, the image titled **George_Robertson_0020.jpg** is displayed. To display images from another directory, just change the folder and image names to ones that correspond to your system setup.

5. Specify the path of the OpenCV XML files that will be used to detect faces in each image. Create a folder named **opencv_xml_files** and save **haarcascade_frontalface.xml** inside that folder.

> **NOTE**
>
> The **haarcascade_frontalface.xml** file is available at https://packt.live/38fDThI.

The code is as follows:

```
haarcascade_frontalface = 'opencv_xml_files/ haarcascade_frontalface.xml'
```

6. Detect the faces in an image using the **detect_face** function. This function takes color images as input and converts them to grayscale. Then, it loads the XML cascade files that will detect faces in the image. Finally, it returns the faces and their coordinates in the input image:

```
def detect_face(input_img):
    image = cv2.cvtColor(input_img, cv2.COLOR_BGR2GRAY)

    face_cascade = cv2.CascadeClassifier(\
                   haarcascade_frontalface)

    faces = face_cascade.detectMultiScale(image, \
            scaleFactor=1.2, minNeighbors=5);

    if (len(faces)==0):
        return -1, -1
    (x, y, w, h) = faces[0]
    return image[y:y+w, x:x+h], faces[0]
```

7. Define the **prepare_training_data** function. This will first initialize two empty lists that will store the detected faces and their respective categories. Then it reads images from directories, shown in *Step 3*. Then, for each image, it tries to detect faces. If a face is detected in the image, then it resizes the face to **(121, 121)** and appends it to the list:

```
def prepare_training_data(trainig_data_folder_path):
    detected_faces = []
    face_labels = []
    traning_image_dirs = os.listdir(training_data_folder_path)
    for dir_name in traning_image_dirs:
        label = int(dir_name)
        training_image_path = training_data_folder_path + \
                        "/" + dir_name
        training_images_names = os.listdir(training_image_path)
        for image_name in training_images_names:
            image_path = training_image_path + "/" + image_name
            image = cv2.imread(image_path)
            face, rect = detect_face(image)
            if face is not -1:
                resized_face = cv2.resize(face, (121,121), \
                                interpolation = cv2.INTER_AREA)
                detected_faces.append(resized_face)
                face_labels.append(label)
    return detected_faces, face_labels
```

> **NOTE**
>
> The Fisherface method expects all images to be of the same size. To achieve that, we have fixed the dimensions of all images at **(121,121)**. We can change this value. We can set the minimum size of the image or we can also take an average of the dimensions.

8. Call the function mentioned in the previous step:

```
detected_faces, face_labels = \
prepare_training_data("dataset/training-data")
```

9. Display the number of training images and labels:

   ```
   print("Total faces: ", len(detected_faces))
   print("Total labels: ", len(face_labels))
   ```

   The preceding code produces the following output:

   ```
   Total faces: 105
   Total labels: 105
   ```

10. Use the Fisherface recognizer module from OpenCV. Currently, the default value of the parameter is selected:

    ```
    fisherfaces_recognizer = cv2.face.FisherFaceRecognizer_create()
    ```

11. You have prepared the training data and initialized a face recognizer. Now, in this step, train the face recognizer. Note that the labels are first converted into a NumPy array before being passed to the recognizer because OpenCV expects labels to be a NumPy array:

    ```
    fisherfaces_recognizer.train(detected_faces, np.array(face_labels))
    ```

12. Define the **draw_rectangle** and **draw_text** functions. These help to draw rectangular boxes around detected faces with the predicted class or category:

    ```
    def draw_rectangle(test_image, rect):
        (x, y, w, h) = rect
        cv2.rectangle(test_image, (x, y), (x+w, y+h), \
                     (0, 255, 0), 2)
    def draw_text(test_image, label_text, x, y):
        cv2.putText(test_image, label_text, (x, y), \
        cv2.FONT_HERSHEY_PLAIN, 1.5, (0,255, 0), 2)
    ```

13. Now, let's predict the category of a new image. The **predict** function takes a new image as input. Then, it passes the image to the face detection function to detect faces in the test image. As discussed earlier, the Fisherface method expects a fixed-size image. So, the function will resize the test image to match the size of the training images. Then, in the next step, it passes the resized image to the face recognizer to identify the category of the image. The **label_text** variable will return the category of the image. Finally, it uses the **draw_rectangle** function to draw a green rectangular box around the face using the four coordinates obtained during face detection. Alongside the rectangular box, the next line uses the **draw_text** function to write the predicted category of the image:

```
def predict(test_image):
    detected_face, rect = detect_face(test_image)
    resized_face = cv2.resize(detected_face, (121,121), \
                              interpolation = cv2.INTER_AREA)
    label = fisherfaces_recognizer.predict(resized_face)
    label_text = tags[label[0]]
    draw_rectangle(test_image, rect)
    draw_text(test_image, label_text, rect[0], rect[1]-5)
    return test_image, label_text
```

14. The training images have the following categories. This is similar to the **tags** array we used in the previous exercise. The basic idea behind using these tags is to show that the faces belong to a specific category/class:

```
tags = ['0', '1', '2', '3', '4']
```

15. Then, we need to test the model with the test images. In this step, the test image is read and converted to grayscale:

> **NOTE**
>
> The image is available at https://packt.live/2Zsw3ig.

```
test_image = cv2.imread('dataset/test-data/2/\
                        George_Robertson_0021.jpg')
```

16. Call the **predict** function to predict the category of the test image:

   ```
   predicted_image, label = predict(test_image)
   ```

17. Finally, let's display the result of the test image. *Step 16* returns the predicted image and its respective category label. Use **matplotlib** to display the images. Note that OpenCV generally represents RGB images in multi-dimensional arrays, but in reverse order, that is, BGR. So, to display the images in the original order, we will use OpenCV's built-in **cv2.COLOR_BGR2RGB** function to convert from BGR to RGB. *Figure 7.11* shows the prediction results:

   ```
   fig = plt.figure()
   ax1 = fig.add_axes((0.1, 0.2, 0.8, 0.7))
   ax1.set_title('actual class: ' + tags[2]+ ' | ' \
                 + 'predicted class: ' + label)
   plt.axis('off')
   plt.imshow(cv2.cvtColor(predicted_image, cv2.COLOR_BGR2RGB))
   plt.show()
   ```

The output is as follows:

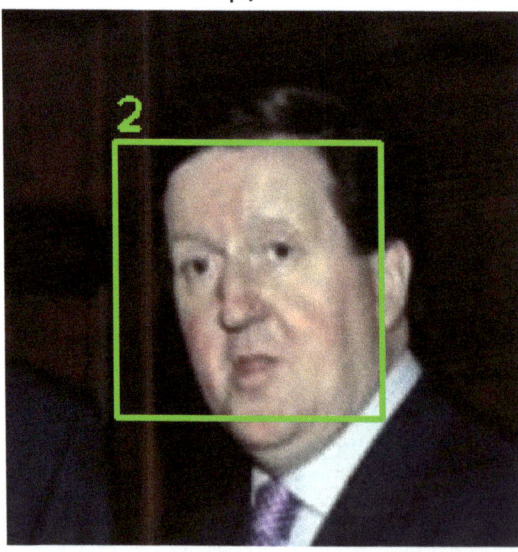

Figure 7.11: Recognition result on test data using the Fisherface method

In this exercise, you have predicted the category of a new image from `class 2`. Similarly, you can predict the categories of different test images from other classes also. We also learned how we can use the Fisherface method for face recognition. One important thing to note here is the ease with which you can shift from the Eigenface recognition method to the Fisherface method by just changing one OpenCV function.

> **NOTE**
>
> The test images of other classes are available at https://packt.live/3gf6hDr.
>
> To access the source code for this specific section, please refer to https://packt.live/38f3ibx.

Now we know that the Fisherface method helps to capture the facial features of all the unique identities separately. So, the facial features of one person will not dominate the others. Despite this, the method still considers varying illumination levels as a feature. We know that illumination changes create incoherence in images, and consequently, the performance of a face recognizer reduces. In the next section, we will discuss another facial recognition algorithm and find out whether it solves the illumination problem.

LOCAL BINARY PATTERNS HISTOGRAMS

We have talked about how the Eigenface and Fisherface methods consider illumination changes as features. We also know that it is not always possible to have ideal lighting conditions. So, we need a method that addresses the limitation of varying illumination. The LBPH method helps to overcome that limitation.

The LBPH algorithm uses a local image descriptor called **Linear Binary Pattern** (**LBP**). LBP describes the contrast information of a pixel with regard to its neighboring pixels. It describes each pixel in an image with a certain binary pattern. It generally applies to grayscale images. The LBP method does not consider a complete image at once. Instead, it focuses on the local structure (patches) of an image by comparing each pixel to its neighboring pixels.

Let's discuss how LBP extracts features from an image. Originally, LBP was defined for a window (filter) of size 3x3, as shown in *Figure 7.12(a)*, but a circular LBP operator can be used to improve performance, shown in *Figure 7.12(b)*. Generally, it takes the center pixel's value as the threshold value for the window and compares this threshold with its neighboring 8 pixels. We can also consider the average pixel value as a threshold:

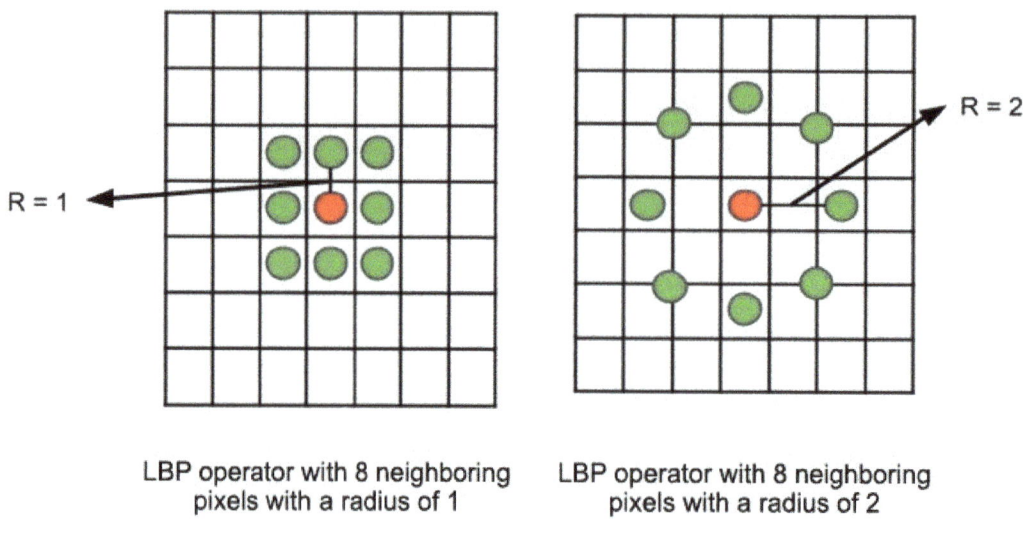

LBP operator with 8 neighboring pixels with a radius of 1

a

LBP operator with 8 neighboring pixels with a radius of 2

b

Figure 7.12: LBP operators with radii of 1 and 2

Now, if the values of adjacent pixels are greater than or equal to the threshold value, then you substitute the pixel value with 1; otherwise, you substitute it with 0, as shown in *Figure 7.13*. As a result, this generates an 8-bit binary number that is then converted to a decimal number to obtain the LBP value of the center pixel:

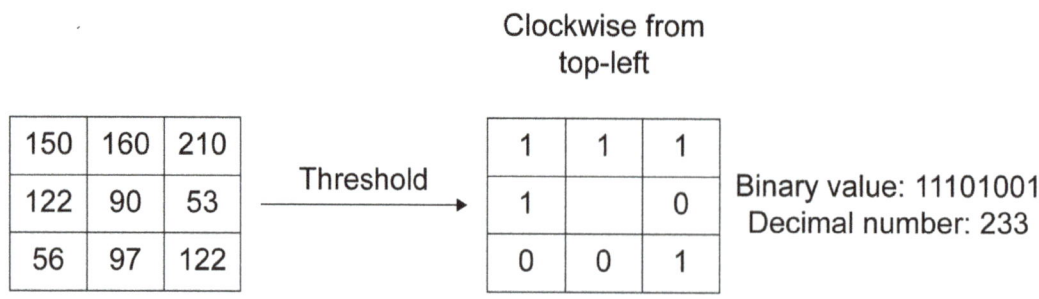

Figure 7.13: Estimation of LBP values

This LBP value is considered as the textual features of that local region. The LBP operator computes the LBP value for each pixel of an image. After calculating the LBP values for each pixel, the LBPH method converts those values into a histogram. Eventually, we will have a histogram for each image in the dataset. So, LBPH uses a combination of LBP and histograms to represent the feature vectors of an image. Note that this local features-based approach provides more robustness against illumination changes.

Until now, we have discussed the underlying concepts of the LBPH method. Now we will build a face recognizer using this method. We will use the LBPH face recognizer module called **cv2.face.LBPHFaceRecognizer_create()** from OpenCV to implement the next exercise. We will feed the images and their respective categories to this module during the training phase, and we will use the trained model to recognize a new face. We will mainly be concerned with three input parameters for this module:

- **radius**: As shown in *Figure 7.12(b)*, the LBPH method can also build circular local binary patterns. The parameter radius determines the positions of adjacent pixels.

- **neighbors**: The number of adjacent pixels to determine the LBP value. We can also create more adjacent pixels using interpolation methods such as bilinear interpolation. Note that the computational cost is proportionate to the number of adjacent pixels.

- **threshold**: Refer to the threshold part of the *Eigenfaces* section of this chapter.

The following functions are used by the **LBPHFaceRecognizer** module to train and test a model respectively:

- **lbphfaces_recognizer.train()**: This function is used to train the model. It takes the previously mentioned parameters as input arguments.

- **lbphfaces_recognizer.predict()**: This function is used to predict the label of a new image. It takes an image as its input argument.

386 | Object Detection and Face Recognition

Let's discuss the major steps involved in the LBPH algorithm. *Figure 7.14* and *Figure 7.15* depict the training steps and test steps in the LBPH algorithm using OpenCV, respectively:

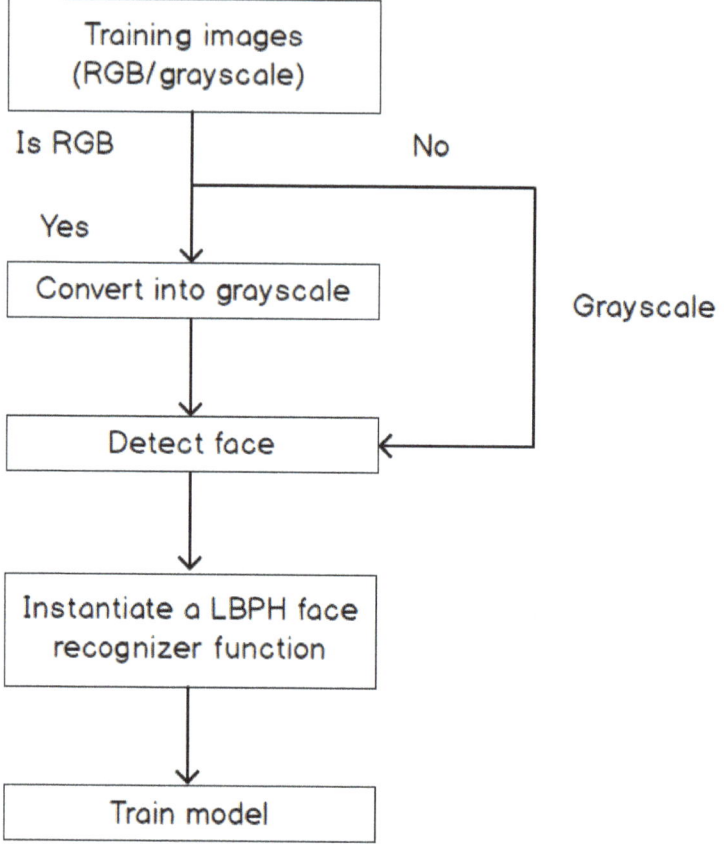

Figure 7.14: Training steps of the LBPH algorithm

As shown in *Figure 7.14*, the training steps first take color or grayscale images as input. If the input is a color image, then the next step converts the input image into grayscale; otherwise, the input image is passed to the face detection function. The LBP algorithm does not necessarily require images to be the same size; hence, a resize function will not be implemented in the next exercise. Then, the next step instantiates the recognizer model using the OpenCV module. Finally, the model is trained using training images:

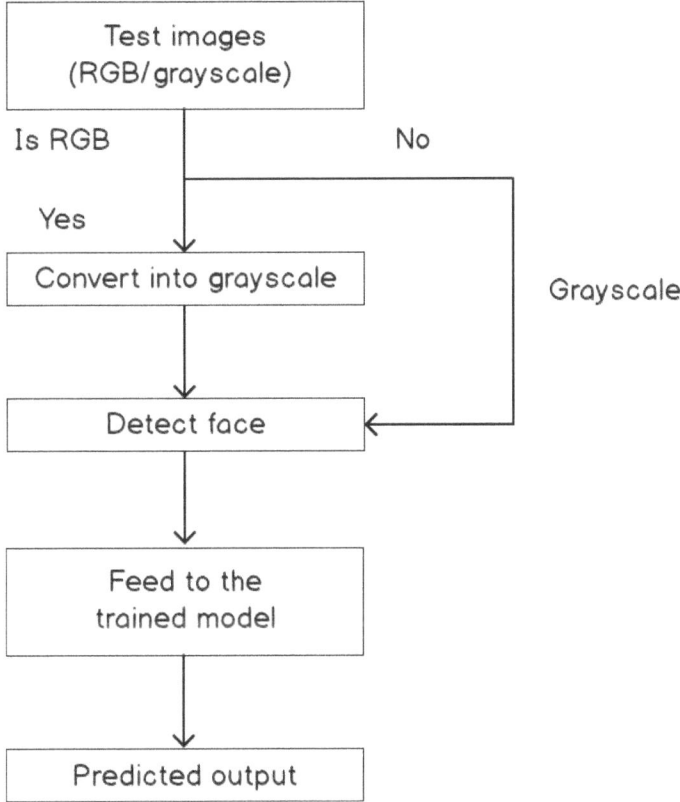

Figure 7.15: Steps during the test phase of the LBPH algorithm

As shown in *Figure 7.15*, the testing starts with a new color or grayscale image being taken as input. If the input is a color image, then the next step converts the input image into grayscale; otherwise, the input image is passed to the face detection function. The LBP algorithm does not necessarily require images to be of the same size; hence, a resize function will not be implemented in the next exercise. Then, the next step feeds the image to the trained model (which was trained during training steps). Finally, it outputs the category of the input image.

EXERCISE 7.03: FACIAL RECOGNITION USING THE LBPH METHOD

In this exercise, we will be performing facial recognition using the LBPH method. The problem statement and the dataset will stay the same as those for the previous two exercises. We will only be changing the facial recognition method being used. This way, you will get hands-on experience of using different techniques for facial recognition:

1. Firstly, open an untitled Jupyter Notebook and name the file **Exercise7.03** or **face_recongition_lbph**.

2. Start by importing all the required libraries for this activity:

```
import cv2
import os
import numpy as np
import matplotlib.pyplot as plt
%matplotlib inline
```

3. Then, specify the path of the dataset. Create a folder named **dataset** and, inside that, create two more folders, **training-data** and **test-data**. Put the training and test images into **training-data** and **test-data**, respectively.

   > **NOTE**
   >
   > The **training-data** and **test-data** folders are available at https://packt.live/3eFDsPr.

   The code is as follows:

   ```
   training_data_folder_path = 'dataset/training-data'
   test_data_folder_path = 'dataset/test-data'
   ```

4. Display a training sample using **matplotlib**. To check whether the images from the dataset are displaying correctly, use the following code:

```
random_image = cv2.imread('dataset/training-data/3\
                /George_W_Bush_0020.jpg')
fig = plt.figure()
ax1 = fig.add_axes((0.1, 0.2, 0.8, 0.7))
ax1.set_title('Image from category 3')
# change category name accordingly
plt.imshow(cv2.cvtColor(random_image, cv2.COLOR_BGR2RGB))
plt.show()
```

The output is as follows:

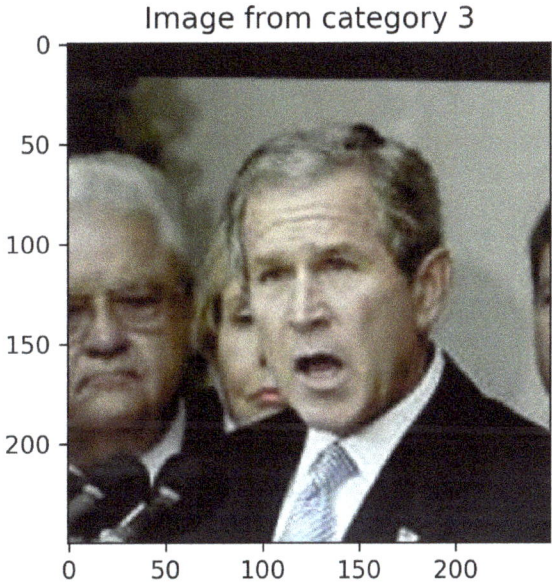

Figure 7.16: An image from the training dataset

In the preceding code, the image titled **George_W_Bush_0020.jpg** is displayed. To display images from another directory, just replace the folder and image names with any other existing folder and image names.

5. Specify the path of the OpenCV XML files that will be used to detect the faces in each image. Create a folder named **opencv_xml_files** and save **lbpcascade_frontalface.xml** inside that folder. Instead of **haarcascade**, this time we will use **lbpcascade**, for LBP-based face detection:

```
lbpcascade_frontalface = 'opencv_xml_files/lbpcascade_frontalface.xml'
```

> **NOTE**
>
> The **lbpcascade_frontalface.xml** file is available at https://packt.live/2CTmHD3.

6. Use the **detect_face** function to detect the faces in an image. It first takes a color image as an input and converts it to grayscale. Then, it loads the XML cascade files that will detect the faces in the image. Finally, it returns the faces and their coordinates in the input image:

```
def detect_face(input_img):
    image = cv2.cvtColor(input_img, cv2.COLOR_BGR2GRAY)
    face_cascade = cv2.CascadeClassifier(lbpcascase_frontalface)
    faces = face_cascade.detectMultiScale(image, \
            scaleFactor=1.2, minNeighbors=5)
    if (len(faces)==0):
        return -1, -1
    (x, y, w, h) = faces[0]
    return image[y:y+w, x:x+h], faces[0]
```

7. Define the **prepare_training_data** function such that it first initializes two empty lists that will store the detected faces and their respective categories, which will be eventually used for the training. Then, it should read the images from the directories shown in *Step 3*. Then, from each image, it tries to detect faces. If a face is detected in the image, then it appends the image to the list:

```
def prepare_training_data(training_data_folder_path):
    detected_faces = []
    face_labels = []
    traning_image_dirs = os.listdir(training_data_folder_path)
    for dir_name in traning_image_dirs:
        label = int(dir_name)
        training_image_path = training_data_folder_path + \
```

```
                              "/" + dir_name
        training_images_names = os.listdir(training_image_path)
        for image_name in training_images_names:
            image_path = training_image_path + "/" + image_name
            image = cv2.imread(image_path)
            face, rect = detect_face(image)
            if face is not -1:
                resized_face = cv2.resize(face, (121,121), \
                                    interpolation = cv2.INTER_AREA)
                detected_faces.append(resized_face)
                face_labels.append(label)
    return detected_faces, face_labels
```

> **NOTE**
>
> The LBPH method does not expect all the training images to be of the same size.

8. Call the function mentioned in the previous step:

   ```
   detected_faces, face_labels = prepare_training_data("dataset/training-data")
   ```

9. Display the number of training images and labels:

   ```
   print("Total faces: ", len(detected_faces))
   print("Total labels: ", len(face_labels))
   ```

   The preceding code produces the following output:

   ```
   Total faces: 93
   Total labels: 93
   ```

10. OpenCV is equipped with face recognizer modules. Use the **LBPHFaceRecognizer** module from OpenCV:

    ```
    lbphfaces_recognizer = \
    cv2.face.LBPHFaceRecognizer_create(radius=1, neighbors=8)
    ```

11. In the previous two steps, we have prepared the training data and initialized a face recognizer. Now, in this step, we will train the face recognizer. Note that the labels are first converted into a NumPy array before being passed into the recognizer because OpenCV expects labels to be a NumPy array:

    ```
    lbphfaces_recognizer.train(detected_faces, np.array(face_labels))
    ```

12. Define the **draw_rectangle** and **draw_text** functions. This helps to draw rectangular boxes around the detected faces with the predicted class or category:

    ```
    def draw_rectangle(test_image, rect):
        (x, y, w, h) = rect
        cv2.rectangle(test_image, (x, y), (x+w, y+h), \
                      (0, 255, 0), 2)
    def draw_text(test_image, label_text, x, y):
        cv2.putText(test_image, label_text, (x, y), \
        cv2.FONT_HERSHEY_PLAIN, 1.5, (0, 255, 0), 2)
    ```

13. Create the **predict** function to predict the category of a new image. The **predict** function takes a new image as an input. Then, it passes the image to the face detection function to detect faces from the test image. Though the LBPH method does not expect a fixed size of images, it is always a good practice to process images of the same size. So, the function will resize the test image to match the size of the training images. Then, in the next step, it passes the resized image to the face recognizer to identify the category of the image. The **label_text** variable will return the category of the image. Finally, it uses the **draw_rectangle** function to draw a green rectangular box around the face using the four coordinates obtained during face detection. Alongside the rectangular box, the next line uses the **draw_text** function to write the predicted category of the image:

    ```
    def predict(test_image):
        face, rect = detect_face(test_image)
        label= lbphfaces_recognizer.predict(face)
        label_text = tags[label[0]]
        draw_rectangle(test_image, rect)
        draw_text(test_image, label_text, rect[0], rect[1]-5)
        return test_image, label_text
    ```

14. The training images have the following categories. This is the same as the **tags** array used in the previous exercise:

    ```
    tags = ['0', '1', '2', '3', '4']
    ```

15. So far, we have trained the recognizer model. Now, we will test the model with some test images. In this step, you will first read a test image:

    ```
    test_image = cv2.imread('dataset/test-data/3/'\
                            'George_W_Bush_0021.jpg')
    ```

16. In this step, call the **predict** function to predict the category of the test image:

    ```
    predicted_image, label = predict(test_image)
    ```

17. Now, we will display the results for the test images. *Step 16* returns the predicted images and their respective category labels. Use **matplotlib** to display the images. Note that OpenCV generally represents RGB images in multi-dimensional array but in reverse order, that is, BGR. So, to display the images in the original order, we will use OpenCV's built-in **cv2.COLOR_BGR2RGB** function to convert BGR to RGB. *Figure 7.17* shows the prediction results:

    ```
    fig = plt.figure()
    ax1 = fig.add_axes((0.1, 0.2, 0.8, 0.7))
    ax1.set_title('actual class: ' + tags[3]+ ' | '\
                  + 'predicted class: ' + label)
    plt.axis('off')
    plt.imshow(cv2.cvtColor(predicted_image, \
               cv2.COLOR_BGR2RGB))
    plt.show()
    ```

The output is as follows:

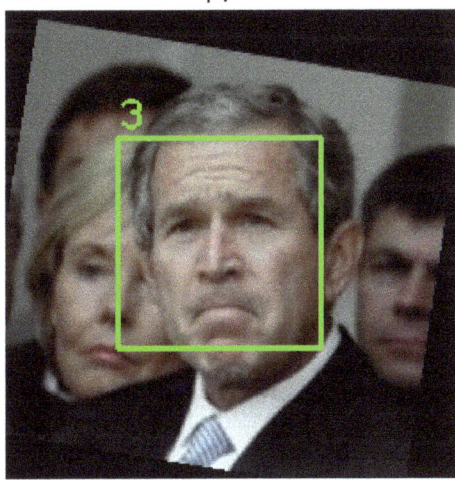

Figure 7.17: Recognition result on the test data using the LBPH method

You have predicted the category of a new image from class **3**. Similarly, you can predict the category of different test images from other classes as well. You can even try varying the parameters that we input in *Step 10* of this exercise. For example, the following figure shows the variation of results when we vary the parameters:

actual class: 0 | predicted class: 4

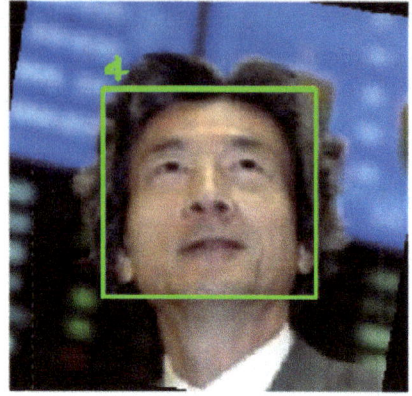

(radius=1, neighbors=8)

actual class: 0 | predicted class: 0

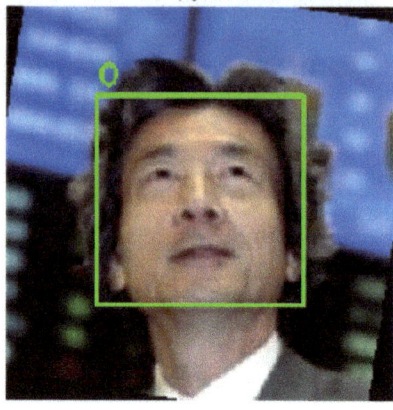

(radius=2, neighbors=12)

Figure 7.18: Recognition results upon varying the parameters

> **NOTE**
>
> The test images of other classes are available at https://packt.live/2ZtQycO.
>
> To access the source code for this specific section, please refer to https://packt.live/2ZtLuoP.

In this exercise, you have learned how to implement a face recognizer model that overcomes the issue of varying illumination levels. Also, note that the examples in *Figure 7.18* show that by varying the parameters of the model, you can enhance the accuracy of the face recognizer model. Initially, if we pass the test image (left) shown in *Figure 7.18* to the trained model, it predicts the wrong category. But once you change the input parameter (that is, once you increase the radius and number of pixels in the face recognizer model) and specify **radius=2** and **neighbors =12** in *Step 10* and train the model again, the model returns the correct result. Hence, you are advised to adjust the parameters as per your requirements.

We have gone through various face recognizer models. We have discussed the pros and cons of each algorithm. We have also discussed how parameter tuning affects our results. We can conclude that these algorithms work well if images are properly pre-processed. If there are varying illumination levels and the positions of faces are beyond a certain angle, then the performance of these algorithms reduces. In the next section, we will discuss another important application of computer vision called object detection. We will also learn how to implement an object detection model using OpenCV.

OBJECT DETECTION

Object detection is a technique that's used for locating and identifying instances of objects in images or videos. Object detection is popular in applications such as self-driving cars, pedestrian detection, object tracking, and many more. Generally, object detection comprises two steps: first, object localization, to locate an object in an image; and second, object classification, which classifies the located object in the appropriate category. There are many algorithms for object detection, but we will discuss one of the more widely used algorithms, called **Single Shot Detector** (**SSD**). Later, we will also implement object detection models using classical algorithms such as the Haar Cascade and LBPH algorithms.

SINGLE SHOT DETECTOR

Single Shot Detector (**SSD**) is a state-of-the-art real-time object detection algorithm that provides much better speeds compared to the **Faster-Regional Convolutional Neural Network** (**RCNN**) algorithm (another popular object detection algorithm). It performs relatively well but takes a different approach. In most of the popular object detection algorithms, such as RCNN or Faster-RCNN, we perform two tasks. First, we generate **Regions of Interest** (**ROIs**) or region proposals (the probable locations of objects), and then we use **Convolutional Neural Networks** (**CNNs**) to classify those regions. SSD does these two steps in a single shot. As it processes the image, it predicts both the bounding (anchor) box and the class. Now, let's learn how SSD achieves this once we have the input image and the ground truth table for the bounding boxes – that is, their coordinates and the objects they contain. Firstly, an image is passed through a CNN (a series of convolutional layers), which generates sets of feature maps at different scales (to include different object size possibilities) – 10 × 10, 5 × 5, and 3 × 3. Now, for each location for the different feature maps, a 3 × 3 filter is used to evaluate a small set of default bounding boxes or anchor boxes. During training, the **Ground Truth Box** (**GTB**) is matched with the **Predicted Box** (**PB**) based on the **Intersection over Union** (**IoU**) metric.

The IoU is calculated as shown here:

```
IoU = (Common area shared by GTB and PB) / (Area of GTB + Area of
PB-Common area shared by GTB and PB)
```

Figure 7.19 (a) is a diagram that describes the IoU. The red and green boxes represent the ground truth and predicted boxes, respectively. The shaded region in the numerator represents the common area (the overlapped region), and the shaded region in the denominator represents the union of the ground truth and predicted boxes. *Figure 7.19 (b)* shows the various possibilities for the IoU. The left-most image is an indicator of a poor IoU and the right-most is an ideal result:

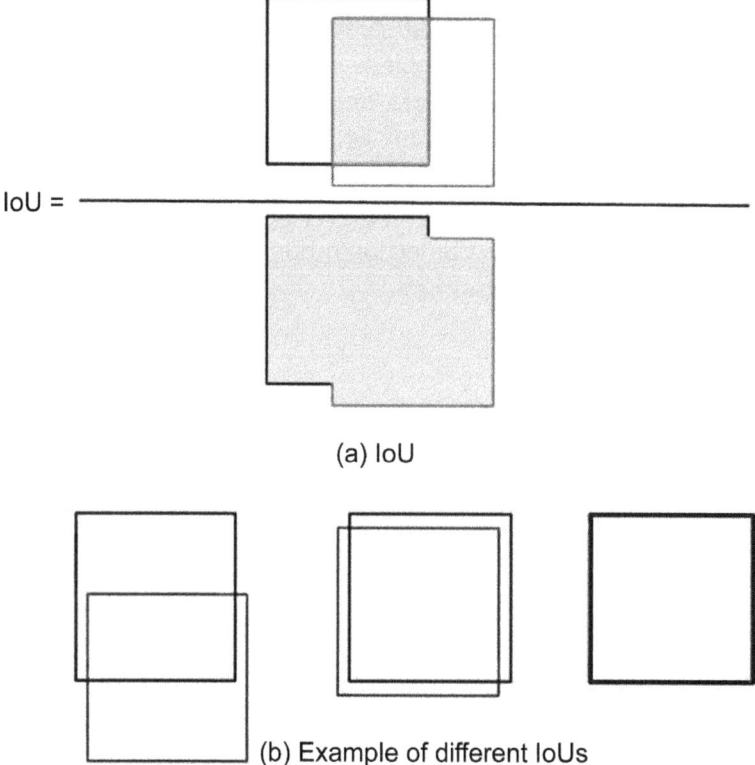

Figure 7.19: An example of non-maximum suppression

The SSD model outputs a large number of bounding boxes with a wide range of IoU values. Out of all those boxes, only the ones with an IoU value of more than 0.5 (or any other threshold) are used for further processing. That is, first, the algorithm detects all positive boxes and then checks for a threshold. Through this approach, a large number of bounding boxes are generated, as we classify and draw bounding boxes for every single position in the image, using multiple shapes, at several different scales. To overcome this, non-maximum suppression (or just **non-max suppression**) is used to combine highly overlapping boxes into one. *Figure 7.20* shows an example of non-max suppression:

Figure 7.20: An example of non-max suppression

To summarize, SSD carries out the region proposal and classification steps simultaneously.

Given *n* classes, each point in the feature map (bounding box) is associated with a **4+n**-dimensional vector that returns **4** on box offset coordinates and *n* class probabilities as output, and the training is carried out by matching this vector with ground truth labels. Look at *Figure 7.21* to understand the SSD framework:

a: Image with ground truth boxes

b: Image with predicted boxes

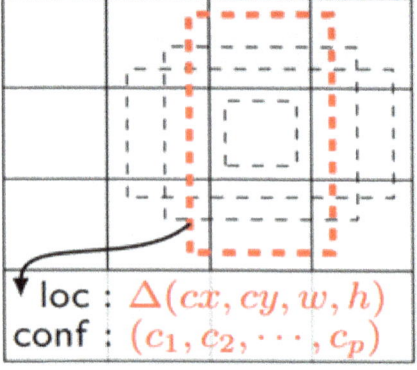

c: 4 x 4 feature map

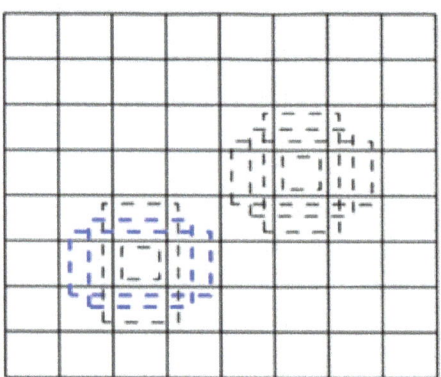

d: 8 x 8 feature map

Figure 7.21: SSD network

To train an object, SSD only requires an input image and GTBs. *Figure 7.21(b)* shows the predicted bounding box with a confidence score. At each location, we convolutionally evaluate a small set (say 4) of bounding boxes of various aspect ratios in several feature maps with different scales (for example, in Figures 7.21 (c) and 7.21(d), we see 4x4 and 8x8 feature maps respectively). Here, the feature maps represent different features of the images, which are outputs of convolutional operations using different filters. For each bounding box, we predict both the shape offsets and the confidence scores (probabilities) for every bounding box, for all object categories, $c_1, c_2, \ldots, c_p$. During training, the bounding boxes are matched to the GTBs. For example, in Figure 7.21(b), we have matched three bounding boxes with bicycle, car, and dog, which we treat as positive. The rest are treated as negatives.

MOBILENET

SSD takes feature maps generated by CNNs. There is a question that can be asked here: if we already have architectures such as VGG (*Very Deep CNNs for Image Recognition, by Simonyan et al., 2015*) or ResNet (*Deep Residual Learning for Image Recognition, by He et al., 2015*), then why consider MobileNet? We know that networks such as VGG and ResNet comprise a large number of layers and parameters and, consequently, this increases the size of the network. But some real-time applications, such as real-time object tracking and self-driving cars, need a faster network. MobileNet is a relatively simple architecture. MobileNet is based on a depth-wise and point-wise convolution architecture. This two-step convolutional architecture reduces the complexity of the convolutional operations of a model.

For more information on MobileNet, please refer to the paper *MobileNets, by Howard et.al, 2017*.

So far, we have discussed the SSD algorithm and the advantages of MobileNet. Next, we will create an object detection model using SSD and MobileNet. Fortunately, OpenCV provides the **dnn** (which stands for **deep neural network**) module to perform object detection with. We will use the Caffe framework to implement object detection. We will follow these steps while implementing the code:

1. Firstly, we will load a pre-trained object detection network from OpenCV's **dnn** module. To implement this, we will use the `cv2.dnn.readNetFromCaffe()` function. This function takes a configuration file and a pre-trained module as inputs.

2. Next, we will pass input images to the network, and the network will return a bounding box with the coordinates of all the instances of objects in each image. To implement this, we will use the following functions:

    `cv2.dnn.blobFromImage()`: This function is used to perform the pre-processing (resizing and normalization) steps. It takes an input image, a scale factor, the required size of the image for the network (the output size of the resized image), and the mean of the image as input arguments. Note that the scale factor is $1/\sigma$, where σ is the standard deviation.

    `net.setInput()`: Since we are using a pre-trained network, this function helps to use the blob created in the last step as input to the network. Finally, we will get the name of the detected object with confidence scores (probabilities). To implement this, we will use the `net.forward()` function. This function helps to pass the input through the network and compute the prediction.

EXERCISE 7.04: OBJECT DETECTION USING MOBILENET SSD

In this exercise, we will implement object detection using MobileNet SSD. We will be using the OpenCV functions from the **dnn** module to perform object detection. Note that we are going to be using a pre-trained MobileNet SSD model implemented in Caffe to perform the inference:

1. Firstly, open an untitled Jupyter Notebook and name the file **Exercise7.04**.

2. Start by importing all the required libraries for this exercise:

```
import cv2
import numpy as np
import matplotlib.pyplot as plt
%matplotlib inline
```

3. Next, specify the path of the pre-trained MobileNet SSD model. Use this model to detect the objects in a new image. Please make sure that you have downloaded the **prototxt** and **caffemodel** files and have saved them in the path where you're running the Jupyter Notebook.

> **NOTE**
>
> The **prototxt** and **caffemodel** files can be downloaded from https://packt.live/2WtqhuM.

The code is as follows:

```
# Load the pre-trained MobileNet SSD model
net = cv2.dnn.readNetFromCaffe('MobileNetSSD_deploy.prototxt.txt', \
                               'MobileNetSSD_deploy.caffemodel')
```

4. The pre-trained model can detect a list of object classes. Let's define those classes in a list:

```
categories = { 0: 'background', 1: 'aeroplane', \
               2: 'bicycle', 3: 'bird', 4: 'boat', \
               5: 'bottle', 6: 'bus', 7: 'car', 8: 'cat', \
               9: 'chair', 10: 'cow', 11: 'diningtable', \
               12: 'dog', 13: 'horse', 14: 'motorbike', \
               15: 'person', 16: 'pottedplant', \
               17: 'sheep', 18: 'sofa', 19: 'train', \
               20: 'tvmonitor'}

# defined in list also
classes = ["background", "aeroplane", "bicycle", "bird", \
           "boat", "bottle", "bus", "car", "cat", "chair", \
           "cow", "diningtable", "dog", "horse", \
           "motorbike", "person", "pottedplant", "sheep", \
           "sofa", "train", "tvmonitor"]
```

5. Now, we will read the input image and construct a blob for the image. Note that MobileNet requires fixed dimensions for all input images, so first resize the image to 300 x 300 pixels and then normalize it.

> **NOTE**
>
> Before proceeding, ensure that you change the paths to the image (highlighted in the following code) based on where they are saved on your system. The images can be downloaded from https://packt.live/3h0nDEm.

The code is as follows:

```
# change image name to check different results
image = cv2.imread('dataset/image_3.jpeg')
(h, w) = image.shape[:2]
blob = cv2.dnn.blobFromImage(cv2.resize(image, (300, 300)), \
                             0.007843, (300, 300), 127.5)
```

6. Feed the scaled image to the network. It will compute the forward pass and predict the objects in the image:

```
net.setInput(blob)
detections = net.forward()
```

7. Select random colors for the bounding boxes. Each time a new box is generated for a different category, it will be in a new color:

```
colors = np.random.uniform(255, 0, size=(len(categories), 3))
```

8. Now, we will iterate over all the detection results and discard any output whose probability is less than **0.2**, then we will create a bounding box with the object name around each detected object:

```
for i in np.arange(0, detections.shape[2]):
    confidence = detections[0, 0, i, 2]
    if confidence > 0.2:
        idx = int(detections[0, 0, i, 1])
        # locate the position of detected object in an image
        box = detections[0, 0, i, 3:7] * np.array([w, h, w, h])
```

```
            (startX, startY, endX, endY) = box.astype("int")
            # label = "{}: {:.2f}%".format(categories[idx], \
            # confidence * 100)
            label = "{}: {:.2f}%".format(classes [idx], \
                    confidence*100)
            # create a rectangular box around the object
            cv2.rectangle(image, (startX, startY), (endX, endY), \
                    COLORS[idx], 2)
            y = startY - 15 if startY - 15>15 else startY + 15
            # along with rectangular box, we will use cv2.putText
            # to write label of the detected object
            cv2.putText(image, label, (startX, y), \
            cv2.FONT_HERSHEY_SIMPLEX, \
            0.5, colors[idx], 2)
```

9. Use the **imshow** function of OpenCV to display the detected objects with boxes and confidence scores on the image. Note that **5000** in **waitkey()** means that the image will be displayed for 5 seconds. We can also display the image using **matplotlib**:

```
cv2.imshow("detection", image)
cv2.waitKey(5000)
cv2.destroyAllWindows()
# using matplotlib
fig = plt.figure()
ax1 = fig.add_axes((0.1, 0.2, 0.8, 0.7))
# ax1.set_title('object detection')
plt.axis("off")
plt.imshow(cv2.cvtColor(image, cv2.COLOR_BGR2RGB))
plt.show()
```

The preceding code produces the following output:

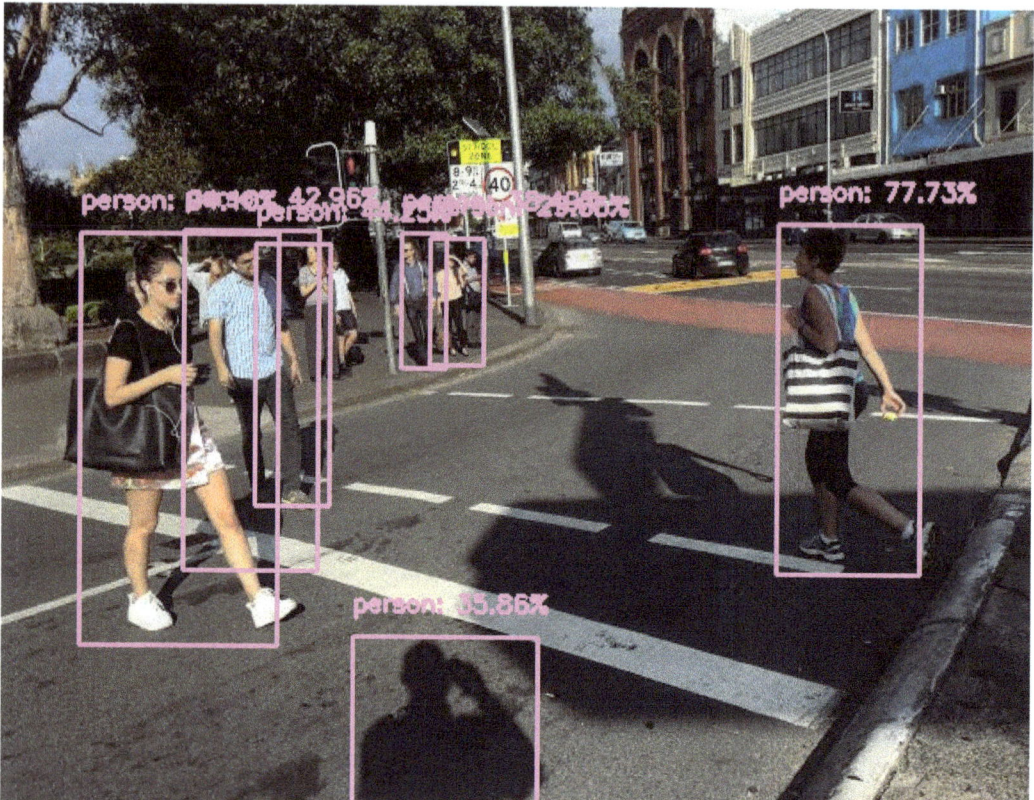

Figure 7.22: Prediction results on the test images using MobileNet SSD

Let's recap what we just did. The first step was to read a color image as an input. Then, the second step was to load the pre-trained model using the OpenCV module and take the image into the model as an input. In the next step, the model output multiple instances of objects with their associated probabilities. The step that followed filtered out the less-probable object instances based on a threshold value. Finally, the last step created rectangular boxes and corresponding probability values for each of the objects detected.

> **NOTE**
>
> To access the source code for this specific section, please refer to https://packt.live/3eQWnYz.

In this exercise, we learned how to implement object detection using MobileNet SSD. Note that in this exercise, we have used just one version of the MobileNet model; we can also try different versions to check the performance differences.

Next, we will look into classical algorithms for object detection. First, we will go through the LBPH-based method, then we will conclude with the Haar-based method.

OBJECT DETECTION USING THE LBPH METHOD

In an earlier section, we discussed the concepts of LBPH and how it works. In this section, we will directly jump into the implementation of object detection using the LBPH method. In the upcoming exercise, we will try to sort images into three categories: car, ship, and bicycle. We will be using scikit-image to implement the LBPH algorithm. Note that OpenCV also implements LBPH (used in *Exercise 7.03, Facial Recognition Using the LBPH Method*), but it is strictly focused on facial recognition. The scikit-image library helps to compute the raw LBPH value for each image. In addition to scikit-image, we will also use the scikit-learn library to build a linear **Support Vector Machine** (**SVM**) classifier.

The `skimage` library uses the `feature.local_binary_pattern()` function to compute the descriptive features of an input image. The function takes an input image, a radius value, and several neighboring pixels (or we can define this as the sample points for a circle of radius `r`) as input arguments. It outputs a two-dimensional array.

In the next exercise, we will build a linear SVM classifier to classify the images into categories. Fortunately, the scikit-learn library provides a module called `svm` to enable us to build a linear SVM classifier. The `svm` module uses the `LinearSVC()` and `LinearSVC().fit()` functions to initialize and train a model, respectively.

The `LinearSVC()` function takes various parameters as input arguments, but, in this exercise, the value of the regularization parameter, **C**, is specific. The values of the remaining parameters are set to the default values.

The `LinearSVC().fit()` function takes an array of input images and their corresponding labels as input arguments.

> **NOTE**
>
> Before we go ahead and start working on the exercise, make sure that you have installed the **skimage**, **imutils**, and **scikit-learn** modules. If you are using Windows, then we recommend using Anaconda. To install **skimage**, you can use `pip install scikit-mage`; for **imutils**, use `pip install imutils`; and for **sklearn**, use `pip install sklearn`.

EXERCISE 7.05: OBJECT DETECTION USING THE LBPH METHOD

In this exercise, we will implement object detection using the LBPH algorithm. We will use images of different vehicles to implement this object detection algorithm. Note that you can use the solution from this exercise to detect various kinds of objects, such as flowers, stationery items, and so on. We will start by using the training dataset to fit the linear SVM-based classifier. We will then use the testing dataset to find out the performance of the classifier on an unseen dataset:

1. Firstly, open an untitled Jupyter Notebook and name the file **Exercise7.05**.

2. Start by importing all the required libraries for this activity:

```
import os
from imutils import paths
import cv2
import numpy as np
import matplotlib.pyplot as plt
from skimage import feature
from sklearn.svm import LinearSVC
```

3. Next, specify the path of the dataset. Create a folder named **dataset** and, inside that, create two more folders, **training** and **test**. Put the training and test images into **training** and **test**, respectively:

> **NOTE**
>
> These folders are available on GitHub at https://packt.live/2YOsbHR.

```
training_data_path = 'dataset/training'
test_data_path = 'dataset/test'
```

4. We know that the LBPH algorithm requires two parameters, a radius value (to determine the neighboring pixels) and a number of pixels. Set these two parameter values, which will be used as inputs to the **skimage** function. To tune the performance, we can vary these values:

```
numPoints = 24
radius = 8
```

5. Next, iterate over all the images and create histogram features for each image. Then, we will store the features with respective classes in empty lists:

```
train_data = []
train_labels = []
eps = 1e-7
for image_path in paths.list_images(training_data_path):

    # read the image, convert it to grayscale
    image = cv2.imread(image_path)
    gray = cv2.cvtColor(image, cv2.COLOR_BGR2GRAY)

    """
    extract LBPH features, method is uniform means feature
    extraction isrotation invariance
    """
    lbp_features = feature.local_binary_pattern(\
                   gray, numPoints, radius, \
                   method="uniform")
```

```
        (hist, _) = np.histogram(lbp_features.ravel(), \
                bins=np.arange(0, numPoints + 3), \
                range=(0, numPoints + 2))
        hist = hist.astype("float")

        # normalize histograms
        hist /= (hist.sum() + eps)

        #get the label from the image path, append in a list
        train_data.append(hist)
        train_labels.append(image_path.split(os.path.sep)[-2])
```

Note that we are using a variable named **eps**, and its value is a very small number. The purpose of using this variable is to avoid division by zero. In this step, we are normalizing the histograms by dividing by the sum of all the histogram values. In this case, if any histogram values are zeros, then the sum would be zero, so **eps** helps to avoid this situation.

6. Use the linear SVM classifier to build and train the model:

```
model = LinearSVC(C=70, random_state=42)
model.fit(train_data, train_labels)
```

This produces the following output:

```
C:\ProgramData\Anaconda3\lib\site-packages\sklearn\svm\_base.py:947: ConvergenceWarning: Liblinear failed
to converge, increase the number of iterations.
  "the number of iterations.", ConvergenceWarning)
LinearSVC(C=70, class_weight=None, dual=True, fit_intercept=True,
          intercept_scaling=1, loss='squared_hinge', max_iter=1000,
          multi_class='ovr', penalty='l2', random_state=42, tol=0.0001,
          verbose=0)
```

Figure 7.23: Configuration of the linear SVM classifier

7. Now, test the performance of the trained model using the test images. This code iterates over each of the test images, then it calculates the LBPH features. Finally, it will pass the extracted features to the model to find the class of the image. In the end, the OpenCV library is used to display the name of the predicted class in a window:

```
test_data = []
prediction_result = []
for image_path in paths.list_images(test_data_path):

    # read the image, convert it to grayscale
    image = cv2.imread(image_path)
    gray = cv2.cvtColor(image, cv2.COLOR_BGR2GRAY)

    # Extract LBPH features
    lbp = feature.local_binary_pattern(gray, numPoints, \
        radius, method="uniform")
    (hist, _) = np.histogram(lbp.ravel(), \
                bins=np.arange(0, numPoints + 3), \
                range=(0, numPoints + 2))
    hist = hist.astype("float")

    # normalize histograms
    hist /= (hist.sum() + eps)
    test_data.append(hist)
    prediction = model.predict(hist.reshape(1, -1))

    # display the image and the prediction
    cv2.putText(image, prediction[0], (10, 30),
    cv2.FONT_HERSHEY_SIMPLEX, 0.5, (0, 0, 255))
    cv2.imshow("predicted image", image)
    cv2.waitKey(5000)
cv2.destroyAllWindows()
```

The preceding code produces the following output:

Figure 7.24: Prediction results on the test images using the LBPH algorithm

> **NOTE**
> All output will be displayed after 5 seconds in a single window.

Object Detection | 411

In this exercise, we first trained a linear SVM classifier and then used it to predict labels for an unseen dataset/testing dataset. The important thing to remember here is that the classifier was trained based on the features obtained from the LBPH algorithm. This resulted in better performance, even using a model as simple as a linear SVM-based classifier.

Figure 7.25 and *Figure 7.26* show a brief overview of the training steps and test steps respectively, used in *Exercise 7.05, Object Detection Using the LBPH Method*:

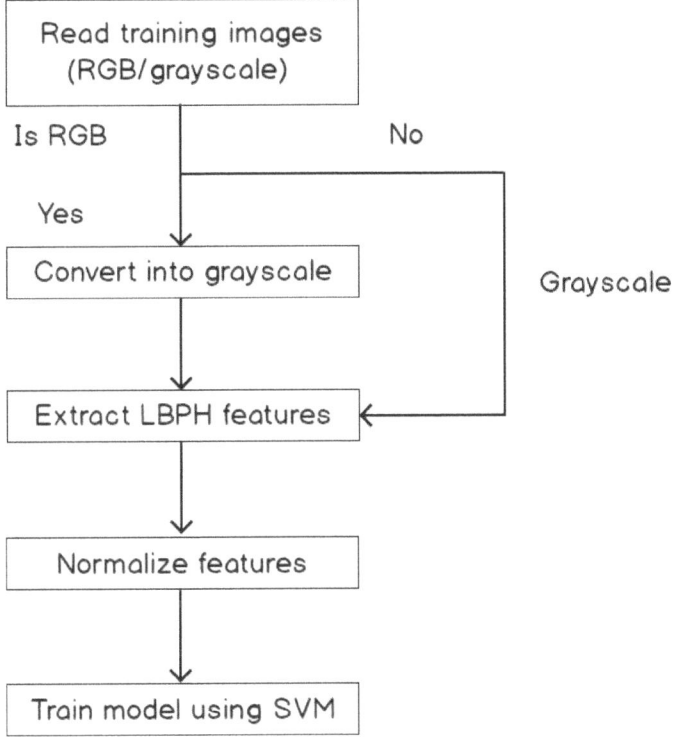

Figure 7.25: Training steps in exercise 7.05

As shown in *Figure 7.25*, the training steps first take color or grayscale images as input. If the input image is a color image, then the next step converts the input image into grayscale; otherwise, the input image is passed to the LBPH feature extraction function. In the next step, the computed features are normalized. Then, the next step instantiates an SVM model. Finally, the model is trained using training images:

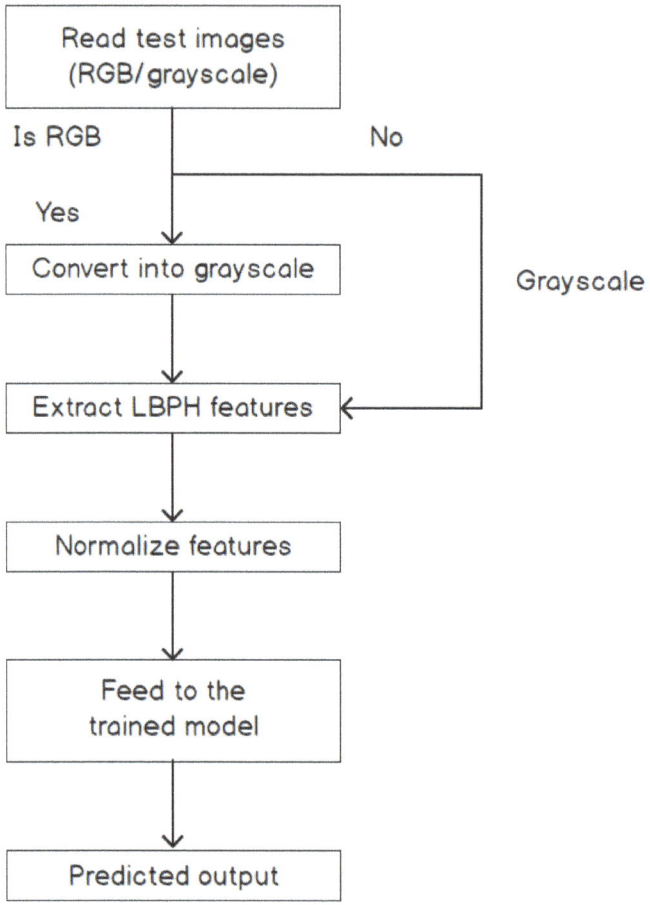

Figure 7.26: Test steps in the exercise 7.05

As shown in *Figure 7.26*, the test step first takes a new color or grayscale image as input. If the input is a color image, then the next step converts the input image into grayscale; otherwise, the input image is passed to the LBPH feature extraction function. In the next step, the computed features are normalized. Then, the next step feeds the extracted features to the trained model (which is trained during the training steps). Finally, it outputs the category of the input image.

In this exercise, you learned how to implement object detection with LBPH. You built an object detection model for three categories. It is recommended that you add more categories and images to check the performance of the algorithm.

> **NOTE**
>
> To access the source code for this specific section, please refer to https://packt.live/3dVIOWA.

Next, we will implement a Haar-based object detection model.

HAAR CASCADES

In *Chapter 5, Face Processing in Image and Video*, we discussed the Haar Cascades classifier. We implemented face detection using it. Now, we will quickly recap the concepts of Haar Cascades and implement object detection using it. In this algorithm, Haar-like feature extraction is carried out by considering a large set of adjacent rectangular regions of different scales, which then move over an image. At any specific region, pixels in adjacent windows are summed up and the difference of these sums is then used to categorize that region of the image. Generally, areas around the eyes are darker than cheeks. So, one example of Haar features for face detection would be a set of rectangular windows for the eye region and the cheek region. Through this approach, many Haar features are generated. But not all the features are relevant. At this step, a well-known boosting technique – **AdaBoost** – comes into the picture, which can reduce the number of features to a great extent and can yield only the relevant features out of a huge pool.

Now that we have revised the concepts of Haar Cascades, let's implement car and pedestrian detection models using it.

EXERCISE 7.06: OBJECT DETECTION USING HAAR-BASED FEATURES

In this exercise, we will perform object detection using Haar-based features. In the previous exercise, we used LBPH features to train a support vector classifier that was then used for object detection. In this exercise, we will carry out object detection using pre-trained Haar Cascade classifiers:

> **NOTE**
>
> Before we go ahead and start working on this, please make sure that you have downloaded the **haarcascade_car.xml** and **haarcascade_fullbody.xml** files. These are available at https://packt.live/3j2rYcc.
>
> We will use the **cv2.VideoCapture()** function to read frames from videos and also from webcams.

1. Firstly, open an untitled Jupyter Notebook and name the file **Exercise7.06** or **object_detection_haar_cascade**.

2. Start by importing all the required libraries for this activity:

    ```
    import cv2
    ```

3. Specify the path of the OpenCV XML files that will be used to detect people and cars in each image. Create a folder named **opencv_xml_files** and save the **haarcascade_car.xml** and **haarcascade_fullbody.xml** files inside that folder:

    ```
    haarcascade_car = 'opencv_xml_files/haarcascade_car.xml'
    haarcascade_fullbody = 'opencv_xml_files/haarcascade_fullbody.xml'
    ```

4. Next, load the XML file that will be used to detect cars. To detect pedestrians, replace **haarcascade_car.xml** with **haarcascade_fullbody.xml**:

    ```
    car_classifier = cv2.CascadeClassifier(haarcascade_car)
    # for pedestrian detection use file
    #'opencv_xml_files/haarcascade_fullbody.xml'
    # classifier = cv2.CascadeClassifier(haarcascade_fullbody)
    ```

5. Read frames from the video and detect the cars in it. Currently, a video named `cars.avi` is being used. *Steps 5 - 6* explain how to read frames using video.

> **NOTE**
>
> `cars.avi` can be downloaded from https://packt.live/3dPxsDk.

The code is as follows:

```
cap = cv2.VideoCapture('cars.avi')
# for pedestrian detection
# cap = cv2.VideoCapture('walking.avi')
count = 0
```

Note that a `count` variable is used in this code. The purpose of this variable to select every *n*th frame in a video. Sometimes, we get a video with 30 or more frames per second. It becomes slightly difficult to see the predicted result for each frame, so instead of detecting objects in each frame, we can apply a detection model on every *n*th frame. Currently, we are detecting every 20th frame, but we can change this to any number.

6. In this step, you will read every 20th frame, until the last frame. For each frame, the image will be converted to grayscale. Then, pre-trained files from *Step 4* will detect the cars in each frame. Next, a rectangular box will be created around the detected cars with labels on them. Finally, the model will display a stream of frames in a window. The window will be closed once the code reaches the last frames of the input video:

```
while cap.isOpened():
    count +=1
    # here 20 means every 20th frame.
    if count%20==0:
        ret, frame = cap.read()
        gray = cv2.cvtColor(frame, cv2.COLOR_BGR2GRAY)
        cars=car_classifier.detectMultiScale(gray,1.1, 3)
        for (x,y,w,h) in cars:
            cv2.rectangle(frame, (x, y), (x+w, y+h), (0, 255, 0), 2)
            cv2.putText(frame, 'car', (x, y-10), \
            cv2.FONT_HERSHEY_SIMPLEX, 0.4, (0, 0, 255))
        cv2.imshow('car', frame)
```

```
            # for pedestrian
            #cv2.imshow('person', frame)
    # press 'q' to destroy window
    if cv2.waitKey(1) & 0xFF == ord('q'):
        break
cap.release()
cv2.destroyAllWindows()
```

Figure 7.27 shows the result of car detection using a Haar Cascades classifier:

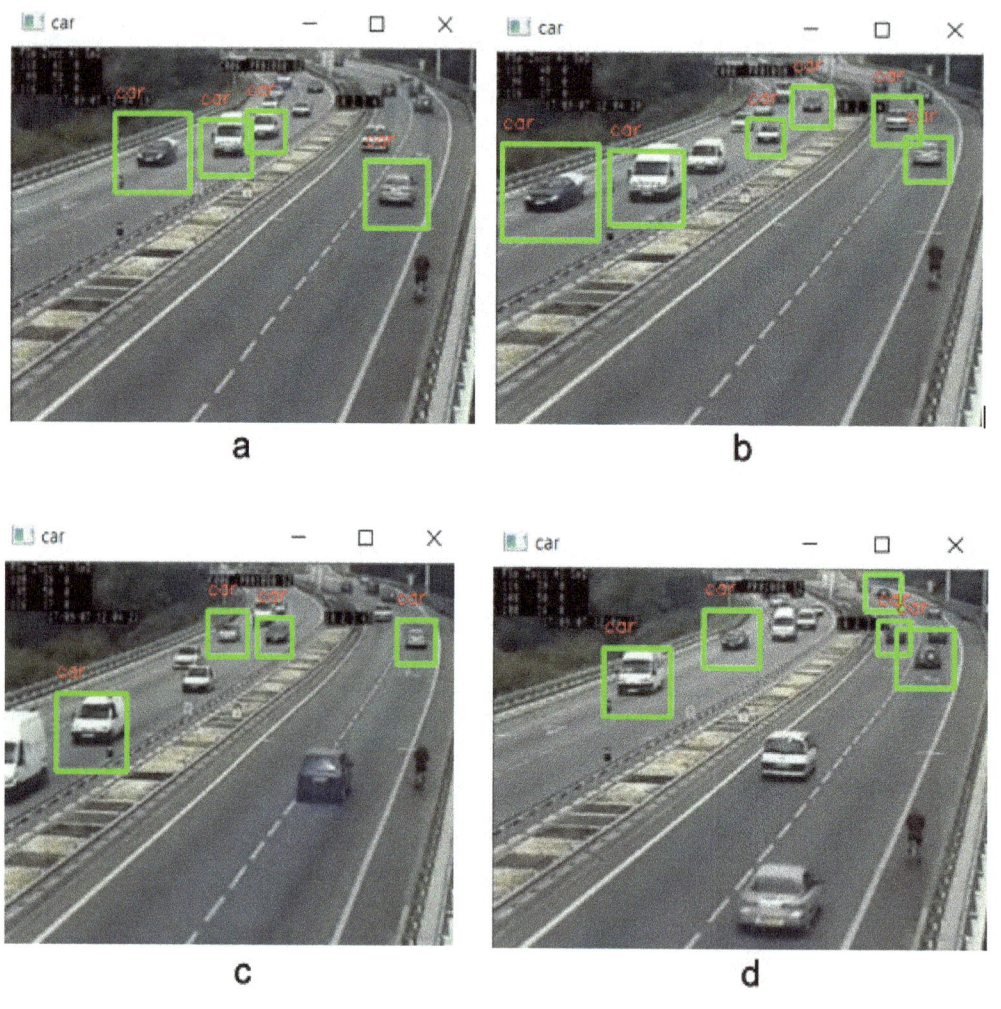

Figure 7.27: Detection results on a few frames of the video using the Haar Cascades algorithm

Object Detection | 417

> **NOTE**
>
> All output will be displayed in a single window.
>
> To access the source code for this specific section, please refer to https://packt.live/2NlOQij.

In this exercise, we used Haar Cascades to detect the cars present in a video frame. This is different from the LBPH method we used in the previous exercise, where we had to train a support vector classifier to detect the objects present in a frame. Here we have used a pre-trained Haar Cascades classifier, and that's why we have effectively only performed the inference step.

Figure 7.28 shows a brief overview of the steps in this exercise:

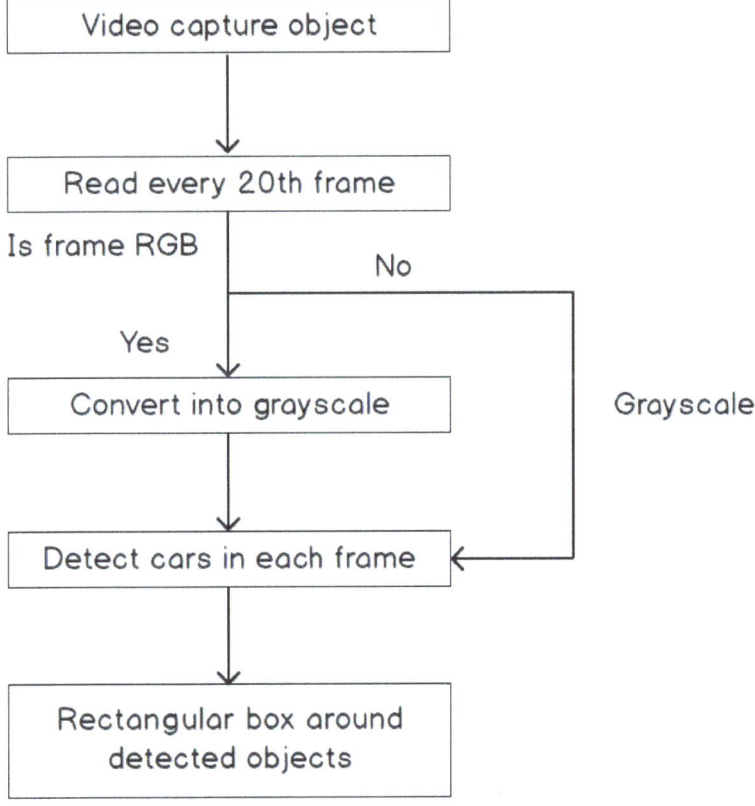

Figure 7.28: Brief overview of steps in exercise 7.06

The first step creates a video capture object and reads an input file. Then, the second step reads every 20th frame. If the input is a color image, then the next step converts the input image into grayscale; otherwise, it passes each frame into the car detection module. Finally, the last step creates a rectangular box with a label for each car detected in the frame.

In this exercise, you have learned how to implement object (car) detection using the Haar Cascades method in a video. By following the instructions in this exercise, you can also detect humans in a video.

ACTIVITY 7.01: OBJECT DETECTION IN A VIDEO USING MOBILENET SSD

In this activity, you will implement an object detection model for a recorded video and for some real-time video. You can use any video to detect object instances in each frame. Other than that, a webcam can be used to detect objects in real-time.

The steps to follow to solve the problem have been listed here:

1. First, open an untitled notebook and create a file named **Activity7.01**. We will type out our code in this file.

2. Import the necessary Python modules.

3. Download the **MobileNetSSD_deploy.prototxt** file, and the pre-trained MobileNet SSD model: **MobileNetSSD_deploy.caffemodel**.

   > **NOTE**
   >
   > The **MobileNetSSD_deploy.prototxt** and **MobileNetSSD_deploy.prototxt** files can be downloaded from https://packt.live/2ZvqpdI.

4. Load both the downloaded files using OpenCV's **dnn** module.

5. Prepare a dictionary of categories that can be detected by the model.

6. Provide the path of the input video to **cv2.VideoCapture(path)** and read each frame.

   > **NOTE**
   >
   > The input video, **walking.avi**, can be downloaded from https://packt.live/3gecS12.

7. Perform the pre-processing steps for each frame, such as resizing and converting into a blob, before feeding it to the model.

8. Store the model output to a variable (say, **detection_result**).

9. For each frame, the model will return detected instances. Iterate over all the detected instances and check whether the confidence score (probability) of the detected object is greater than the threshold (say, **0.2** or **0.3**).

10. If the confidence of detection is greater than the threshold, then get the coordinates and category ID of the object from the **detection_result** variable.

11. Note that in the pre-processing step, each image is resized, so to get the locations of the objects in the original frame, scale the coordinates.

12. Check whether the category ID is present in the dictionary created in *Step 5*. If it is, then create a rectangular box around the detected object and label it.

13. Display the result in the **cv2** window.

14. Once these steps are done, replace the path with 0 in **cv2.VideoCapture()**. It will start using a webcam. Bring your face or any other object in front of the webcam. It will start detecting objects in real-time.

15. It is advised to test the results with various input videos.

Once all the steps are completed successfully, you should be able to get the images shown in *Figure 7.29* and *Figure 7.30*. *Figure 7.29* shows the prediction results for one of the frames of the input video:

7.29: Sample output of the prediction results for one of the frames of the input video

Figure 7.30 is a sample of how the result would look if you were to use a webcam. For representation purposes, an image of a train was placed in front of the webcam. You are advised to bring any object in front of the webcam from the list that we created in *Step 4* of *Exercise 7.04, Object Detection Using MobileNet SSD*:

7.30: Sample output of the prediction results using a webcam

> **NOTE**
>
> The solution for this activity can be found on page 518.

ACTIVITY 7.02: REAL-TIME FACIAL RECOGNITION USING LBPH

In this activity, you will implement a real-time facial recognition model. The objective is to use a camera to recognize a person in real-time. The solution can be extended to real-world problems. To start with, you can use real-time facial recognition to build an attendance system for offices or schools.

The steps to follow to solve the problem have been listed here:

1. First, open an untitled notebook and create a file named **Activity7.02**.

2. Import the necessary Python modules.

3. Download the **lbpcascade_frontalface.xml** file. It will be used for face detection.

   > **NOTE**
   >
   > The **lbpcascade_frontalface.xml** file can be downloaded from https://packt.live/3ifLn9j.

4. Use the webcam to detect faces and store them in a folder. If you like, you can start by capturing your own face images and storing them in the folder.

5. Read the files from the folder and create a set of training images with their respective labels.

6. Use the face recognizer module from OpenCV to build a recognizer model.

7. Train the model on the training data.

8. Again, use the **webcam** and **haarcascade** XML files to detect a face and pass this face to the recognizer model and check whether the trained model can recognize the new face correctly.

9. Increase the number of unique faces (categories) to check the performance of the model.

> **NOTE**
>
> As mentioned in *step 4*, this activity will access your webcam and store the faces of people who are in front of the camera.

Once all the steps are done successfully, you should be able to get the image of a person and their name. *Figure 7.31* is a sample representation of how the output should look. So, instead of an emoji, you will see a real face and their assigned name:

Figure 7.31: Example output for this activity

> **NOTE**
>
> The solution for this activity can be found on page 523.

In this activity, you have learned how to build a real-time facial recognition model. You can extend this solution to build an attendance system in an office or to monitor areas around your home (to check whether unknown people are roaming).

SUMMARY

In this chapter, we discussed the concepts of facial recognition and object detection. In the first part, we discussed three popular classical algorithms for facial recognition. We started with the Eigenface method, and then we discussed its limitations. Then, we moved on to the next method, called Fisherface, and discussed how it solves the limitations of the Eigenface method. We looked into the limitations of Fisherface as well. Finally, we concluded our study of facial recognition with a look at the LBPH method. We have also implemented three different face recognizers using these three methods. We observed that these methods work well if the illumination of images is uniform, and if the face in the image is facing the camera head-on rather than at an angle.

In the second part, we discussed three different object detection algorithms. We started with a recently developed algorithm called MobileNet SSD. In this section, we discussed how SSD performs object detection in a single shot, which makes this algorithm faster than some of its competitors. Then, we discussed how the low complexity of MobileNet makes it a preferred choice for real-time use cases. Moving on, we discussed classical approaches to object detection, such as the LBPH and Haar Cascade methods. We concluded our look at object detection by implementing object detection models using these two classical methods.

In the next chapter, we will discuss Intel's OpenVINO toolkit, a recent addition to the OpenCV module. We will start with an introduction to OpenVINO, where we will discuss the various components of OpenVINO and how to combine it with OpenCV at various steps – image pre-processing, model optimizing, inference engines, and so on. We will also use some pre-trained models present in ModelZoo to build some applications.

8
OPENVINO WITH OPENCV

OVERVIEW

This chapter introduces a very important new toolkit – the OpenVINO toolkit by Intel. The OpenVINO toolkit is a comparatively new name in the computer vision domain, but it's so powerful that it cannot be ignored. We will start by understanding what the OpenVINO toolkit is, examining its prerequisites and components, and exploring how we can use them individually and alongside OpenCV. We will not be building any end-to-end applications, but we will explore the computer vision application pipeline. Specifically, we will focus on OpenVINO's Model Zoo, inference engine, and model optimizer. By doing so, you will be able to apply this knowledge in real life and increase the performance of your computer vision applications.

By the end of this chapter, you will be able to use the individual components of the OpenVINO toolkit for your deep learning and computer vision tasks.

INTRODUCTION

In the previous chapters, we went over commonly used computer vision applications, including object detection. Let's start this chapter by talking about how real-life **artificial intelligence** (**AI**) (and specifically, computer vision) applications are deployed and used. First, what do we mean by deployment? **Deployment** is typically the step that comes after a certain program or software has been developed and thoroughly tested. It is the first step in which access to the application is provided to consumers; for example, let's say we are developing a computer vision application for adding an emoji filter (recall that we developed something similar in *Chapter 5, Face Processing in Image and Video*). If you think about the various steps the program will go through, you will see that it first starts off with idea development, which, after further iterations of refinement, goes to the development stage, where the actual code for the software is written. Of course, the software can't be ready for use by customers just after the code has been written. The developed software is tested for quality and performance, and after multiple iterations, it moves on to the deployment step, where the program is either distributed as open-source or is sold as a product. In our emoji filter example, this would mean making the program available as an application on, say, the Google Play store. The reason for this discussion is that it is important to understand the thought process that goes into application development and deployment. This process must take into account the client requirements, the performance specification required to make the program work, and a lot of other considerations.

Let's think of a simple scenario. You have one mobile application that can apply emoji filters but takes 3 GB of space, and then another application is just 30 MB in size. Which one would you prefer to install? Typically, you would go with the second application. Clearly, size is of importance. Now, think about a computer game. Games that have amazing graphics are generally preferred, but what's their use if you don't have a powerful enough graphics card or the RAM space? So, computational requirements are also important. The one thing that is common in both examples is *where* the computation is taking place. In the case of a mobile application, it is your mobile that will be responsible for applying the filters, while in the case of a computer game, your system will be responsible for all the computation. This means that as a developer, you must understand the limitations set by a customer's system. Now, think about what the ideal case would be. What if we could reduce the computation and space requirement without losing any significant performance? That's what we will be looking at in this chapter. The part about computation happening on your system and not in some cloud is called an **edge application**.

These applications are commonly used in self-driving cars, mobile applications, **Internet of Things (IoT)** devices, and so on because every network packet sent and received costs time and money. Even cloud computation requires a significant amount of money. So, by using an edge application, we not only save a huge amount of money, but we also save unnecessary time spent on to-and-fro communication between the client's device and the cloud system. We are also able to make sure that data privacy is maintained as the data is not stored on any cloud instance or server.

Let's try to understand an important product/package that comes in handy in edge applications – the OpenVino toolkit.

EXPLORING THE OPENVINO TOOLKIT

The **OpenVINO toolkit** is developed and packaged by Intel and is one of the most important packages for optimizing model performance. This makes the product useful in edge applications. It should not come as a surprise that Intel is the key player behind this toolkit since Intel is well-known for developing highly optimized libraries and tools for computation, such as `OpenCV*`, `OpenVX*`, `Python*` (this is different from Python and is an Intel distribution of accelerated Python), Intel MKL, and many more.

> **NOTE**
>
> These asterisks (*) imply that we are talking about highly optimized versions of the corresponding tool or library.

The OpenVINO toolkit specifically focuses on two things – **interoperability** and **performance optimization**. Let's talk about interoperability first, referencing deep learning for a moment to understand this term. We have several types of deep learning libraries – PyTorch, TensorFlow, and Caffe, to name a few. We spoke about a Caffe model in *Chapter 6, Object Tracking*, when we looked at the GOTURN object tracker. Now, the frustrating part about the deep learning domain is that all these key players have their own model extensions. A Caffe model will have `prototxt` and `caffemodel` files. PyTorch saves model weights in `.pt` or `.pth` files. TensorFlow 2.0 saves model weights in a `.h5` file. Now, imagine that you are developing a computer vision application. Of course, your first choice will be OpenCV, but since OpenCV cannot train a deep learning model, you first have to use a deep learning library to train the model and then carry out inference using OpenCV.

This poses a challenge because every library has its own model extensions, deep learning architecture layer names, and other functionalities. OpenCV developers will end up spending a lot of time providing support for each of these models and there might still be some bugs left unsolved, or some functionality missed. This raises the need for interoperability. A model trained in TensorFlow cannot be opened in PyTorch and thus we cannot use the strength of PyTorch. However, interoperability provides one model extension that every library will implement, and that way, any deep learning library can load it. The model extension that we are talking about here is called **Open Neural Network Exchange** (**ONNX**). ONNX allows libraries such as OpenCV to focus more on providing support for ONNX models rather than providing support for every model extension that exists. The OpenVINO toolkit comes with its own set of utility functions that let us convert a one-model file into an ONNX model that can then be directly used in our edge application.

The second thing that OpenVINO focuses on is model performance. OpenVINO provides support for most Intel architectures, which means that OpenVINO has optimized model performance not only for the software part but also for the hardware part. OpenVINO comes with a component called **Model Optimizer**, which is responsible for converting an input model file into an **Extensible Markup Language** (**XML**) file and a binary (`.bin`) file. These files are much smaller in size and are highly optimized. This enables a developer to ship a model with similar performance in a much smaller-sized file on any device – for example, a 5 MB model file rather than an unoptimized 50 MB model file on your mobile.

Now, let's start talking about the various components of the OpenVINO toolkit and their roles.

COMPONENTS OF THE OPENVINO TOOLKIT

While OpenVINO comes with a huge set of utilities, we will focus only on the following components:

- Model Optimizer
- Inference Engine

We will also focus on Model Zoo. While **Model Zoo** is not a component of the `OpenVINO` toolkit, it is a public repository where a lot of models trained for different computer vision and deep learning applications are shared by users, as well as Intel, for others.

Let's have a look at a typical edge application pipeline and see how `OpenVINO` will fit into it:

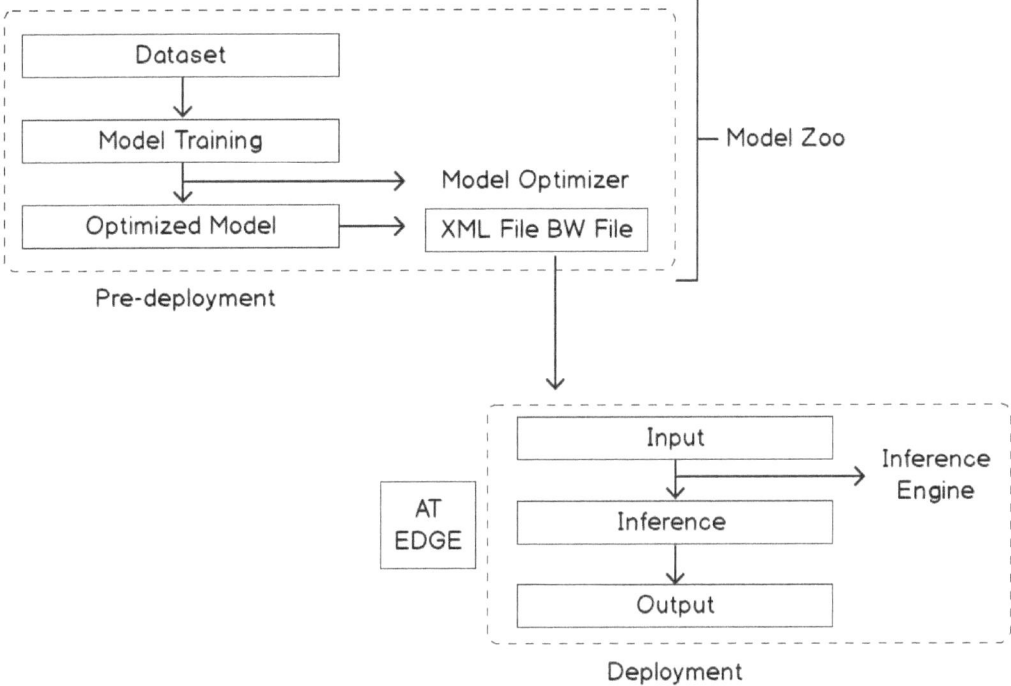

Figure 8.1: Edge application pipeline

During pre-deployment, we have to train our model on a given dataset. The first thing to do when you get a case study to work on is to check whether a similar model exists in Model Zoo. If the model is available, it's better to use the pre-trained model and fine-tune it for your use case – you should only try to train your own model if you can't find the relevant model in Model Zoo. You should also make sure that you check the license of the pre-trained model before using it in any commercial application. There is no need to reinvent the wheel. If you are training your model, you can optimize the model using **Model Optimizer**, which will give you an XML file and a `.bin` file.

> **NOTE**
> You will need both files for the deployment process.

On the deployment or post-deployment side, we have model inference. Model inference means that you take input and use your trained model to obtain an output. At this step, we can take the help of **Inference Engine** to perform inference, provided we are dealing with the model files that were obtained from Model Optimizer.

Now, before we jump into using OpenVINO with OpenCV, let's have a look at how to install the OpenVINO toolkit.

INSTALLING OPENVINO FOR UBUNTU

The steps to install OpenVINO are very different based on your operating system. Here, we will take a look at the steps for installing OpenVINO on Ubuntu.

Windows users can refer to https://docs.openvinotoolkit.org/latest/_docs_install_guides_installing_openvino_windows.html for installation steps. Similarly, macOS users can refer to https://docs.openvinotoolkit.org/latest/_docs_install_guides_installing_openvino_macos.html. Let's take a look at how to install OpenVINO on Ubuntu systems:

1. First, you will need to download the OpenVINO installer from https://software.intel.com/en-us/openvino-toolkit/choose-download/free-download-linux.

2. After installation, you will have to register yourself on the Intel website. Once you have registered, you will be directed to a page that looks as follows:

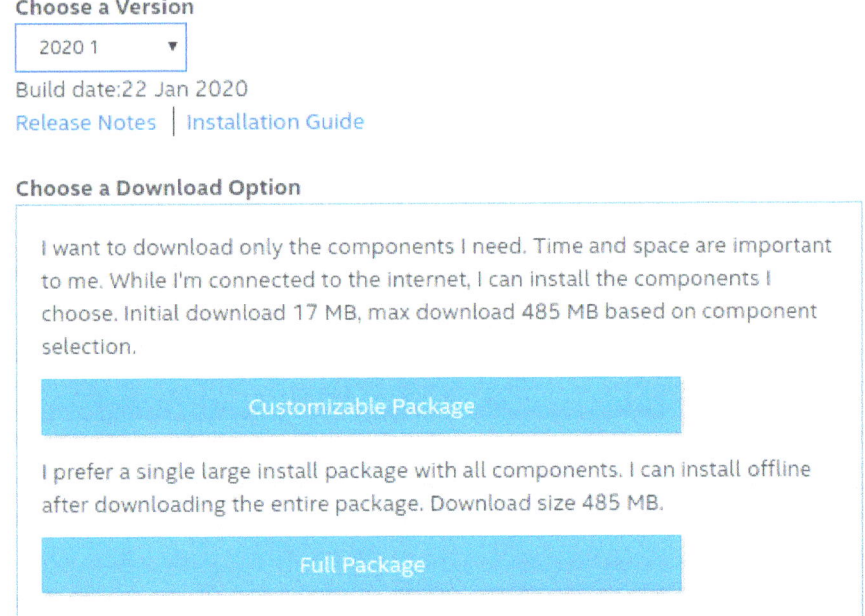

Figure 8.2: OpenVINO download page

3. Now, if you have a steady and high-speed internet connection, you can download **Full Package**; otherwise, go for **Customizable Package**. Here, we will be going for **Full Package**. Make sure you choose **2020 1** as your OpenVINO version, as shown in the preceding screenshot. This 485 MB `tgz` file will take some time to download. The downloaded file's name should be `l_openvino_toolkit_p_2020.1.023.tgz`.

4. This file will be downloaded in your **Downloads** folder by default. Currently, the path of the system is `~/Downloads`. You must move it to a new directory in your `$HOME` directory – **openvino**. This means that the new location of the zipped file will be `~/openvino`.

5. The following command will unzip the installer:

   ```
   tar -xvzf l_openvino_toolkit_p_2020.1.023.tgz
   ```

6. You should be able to see a new directory – **l_openvino_toolkit_p_2020.1.023**. You will need to navigate to that directory and start the installer:

   ```
   cd l_openvino_toolkit_p_2020.1.023/
   sudo ./install_GUI.sh
   ```

   You should see the following window:

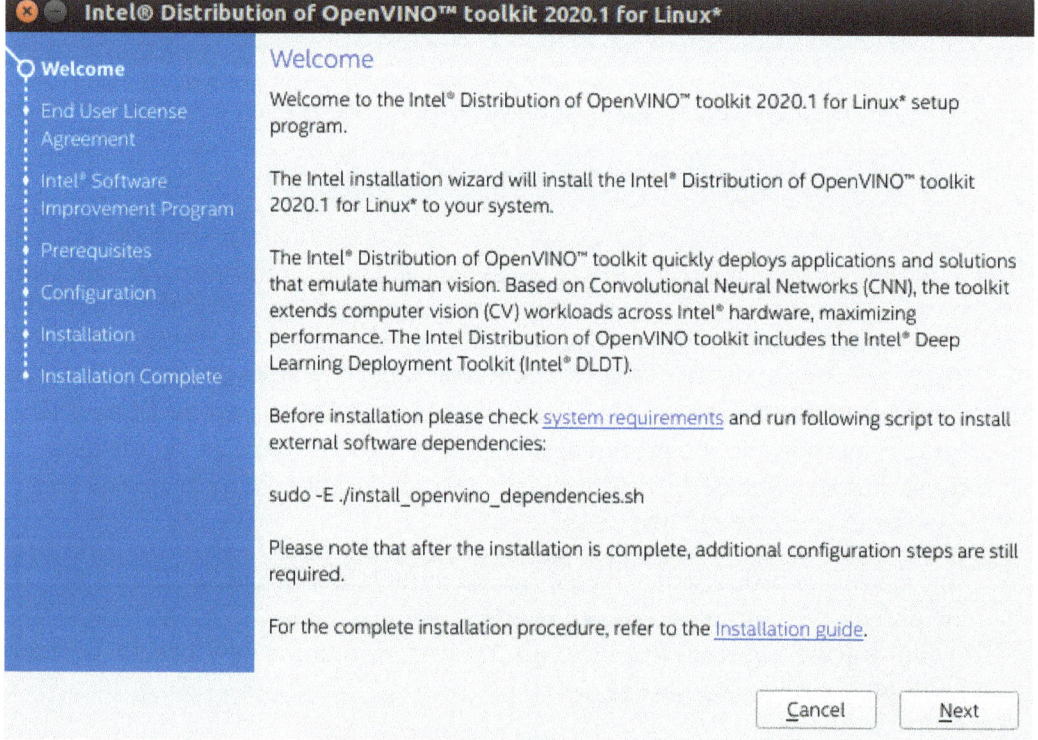

Figure 8.3: OpenVINO GUI installer

7. Click **Next**, click the checkbox corresponding to **I accept the terms of the license agreement**, and then click **Next** again.

8. Next, depending on your choice, select whether you consent to your information being collected and then click **Next**.

9. The OpenVINO installer will then check whether the prerequisites are present on your system and display a message accordingly. Click **Next** and finally click **Install**.

10. The installation might take a few minutes, depending on your system specifications. You should then see the following window. Click **Finish**:

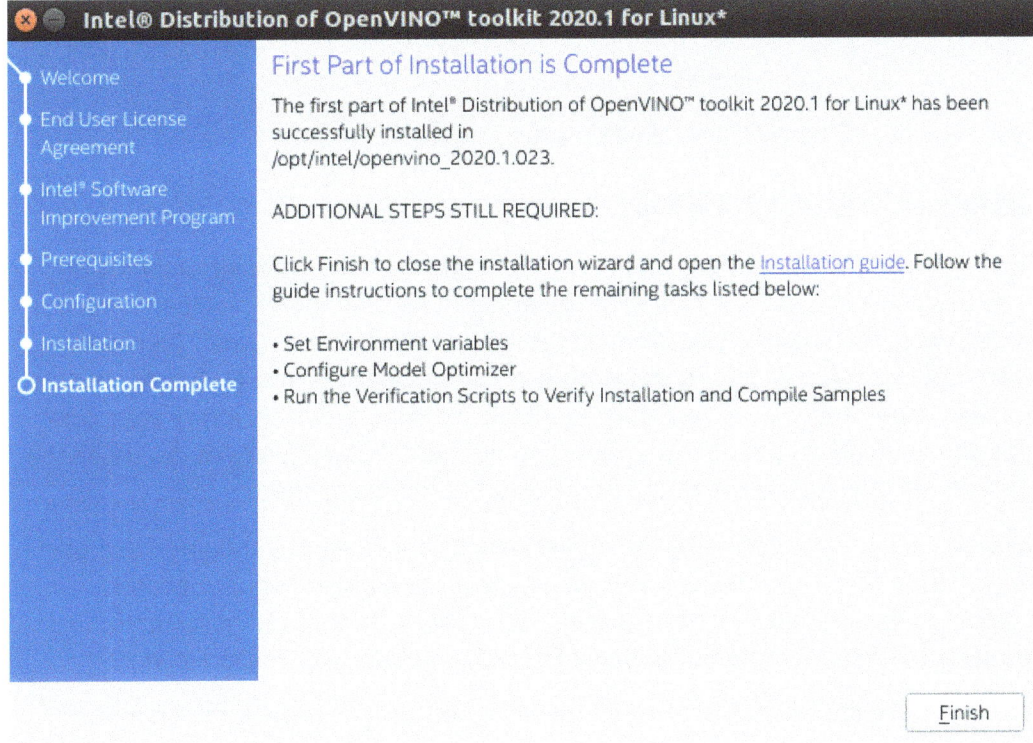

Figure 8.4: OpenVINO GUI installer (Finish window)

11. Next, go to OpenVINO's `install_dependencies` folder and install the software dependencies. Enter **y** whenever you're prompted:

```
cd /opt/intel/openvino/install_dependencies
sudo -E ./install_openvino_dependencies.sh
```

12. Finally, we will need to set our environment variables. To do that, you can add the following line to your `~/.bashrc` file:

```
source /opt/intel/openvino/bin/setupvars.sh
```

13. Close your current Terminal and open a new one. You should see the following at the top of your Terminal:

```
[setupvars.sh] OpenVINO environment initialized
```

14. Now that we have OpenVINO installed, let's configure certain components of OpenVINO, starting with Model Optimizer:

```
cd /opt/intel/openvino/deployment_tools/model_optimizer/install_prerequisites
sudo ./install_prerequisites.sh
```

Now, you have OpenVINO installed on your system.

OPENVINO AS A BACKEND IN OPENCV

The first topic that we are going to cover with regards to the OpenVINO toolkit is how to use it with OpenCV. First, let's understand the scenario here. OpenCV comes with a deep learning module, as we discussed in the previous chapter. Commonly, the module is referred to as **dnn module**, short for **Deep Neural Network module**. It's also important to note here that OpenCV is a computer vision library at its core and can only carry out model inference and not model training. We've discussed the difference between the two terms previously, but let's have another look.

Model training involves finding the model weights and biases by training the model architecture on a given dataset. This usually requires a huge amount of data, computation, and time. We have specific Python libraries that can help with model training, the most common ones being PyTorch and Keras (or TensorFlow).

Model inference comes into the picture when you already have a trained model and you want to put it to use. Imagine that you've trained a model to carry out image classification. Training the model would require feeding in the dataset of images and the image classes. Once the model is ready, we want to give it an image as input and get the class of the image as the output. This is called model inference.

Model inference is less expensive than model training in terms of computation, memory requirements, and time. Typically, model training requires GPUs, whereas model inference can be carried out on a CPU (or even a Raspberry Pi) or the Movidus Neural Compute Stick 2.

Even though model inference is computationally economical, the number of times it will need to be carried out in a computer vision application is huge. Imagine that you are trying to detect and recognize the number plate text of all the vehicles passing through a gate. This is a typical **Optical Character Recognition** (**OCR**) problem. This process needs to be carried out on every frame of the video, in which all frames need to be fed to the model trained for OCR. This means that even a small amount of performance optimization in model inference can be very significant. That's exactly where OpenVINO comes into the picture.

OpenVINO serves as a backend for OpenCV's **dnn** module. With OpenVINO, you can set target hardware where you would like to perform inference, for example, Intel CPU, Intel GPU, Intel NCS2, FPGA, and so on. This is the main purpose of using OpenVINO as a backend. Otherwise, OpenCV will perform the inference but using only the CPU. On top of that, OpenVINO uses highly optimized libraries such as Intel MKL for CPU, ClDNN for GPU, and so on. By that, we mean that it is responsible for most inference-related computation and since we already know that OpenVINO is highly optimized (and also has several utilities for OpenCV* – Intel's optimized version of OpenCV), that makes it the perfect candidate for performance optimization in the model inference step.

Now, let's talk about how easy it is to specify OpenVINO as the backend. It's all about the following line of code:

```
setPreferableBackend(cv2.dnn.DNN_BACKEND_INFERENCE_ENGINE)
```

This can be used with the following two-part process:

1. Load the model using OpenCV's relevant function. For example, if you want to load a Caffe model, you would use the **cv2.dnn.readNetFromCaffe** function. If you want to load an ONNX model, you would use **cv2.dnn.readNetFromONNX**, and so on. The most important one for us in this chapter is **cv2.dnn.readNetFromModelOptimizer**, which can be used to load model files exported by OpenVINO's Model Optimizer. Let's store the output of these functions (the network) in the variable net.

2. Next, specify the backend for computation for the network (**net**) using the following command:

```
net.setPreferableBackend(BACKEND_CODE)
```

There are four backends that are available in OpenCV, which are detailed in the following figure:

Backend Name	Description	BACKEND_CODE
Halide backend	It is typically not used but is still there as an option.	cv2.dnn.DNN_BACKEND_HALIDE
OpenCV backend	This is the backend provided by OpenCV itself.	cv2.dnn.DNN_BACKEND_OPENCV
Inference Engine backend	This is OpenVINO's Inference Engine backend.	cv2.dnn.DNN_BACKEND_INFERENCE_ENGINE
Default backend	If OpenVINO is present in the system (with OpenCV), then Inference Engine is used as the default backend; otherwise, the OpenCV backend is used.	cv2.dnn.DNN_BACKEND_DEFAULT

Figure 8.5: OpenCV supported backends

That's all you need to do to use OpenVINO's Inference Engine as the backend. At this point, it is recommended to specify the backend as Inference Engine in all the **dnn** module code that you have used in previous chapters and compare the performance of the code.

THE NEED FOR PRE-TRAINED MODELS

Before we jump into the details of Model Zoo, which we discussed briefly in the *Components of the OpenVINO Toolkit* section, let's establish the need for pre-trained models.

Previously, we discussed that there are primarily two factors that a model training process requires:

- Data
- Computation

Of course, it needs time as well, but let's not consider that right now. Let's start with understanding data by looking at an object detection problem. Typically, for an object detection problem, model training requires data that is annotated (labeled) in such a way that every object in the image has a bounding box around it. Now, we are not talking about a couple of thousand images here – we are talking about 0.5 million images.

Have a look at **ImageNet Large Scale Visual Recognition Challenge** (**ILSVC**) 2014 here: http://image-net.org/challenges/LSVRC/2014/. There were 0.5 million images in the training set, 20,000 images in the validation set, and around 40,000 images in the testing set. Imagine the time it would have taken to manually annotate that many images. There is only one best-case scenario when it comes to finding data for your deep learning problem – you find the dataset online and procure the necessary permissions to use it. On the other hand, the worst-case scenario would mean that you don't have any such dataset or don't have the right to use it. Now, imagine hiring a team or service to collect and label 0.5 million images. Even if we're talking about a standard service that charges 0.01 USD for every image labeled, then labeling 0.5 million images would require 5,000 USD, and this doesn't include the cost of collecting the images.

Now, let's talk about computation. If you do not have a high-end GPU with you, you will need to use cloud services to train your model. This can require anything between 1,000 and 50,000 USD. And if you are looking to train a **State-of-the-Art** (**SOTA**) model, have a look here to find out the pricing for other SOTA AI models: https://medium.com/syncedreview/the-staggering-cost-of-training-sota-ai-models-e329e80fa82. SOTA models are models that are known to provide the best performance in terms of accuracy, the time required for inference, and so on.

Not everyone can afford to spend several thousands of dollars to just train a model, which is where pre-trained models come into the picture. Pre-trained models can be used directly in the best-case scenario, or we can use **transfer learning** to fine-tune them for our specific case study. Transfer learning means that you freeze most of the starting layers of your model architecture and just fine-tune the last few layers to make it work. This involves significantly lower costs and a smaller amount of data for training.

OPENVINO MODEL ZOO

Now that we understand the importance of pre-trained models, let's see where we can find them for OpenVINO. We already know that OpenVINO's pre-trained models consist of an XML file and a `.bin` file. The pre-trained models can be found at Model Zoo here: https://github.com/opencv/open_model_zoo.

We will use the downloader script provided by OpenVINO to download these pre-trained models, which can be found at the following path on your system (provided you successfully installed it by following the steps provided in the *Installing OpenVINO for Ubuntu* section):

```
/opt/intel/openvino/deployment_tools/open_model_zoo/tools/downloader/download.py
```

The first step to using these pre-trained models is to figure out which one you need. A few of the interesting models to point out here are the Pedestrian and vehicle detection models (we will be using them here), age and gender recognition, license plate recognition, the road segmentation model, and the human pose estimation model. Let's say we are carrying out a pedestrian detection model. In such a case, we can use the pedestrian and vehicle detection model, that is, **pedestrian-and-vehicle-detector-adas-0001**. Before we begin using it, let's learn how to download this model.

> **NOTE**
>
> We will be using the **os** module in the upcoming exercise. This module provides operating system-level functionalities that can be used to list the files in a directory, create a new directory, change the current directory, create a new environment variable, and so on. The **os.environ** function for setting an environment variable is used so that we can call it in a bash command (think of it as a command on Command Prompt).
>
> We will use the **os.environ.pop** function to remove an environment variable. We will also use the **os.listdir** function to list the files present in a directory.

EXERCISE 8.01: DOWNLOADING THE PEDESTRIAN AND VEHICLE DETECTION MODEL

In this exercise, we will use the downloader script provided by OpenVINO to download the vehicle and pedestrian detection model – **pedestrian-and-vehicle-detector-adas-0001**. Follow these steps to complete this exercise:

1. Create a new Jupyter notebook and name it **Exercise8.01.ipynb**. We will be writing our code in this notebook.

2. First, let's import the **os** module:

```
import os
```

3. Now, let's start specifying some variables. We will start off with the location of the **downloader.py** file:

```
# Specify the path to the downloader script
DOWNLOADER_SCRIPT = "/opt/intel/openvino/deployment_tools/"\
                    "open_model_zoo/tools/downloader/"\
                    "downloader.py"
```

4. Next, we will store the model name in the **MODEL_NAME** variable:

```
# Specify the model name which we want to download
MODEL_NAME = "pedestrian-and-vehicle-detector-adas-0001"
```

5. We will also need the directory where we are going to store the model files. You can change this path according to your system:

```
# Specify where you want to store the files
OUTPUT_DIR = "/home/hp/workfolder/"\
             "The-Computer-Vision-Workshop/Chapter08/data/"
```

> **NOTE**
>
> The path (highlighted) should be specified based on the location of the folder on your system.

6. The model that we are dealing with comes with three precisions – **FP16**, **FP32**, and **INT8**. These values dictate how the calculations are going to be done and how precise the results will be. **FP16** means that the values will be stored in a 16-bit floating-point format, whereas **FP32** means that the values will be stored in a 32-bit floating-point format. **INT8** means that we will be storing all the integer variables using 8 bits. The higher the precision, the more space it will occupy. Let's go for **FP16** precision:

```
# Specify the precision
PRECISION = "FP16"
```

7. Now, we can use the values defined so far in this exercise to create the command that we are going to run to download the model files:

```
# Specify the command we need to run
DOWNLOAD_COMMAND = "{} --name {} --precisions {} -o {}"\
    .format(DOWNLOADER_SCRIPT, MODEL_NAME, PRECISION, OUTPUT_DIR)
```

8. The command that we are trying to run is a bash command and to run it directly, we will store the command in an environment variable. To create an environment variable, run the following command:

```
# Create environment variable
os.environ["DOWNLOAD_COMMAND"] = DOWNLOAD_COMMAND
```

9. Since we have the command in an environment variable, we can directly call it, as follows:

```
%%bash
$DOWNLOAD_COMMAND
```

The output of this command is as follows:

```
################|| Downloading models ||################

========== Downloading /home/hp/workfolder/The-Computer-Vision-Workshop/Chapter08/data/intel/pedestrian-and-vehicl
e-detector-adas-0001/FP16/pedestrian-and-vehicle-detector-adas-0001.xml
... 100%, 239 KB, 366 KB/s, 0 seconds passed

========== Downloading /home/hp/workfolder/The-Computer-Vision-Workshop/Chapter08/data/intel/pedestrian-and-vehicl
e-detector-adas-0001/FP16/pedestrian-and-vehicle-detector-adas-0001.bin
... 31%, 1024 KB, 816 KB/s, 1 seconds passed
... 63%, 2048 KB, 818 KB/s, 2 seconds passed
... 95%, 3072 KB, 809 KB/s, 3 seconds passed
... 100%, 3222 KB, 814 KB/s, 3 seconds passed

################|| Post-processing ||################
```

Figure 8.6: Downloading the model files

10. Now that we have downloaded the model, let's unset the environment variable:

```
# Remove the environment variable
os.environ.pop("DOWNLOAD_COMMAND")
```

The output is as follows:

```
'/opt/intel/openvino/deployment_tools/open_model_zoo/tools/
downloader/downloader.py --name pedestrian-and-vehicle-detector-
adas-0001 --precisions FP16 -o /home/hp/workfolder/
The-Computer-Vision-Workshop/Chapter08/data/'
```

> **NOTE**
>
> The path shown in the preceding output will differ, based on the location of the folder on your system.

11. Now, let's check whether the file was downloaded:

    ```
    # List files present in the output directory
    os.listdir(OUTPUT_DIR)
    ```

    The output is as follows:

    ```
    ['mobilenet_v2.caffemodel',
     'intel',
     'mobilenet_v2.prototxt',
     'pedestrians.jpg']
    ```

12. We can see that there is a new folder, **intel**, present in the output directory. Let's list the files present in the **intel** folder:

    ```
    # List files in intel folder
    os.listdir("{}/intel".format(OUTPUT_DIR))
    ```

    The output is as follows:

    ```
    ['pedestrian-and-vehicle-detector-adas-0001']
    ```

13. There is a new directory named after our model. Let's check the files in that directory:

    ```
    # Go to the model directory
    os.listdir("{}/intel/{}".format(OUTPUT_DIR,MODEL_NAME))
    ```

    This will return **['FP16']**.

14. There is one more directory here – **FP16**, which was the precision we specified. Let's check the files in this directory:

    ```
    # Go to the precisions directory
    os.listdir("{}/intel/{}/{}"\
    .format(OUTPUT_DIR,MODEL_NAME,PRECISION))
    ```

    The output is as follows:

    ```
    ['pedestrian-and-vehicle-detector-adas-0001.xml',
     'pedestrian-and-vehicle-detector-adas-0001.bin']
    ```

    We can see that these files are found in the directory – **pedestrian-and-vehicle-detector-adas-0001.xml** and **pedestrian-and-vehicle-detector-adas-0001.bin**.

In this exercise, we learned how to use the downloader script provided by OpenVINO to download a pedestrian and vehicle detection model.

> **NOTE**
>
> To access the source code for this specific section, please refer to https://packt.live/2CZeVaY.

MODEL SPECIFICATIONS

Now that we have found the model we are going to use, let's understand what this model needs as input and what it gives as output. For that, we will go over the documentation for the model – https://docs.openvinotoolkit.org/latest/_models_intel_pedestrian_and_vehicle_detector_adas_0001_description_pedestrian_and_vehicle_detector_adas_0001.html.

As we can see, this model is based on MobileNet v1.0, which is a model that's commonly used as a backbone in the **Single Shot Detector** (**SSD**) model. First, let's have a look at the model specifications. This provides details about the framework that was used for model training, as well as the model's performance:

Specification

METRIC	VALUE
AP for pedestrians	88%
AP for vehicles	90%
Target pedestrian size	60x120 pixels
Target vehicle size	40x30 pixels
GFLOPS	3.974
MParams	1.650
Source framework	Caffe*

Figure 8.7: Model specifications

As we can see, the model was trained using Caffe* (an Intel-optimized version of Caffe) and has an average precision of **90%** for vehicles and **88%** for pedestrians. It's not very high but will work for our example.

Next, we need to have a look at the inputs and outputs, beginning with the inputs:

Inputs

1. name: "input" , shape: [1x3x384x672] - An input image in the format [BxCxHxW], where:
   - B - batch size
   - C - number of channels
   - H - image height
   - W - image width. Expected color order is BGR.

Figure 8.8: Model inputs

We can see that the model takes in an input image with **[1×3×384×672]** as its model input shape. Let's try to understand this format using the following diagram:

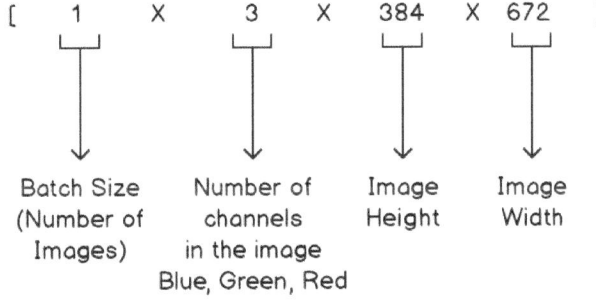

Figure 8.9: Model input shape

From this information, we can conclude:

- We need to pass only one image as input.
- The image should be in BGR color mode (and not the standard RGB mode).
- The image must be resized to have a size of **384×672**.
- The image must be reshaped to have the shape **[1×3×384×672]**.

You need to understand the difference between shape and size here. Size only gives information about the height and width of the image. Shape, on the other hand, also gives information about the number of channels in the image. Resizing can be done using OpenCV's `resize` function, while reshaping can be done using NumPy's `reshape` function.

Therefore, an RGB image and a grayscale version of the same image will have the same size but a different shape.

Next, let's look at the outputs given by the model:

Outputs

1. The net outputs blob with shape: [1, 1, N, 7], where N is the number of detected bounding boxes. For each detection, the description has the format: [image_id, label, conf, x_min, y_min, x_max, y_max]
   - image_id - ID of the image in the batch
   - label - predicted class ID
   - conf - confidence for the predicted class
   - (x_min, y_min) - coordinates of the top left bounding box corner
   - (x_max, y_max) - coordinates of the bottom right bounding box corner.

Figure 8.10: Model output

We can see that the model output has a shape of [1× 1× N×7]. The first two values are used only for providing a -D input (you can see that it's 4D since it has four values in the shape). They do not carry any physical significance. Let's focus on the last two values—N and 7. N tells you how many bounding boxes were detected in the input image, with each corresponding to objects (vehicles and pedestrians). For each object detected, seven values are returned. These are listed as follows:

- **image_id**: This is used to provide a specific ID for an image. It becomes important when we want to find out the output for a specific image and when we have provided a lot of images as input. This is not important for us in the current scenario and can be ignored.

- **label**: This is the label of the class. Since we are only dealing with a two-class problem – vehicle and pedestrian – the object can belong to either of these two categories.

- **`conf`**: This tells us with what confidence the model can say that the object belongs to the class (label). The higher this value is, the surer you can be that the object belongs to the specified class.
- **`x_min`**: This is the **x** coordinate of the top-left corner of the bounding box around the object.
- **`y_min`**: This is the **y** coordinate of the top-left corner of the bounding box around the object.
- **`x_max`**: This is the **x** coordinate of the bottom-right corner of the bounding box.
- **`y_max`**: This is the **y** coordinate of the bottom-right corner of the bounding box.

Now that we have understood the meaning of the input and output that's obtained from pre-trained models, we can use the preceding output to draw bounding boxes around all the objects present in the image using OpenCV. We'll learn how to do that in the next section.

IMAGE TRANSFORMS USING OPENCV

In this section, we will have a look at how we can use OpenCV to make an image input-worthy for the pedestrian and vehicle detection model that we just looked at.

Let's start by understanding the functions we will be using:

- For resizing, we will use the `cv2.resize` function.

  OpenCV's `cv2.resize` function takes two important arguments: the image that we want to resize and the new size of the image – `(width, height)`.

- For reshaping, we will use NumPy's `reshape` function.

  For NumPy's `reshape` function, we just have to provide the new shape of the image – `[Number of images, Number of channels in the image, Image height, Image width]`. In our case, it will become `[1, 3, 384, 672]`.

We will use these functions in the following exercise.

EXERCISE 8.02: IMAGE PREPROCESSING USING OPENCV

In this exercise, we will learn how to preprocess an image using OpenCV and NumPy so that we can then pass the revised image to the vehicle and pedestrian detection model. This is a very important step because the pre-trained models will not work if the dimensions of the input image do not match the size of the images they were trained for. This serves as the first step in inference.

> **NOTE**
>
> The image can be downloaded from https://packt.live/2YT6feV.

Perform the following steps:

1. Open a new Jupyter notebook and name it **Exercise8.02.ipynb**. We will be writing our code in this notebook.

2. First, let's import the required modules – OpenCV and NumPy:

```
# Load modules
import numpy as np
import cv2
```

3. Next, let's load the image:

```
# Load the image
img = cv2.imread("../data/pedestrians.jpg")
```

> **NOTE**
>
> Before proceeding, ensure that you change the path to the image (highlighted) based on where the image is saved on your system.

The image will look as follows:

Figure 8.11: Image of pedestrians and vehicles

4. First, let's establish the shape of the image:

    ```
    # Print image shape
    img.shape
    ```

    This will return **(847, 1280, 3)**. Here, **847** pixels is the height of the image, **1280** pixels is the width of the image, and **3** refers to the three channels present in the image—blue, green, and red.

5. Now, let's print the shape in a more readable format:

    ```
    print("Image height = {}".format(img.shape[0]))
    print("Image width = {}".format(img.shape[1]))
    print("Number of channels = {}".format(img.shape[2]))
    ```

    The output is as follows:

    ```
    Image height = 847
    Image width = 1280
    Number of channels = 3
    ```

6. First, let's resize the image. For that, we will define the target image's width and height:

    ```
    # Target image size
    targetWidth = 672
    targetHeight = 384
    ```

7. Next, let's resize the image using the **cv2.resize** function:

   ```
   # Resize image
   resizedImage = cv2.resize(img, (targetWidth, targetHeight))
   ```

8. Let's check the shape of the resized image:

   ```
   print("Image height = {}".format(resizedImage.shape[0]))
   print("Image width = {}".format(resizedImage.shape[1]))
   print("Number of channels = {}".format(resizedImage.shape[2]))
   ```

   The output is as follows:

   ```
   Image height = 384
   Image width = 672
   Number of channels = 3
   ```

9. Now that we have resized the image, let's reshape it using NumPy:

   ```
   # Reshape image
   reshapedImage = resizedImage.reshape(1,3,384,672)
   ```

10. Finally, we will print the image shape to verify whether the image has the correct shape or not:

    ```
    # Image shape
    reshapedImage.shape
    ```

    The output is as follows:

    ```
    (1, 3, 384, 672)
    ```

We have preprocessed the image using OpenCV and NumPy, and now, it is ready to be provided as input to the model. Notice how we have modified the image shape so that it's 4D data. This is important because the model takes 4D data as input.

> **NOTE**
>
> To access the source code for this specific section, please refer to https://packt.live/3iioAti.

Now, before we try to understand how to use this for model inference, let's have a look at another component of the OpenVINO toolkit – **Model Optimizer**.

MODEL CONVERSION USING MODEL OPTIMIZER

We have already discussed in detail why we need OpenVINO's optimization power. Now, let's understand how Model Optimizer helps with that. Generally speaking, OpenVINO's optimization takes place in two steps: one at the Model Optimizer stage and one at the Inference Engine stage. We will investigate the Inference Engine in the next section, but for now, let's focus on Model Optimizer.

Model Optimizer carries out two main operations – **quantization** and **fusion**.

Let's start by understanding quantization. As you may recall, we talked about precision in *Exercise 8.01*, *Downloading the Pedestrian and Vehicle Detection Model*, when we said that **FP16** means that the number representation is going to use 16 bits of memory and will be represented in floating-point format. **Quantization** is closely related to that. To understand it, let's look at an example. Let's say we want to represent a number, **0.578156**, in a system. If we say that we can only use up to two decimal places, then the number will be stored as **0.57** or **0.58**. If, on the other hand, we say that we can only store integers, then the number will be stored as **0** or **1**, depending on how the system is set in place.

If you compare both the preceding outputs, you will notice that by reducing the number of decimal places we can use, we have reduced the precision of the number. A number **1** in the second decimal representation system can mean anything stored between **0.5** and **1.5**, whereas if we use the first decimal representation system, a number **1** can mean anything between **1.000** and **1.005**. Notice how we have managed to get a more precise value by increasing the number of decimal places we can use. We can directly use this analogy to understand the difference between **FP16** and **FP32**.

Quantization compresses the model size and reduces the inference time by reducing the precision of the weights and biases. Let's consider that, originally, all the weights and biases in a model were stored in **FP64** format and there were 1 million such weights and biases. Such a model would need 64×10^6 bits (or around 8 MB space). If we reduce the precision of all these parameters to **FP16**, we will end up reducing the model size to just 2 MB.

Another way that Model Optimizer improves the model is by fusion, which refers to joining multiple consequential layers into one operation or layer. For example, convolution, batch normalization, and ReLU operations can be fused into one operation (**ConvBNReLU**). That way, we are not only reducing the model size, but we are helping with the model performance.

Model Optimizer uses something called **Intermediate Representation** (**IR**). This refers to the XML and `.bin` files that we saw in *Exercise 8.01, Downloading the Pedestrian and Vehicle Detection Model*. Model Optimizer can convert the models trained in frameworks such as Caffe, TensorFlow, PyTorch (ONNX models), and so on into IR form after optimizing the model using fusion and quantization.

The XML file is responsible for storing the model architecture (layers, number of neurons, number of connections, input and output dimensions, and many other parameters). The following is a screenshot of the XML file for the pedestrian and vehicle detection model:

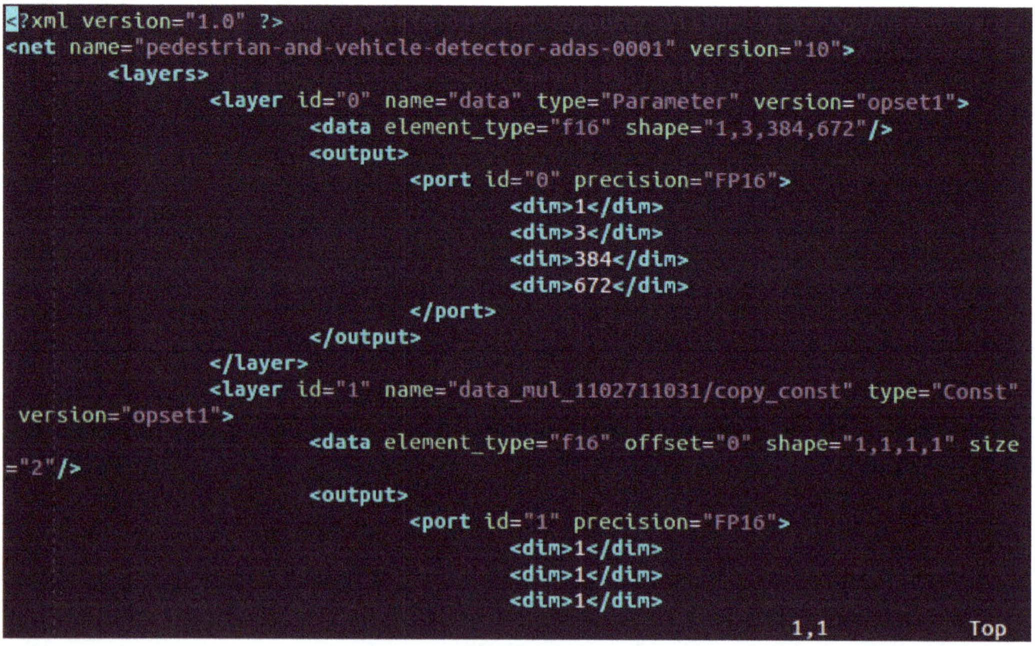

```
<?xml version="1.0" ?>
<net name="pedestrian-and-vehicle-detector-adas-0001" version="10">
        <layers>
                <layer id="0" name="data" type="Parameter" version="opset1">
                        <data element_type="f16" shape="1,3,384,672"/>
                        <output>
                                <port id="0" precision="FP16">
                                        <dim>1</dim>
                                        <dim>3</dim>
                                        <dim>384</dim>
                                        <dim>672</dim>
                                </port>
                        </output>
                </layer>
                <layer id="1" name="data_mul_1102711031/copy_const" type="Const"
 version="opset1">
                        <data element_type="f16" offset="0" shape="1,1,1,1" size
="2"/>
                        <output>
                                <port id="1" precision="FP16">
                                        <dim>1</dim>
                                        <dim>1</dim>
                                        <dim>1</dim>
                                                                        1,1           Top
```

Figure 8.12: Model XML file

The `.bin` file, on the other hand, is responsible for storing the weights and biases in binary format.

Now, because we are looking at two different files here – one with the model architecture and one with the parameters – we will need both files to run any kind of model inference. We saw a similar thing when we talked about the GOTURN object tracker (which is a Caffe model) in *Chapter 6, Object Tracking*.

Let's see how we can convert models from one framework into IR form using Model Optimizer. This is done using the Model Optimizer script, **mo.py**, present in the **/opt/intel/openvino/deployment_tools/model_optimizer** directory. Let's take the example of a Caffe model for MobileNet v2 since we have already seen it before in this chapter (recall that the pedestrian and vehicle detection model was based on MobileNet). You can download the files from these links:

- **MobileNet v2.0 caffemodel**: https://packt.live/3eKLk2Z.
- **MobileNet v2.0 prototxt**: https://packt.live/2BrzgVX.

All we need to do now is use the **mo.py** script and specify the path of **caffemodel** using **--input_model** and the path of the **prototxt** file using **--input_proto**:

```
python3 /opt/intel/openvino/deployment_tools/model_optimizer/mo.py
--input_model mobilenet_v2.caffemodel --input_proto mobilenet_v2.prototxt
```

If you look carefully at the output of the preceding command, you will see some optimization configurations, as shown in the following screenshot:

```
- Precision of IR:               FP32
- Enable fusing:                 True
- Enable grouped convolutions fusing:   True
```

Figure 8.13: Model optimizations

Once the model conversion is complete, you should see the following output. This not only shows the success message but also specifies the location of the XML and **.bin** files:

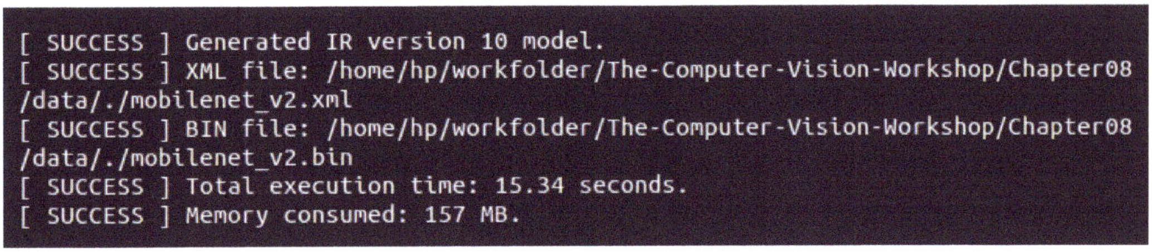

```
[ SUCCESS ] Generated IR version 10 model.
[ SUCCESS ] XML file: /home/hp/workfolder/The-Computer-Vision-Workshop/Chapter08/data/./mobilenet_v2.xml
[ SUCCESS ] BIN file: /home/hp/workfolder/The-Computer-Vision-Workshop/Chapter08/data/./mobilenet_v2.bin
[ SUCCESS ] Total execution time: 15.34 seconds.
[ SUCCESS ] Memory consumed: 157 MB.
```

Figure 8.14: Success message displayed by Model Optimizer

> **NOTE**
>
> You can try converting models from other frameworks using the Model Optimizer script. For more details, you can refer to the documentation here: https://docs.openvinotoolkit.org/latest/_docs_MO_DG_Deep_Learning_Model_Optimizer_DevGuide.html.

Now, it is time to have a look at the last component of the OpenVINO toolkit – Inference Engine.

INTRODUCTION TO OPENVINO'S INFERENCE ENGINE

In the previous section, we discussed the optimizations applied by the OpenVINO toolkit and we said that one such optimization is used at the Inference Engine step. We were referring to hardware optimization. While the optimizations used by Model Optimizer focus more on parameter- and model architecture-based optimization, the Inference Engine uses Intel's processor strength to its benefit and can give better performance.

But that's not the main role of the Inference Engine. The main role of the Inference Engine, as can be inferred from its name, is to perform model inference using the IR of the model.

There are two methods of using the Inference Engine. One is to directly import OpenVINO's components into Python code as modules and then code everything, but unfortunately, that's a slightly more complicated approach. The second, easier, approach is issuing OpenCV with OpenVINO. We have already seen how we can specify OpenVINO's Inference Engine as the backend for OpenCV's **dnn** module, and how to use Model Optimizer to convert a model into IR. Now, let's see how to connect all the dots:

1. We will start by loading the neural network using OpenCV's **cv2.dnn.readNet** function. You will have to pass two arguments to the function. The first will be the path to the **.bin** file, while the second will be the path to the XML file.

2. Next, we can either use the preprocessing technique that we discussed in *Exercise 8.02, Image Preprocessing Using OpenCV*, or use the **cv2.dnn.blobFromImage** function.

3. Next, we will use the **setInput** function to set the preprocessed image we obtained in the preceding step as the input to the loaded network.

4. All that's left to do now is carry out forward propagation or model inference to get the results. This can be done using the OpenCV dnn module's `forward()` function.

5. Finally, we will display the results obtained using model inference with the help of OpenCV functions.

We will go over an exercise to understand these steps in more detail, but first, let's look a bit deeper into the `cv2.dnn.blobFromImage` function. This function provides the preprocessing required to feed an image to a neural network. This can include resizing the image, cropping the image, swapping the channels, or subtracting the mean from the entire image. The steps that need to be carried out depend on the model that we are using. This is because when the model is trained, a very specific image size is used. Similarly, if the model was trained using RGB images, then during the model inference step, an RGB image must be fed in and so on. You can, of course, still use the code that we discussed in *Exercise 8.02*, *Image Preprocessing Using OpenCV*, or as a recommended approach, use the `cv2.dnn.blobFromImage` function for preprocessing.

EXERCISE 8.03: VEHICLE AND PEDESTRIAN DETECTION

Let's carry on with the vehicle and pedestrian detection problem that we were discussing previously. We carried out preprocessing in the previous exercise, so in this exercise, we will use OpenCV's `cv2.dnn.blobFromImage` function. This exercise will give you some hands-on experience on using this function. This serves as an alternative to the image preprocessing process that we saw earlier, in *Exercise 8.02*, *Image Preprocessing Using OpenCV*.

> **NOTE**
>
> The image can be downloaded from https://packt.live/2YT6feV.

Perform the following steps:

1. Open a new Jupyter notebook and name it `Exercise8.03.ipynb`.

2. First, let's import the required modules:

```
# Load the required modules
import cv2
import numpy as np
```

3. Next, let's specify the path to the XML and `.bin` model files (the intermediate representations).

   > **NOTE**
   >
   > Before proceeding, ensure that you change the paths to the `.bin` files (highlighted) based on where they are saved on your system.

   The code is as follows:

   ```
   # Specify the path to IR
   xmlFile = "../data/intel/"\
             "pedestrian-and-vehicle-detector-adas-0001/"\
             "FP16/pedestrian-and-vehicle-detector-adas-0001.xml"
   binFile = "../data/intel/"\
             "pedestrian-and-vehicle-detector-adas-0001/"\
             "FP16/pedestrian-and-vehicle-detector-adas-0001.bin"
   ```

4. Next, we will load the image. We will be using the same image that we used in *Exercise 8.02, Image Preprocessing Using OpenCV*:

   > **NOTE**
   >
   > Before proceeding, ensure that you change the path to the image (highlighted) based on where the image is saved on your system.

   The output is as follows:

   ```
   # Image path
   imgPath = "../data/pedestrians.jpg"
   # Read image
   img = cv2.imread(imgPath)
   ```

5. Now, we will load the model using the `cv2.readNet` function:

   ```
   # Load the network
   net = cv2.dnn.readNet(xmlFile,binFile)
   ```

6. Next comes the image preprocessing part. We know that we want to resize and reshape the image. The reshaping process is taken care of by the **cv2.dnn.blobFromImage** function by default, so we just have to specify the desired size of the image after preprocessing. We can verify the shape of the output image as follows:

```
# Pre-process the image
# We want to resize the image to 384x672
blob = cv2.dnn.blobFromImage(img, size=(672,384))
# Check the blob shape
blob.shape
```

The shape of the **blob** will be as follows:

```
(1, 3, 384, 672)
```

7. We will then proceed with the steps we discussed previously. First, we'll set the image as the network input and then carry out model inference:

```
# Set the image as network input
net.setInput(blob)
# Carry a forward propagation
out = net.forward()
```

8. We can then use the **shape** of the output:

```
# Check output shape
out.shape
```

The output is as follows:

```
(1, 1, 200, 7)
```

9. Compare the output with the output of the model that we saw in *Figure 8.10*:

```
# Compare this with [1,1,N,7]
print("Number of objects found = {}".format(out.shape[2]))
```

This will return the following output:

```
Number of objects found = 200
```

10. Notice that we found **200** objects and that for each of the **200** objects, we have **7** elements describing the confidence, label, and the bounding box coordinates. For ease of coding, let's reshape the **output** array:

```
# Reshape the output
detection = out.reshape(-1,7)
detection.shape
```

The shape is **(200, 7)**.

11. Now, we will iterate over each object and extract its label. If **label** is **1** (vehicle), then we will use a red bounding box; otherwise, we will use a green bounding box.

> **NOTE**
>
> The following code has to be executed in a single cell in the Jupyter notebook.

The code is as follows:

```
for detectedObject in detection:
    # Find label
    label = int(detectedObject[1])
    # Choose color of bounding box
    if label == 1:
        # Green color
        color = (0,255,0)
    else:
        # Red color
        color = (0,0,255)
```

12. Next, let's also extract the confidence for the object that was detected:

```
# Find confidence
confidence = float(detectedObject[2])
```

13. Next, we will find out what the bounding box coordinates for the object are:

```
# Bounding box coordinates
xmin = int(detectedObject[3] * img.shape[1])
ymin = int(detectedObject[4] * img.shape[0])
```

```
xmax = int(detectedObject[5] * img.shape[1])
ymax = int(detectedObject[6] * img.shape[0])
```

Now comes the interesting part. If we were to plot a bounding box around all **200** detected objects, we would end up with something like the following. To generate this image, you can change **0.50** in the code in the next step to **0**. This will plot all the bounding boxes present in the image, irrespective of the confidence:

Figure 8.15: Output image if we use all the detected objects

14. The reason we obtained the preceding image is because we used objects for which the model was not confident whether it was an object or not. To resolve this issue, we use a threshold confidence of **50%**. This means that for an object to be present in a box, the model should be at least **50%** sure that there is an object in that box:

```
# Plot bounding box only if there is at least
# 50% confidence
if confidence >= 0.50:
    cv2.rectangle(img, (xmin, ymin), (xmax, ymax), \
                  color = color)
```

15. We can then display and save the output image:

```
# Display image
cv2.imshow("Output Image",img)
cv2.imwrite("image.png",img)
cv2.waitKey(0)
cv2.destroyAllWindows()
```

The output is as follows:

Figure 8.16: Output image after considering the 50% confidence threshold

In this exercise, we learned how we can use OpenVINO's Inference Engine along with a pre-trained model obtained from Model Zoo to carry out pedestrian detection on an input image. We also learned how to use the model output to plot bounding boxes on the image.

> **NOTE**
>
> To access the source code for this specific section, please refer to https://packt.live/2AiR7h9.

Now that we have understood how to use OpenVINO's Inference Engine with OpenCV, let's try out one more problem.

ACTIVITY 8.01: FACE DETECTION USING OPENVINO AND OPENCV

We studied face detection in *Chapter 5, Face Processing in Image and Video*, when we used Haar Cascades. Now that we are aware of various deep learning techniques and how to use them in OpenCV, let's use deep learning and see whether there is an improvement in the results or not. First, we will carry out face detection using a pre-trained model with OpenVINO and then compare the results obtained by carrying out face detection using Haar Cascades.

The entire activity can be broken down into two parts:

- Face detection using a pre-trained model
- Face detection using Haar Cascades

We will be using the **face-detection-retail-0004** model (refer to https://docs.openvinotoolkit.org/latest/_models_intel_face_detection_retail_0004_description_face_detection_retail_0004.html).

We will use the following image for this activity:

Figure 8.17: People in a crowd

> **NOTE**
>
> The image can be downloaded from https://packt.live/3ik0EG8.

The following steps will help you complete this activity:

1. Open a new Jupyter notebook and name it `Activity8.01.ipynb`. We will be writing our code in this notebook.

2. First, start by loading the required libraries – OpenCV, NumPy, and OS.

3. Next, use *Exercise 8.01, Downloading the Pedestrian and Vehicle Detection Model*, as a reference, and download the XML and `.bin` files. Use 16-bit floating-point precision while downloading the model.

4. Next, use *Exercise 08.03, Vehicle and Pedestrian Detection*, as a reference and specify the path to the IR.

5. Now, you will have to read the `crowd.jpg` image. Make sure that you create a copy of the image after reading it since you will be drawing the bounding boxes on the first image. The second image will be used to draw the results that are obtained using Haar Cascades.

6. Go through the documentation page of the model (https://docs.openvinotoolkit.org/latest/_models_intel_face_detection_retail_0004_description_face_detection_retail_0004.html) and figure out the input dimensions of the image that are required. Accordingly, you will have to modify the parameters for the `cv2.dnn.blobFromImage` function.

7. Once you have obtained the preprocessed image, pass it through the model. While drawing the bounding boxes, use **15%** as the confidence threshold. You can also play around with this value to see if you can get better results for some other value:

Figure 8.18: Faces detected using the deep learning model

8. Now, you will use Haar Cascades for face detection. You can use *Exercise 5.01*, *Face Detection Using Haar Cascades*, from *Chapter 5*, *Face Processing in Image and Video*, as a reference for this part. First, load the frontal face Haar Cascade model.

9. Next, convert the image (*from here onward, image means the copy of the image created previously*) into grayscale since Haar Cascades assumes the input image is a grayscale image.

10. Now, perform multi-scale detection on the image using Cascade classifiers and draw the bounding boxes you obtained from performing detection:

Figure 8.19: Face detection result using Haar Cascades

It's very interesting to note that there was not much difference in the results obtained using Haar Cascades and the deep learning model. We had to even consider bounding boxes with confidence as low as **15%** as correct to get more results. By completing this activity successfully, you will realize a very important point: not everything requires a deep learning solution. Sometimes, the easiest solutions are the most correct ones. Although simplicity can be an effective strategy, in this case, the deep learning model did give a more accurate solution.

> **NOTE**
>
> The solution for this activity can be found on page 529.

You can play around with the values of the parameters in the Haar Cascade solution to obtain various results. The main idea is to show that, sometimes, a simple solution is the best solution.

SUMMARY

This brings us to the end of this chapter. Let's quickly summarize what we discussed in this chapter. We started by understanding the need for optimization in computer vision and how the OpenVINO toolkit fits this picture. We then went over the various components of the toolkit, understanding the importance of each component. We also saw how easy it is to use OpenVINO with OpenCV. We then saw how pre-trained models can save us a huge amount of money and time. We also discussed how these pre-trained models can be downloaded and used. Finally, we looked at a couple of examples of how the various components of the OpenVINO toolkit can be used with OpenCV to give amazing results in face detection and vehicle and pedestrian detection problems.

APPENDIX

CHAPTER 1: BASICS OF IMAGE PROCESSING

ACTIVITY 1.01: MIRROR EFFECT WITH A TWIST

Solution:

1. Let's start by importing the necessary libraries:

   ```
   # Load modules
   import cv2
   import numpy as np
   import matplotlib.pyplot as plt
   ```

2. Next, let's specify the magic command for displaying images in a notebook:

   ```
   %matplotlib inline
   ```

3. Now, we can load the image and display it using Matplotlib.

   > **NOTE**
   >
   > Before proceeding, ensure that you can change the path to the images (highlighted) based on where the image is saved in your system.

   The code is as follows:

   ```
   # Load image
   img = cv2.imread("../data/lion.jpg")
   plt.imshow(img[:,:,::-1])
   plt.show()
   ```

The output is as follows:

Figure 1.42: Image that we are going to use in this activity

4. Next, let's display the image's shape:

```
# Get image shape
img.shape
```

The output is **(407, 640, 3)**.

5. Convert the image into HSV:

```
# Convert image to HSV
imgHSV = cv2.cvtColor(img, cv2.COLOR_BGR2HSV)
plt.imshow(imgHSV[:,:,::-1])
plt.show()
```

The output is as follows:

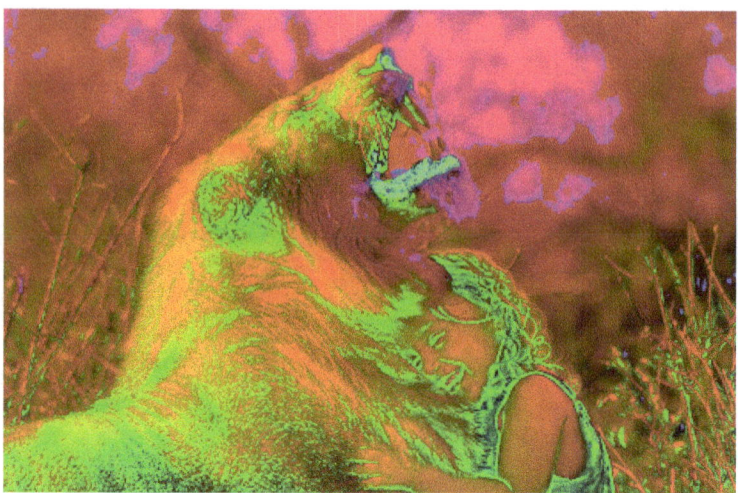

Figure 1.43: Image in the HSV color space

6. We have to split the three channels of the HSV image and display the value channel only:

```
# Obtain the Hue channel
hue, sat, val = cv2.split(imgHSV)
plt.imshow(val, cmap="gray")
plt.show()
```

The output is as follows:

Figure 1.44: Value channel of the HSV color space of the image

7. Creating a negative out of the value channel is very simple. We just have to use **255-val**:

```
val = 255-val
plt.imshow(val, cmap="gray")
plt.show()
```

The output is as follows:

Figure 1.45: Negative of the value channel of the image

8. Now, we can merge the **val** channel with itself:

```
# Merge the val channels
imgHSV = cv2.merge((val,val,val))
plt.imshow(imgHSV[:,:,::-1])
plt.show()
```

The output is as follows:

Figure 1.46: Value channel merged with itself to create a three-channel image

9. Now, we can flip the image horizontally. We just have to reverse the order of the columns:

```
# Flip the image horizontally
imgHSV = imgHSV[:,::-1,:]
plt.imshow(imgHSV[:,:,::-1])
plt.show()
```

The output is as follows:

Figure 1.47: Horizontally flipped negative image

10. Next, we will create the new image filled with zeros:

```
# Create a new image
imgNew = np.zeros((407,640*2,3),dtype=np.uint8)
plt.imshow(imgNew[:,:,::-1])
plt.show()
```

The output is as follows:

Figure 1.48: Black image created using the np.zeros function

11. First, we will copy the original image onto the left half:

```
# Copy img to left half of new image
imgNew[:,:640,:] = img
plt.imshow(imgNew[:,:,::-1])
plt.show()
```

The output is as follows:

Figure 1.49: Original image copied onto the left half

12. Copy the horizontally flipped negative to the right half:

```
# Copy imgHSV to right half of new image
imgNew[:,640:,:] = imgHSV
plt.imshow(imgNew[:,:,::-1])
plt.show()
```

The output is as follows:

Figure 1.50: Horizontally flipped negative image added to the right half

13. Finally, let's save the output image:

```
cv2.imwrite("mirror.png",imgNew)
```

This will return **True**.

> **NOTE**
>
> To access the source code for this specific section, please refer to https://packt.live/38eMp0C.

CHAPTER 2: COMMON OPERATIONS WHEN WORKING WITH IMAGES

ACTIVITY 2.01: MASKING USING BINARY IMAGES

Solution:

1. Create a new notebook – `Activity2.01.ipynb`. We will be writing our code in this notebook.

2. Import the required libraries:

```
# Import modules
import cv2
import numpy as np
import matplotlib.pyplot as plt
%matplotlib inline
```

3. Read the image of the disk and convert it to grayscale.

   > **NOTE**
   > Before proceeding, be sure to change the path to the image (highlighted) based on where the image is saved in your system.

   The code for this is as follows:

```
img = cv2.imread("../data/recording.jpg")
img = cv2.cvtColor(img, cv2.COLOR_BGR2GRAY)
```

4. Display the grayscale image using Matplotlib:

```
plt.imshow(img, cmap='gray')
plt.show()
```

The output is as follows. The X and Y axes refer to the width and height of the image, respectively:

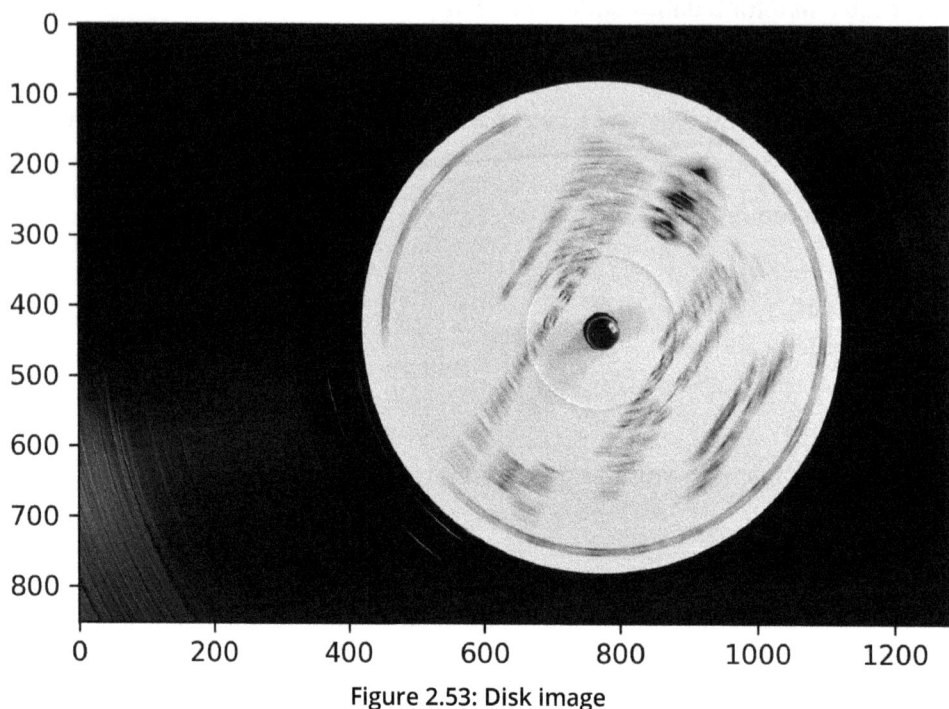

Figure 2.53: Disk image

5. Threshold this image with a threshold of 150 and a maximum value of 255:

```
# Set threshold and maximum value
thresh = 150
maxValue = 255
# Binary threshold
th, dst = cv2.threshold(img, thresh, maxValue, \
        cv2.THRESH_BINARY)
```

6. Display the resultant image, as follows:

```
plt.imshow(dst, cmap='gray')
plt.show()
```

The output is as follows. The X and Y axes refer to the width and height of the image, respectively:

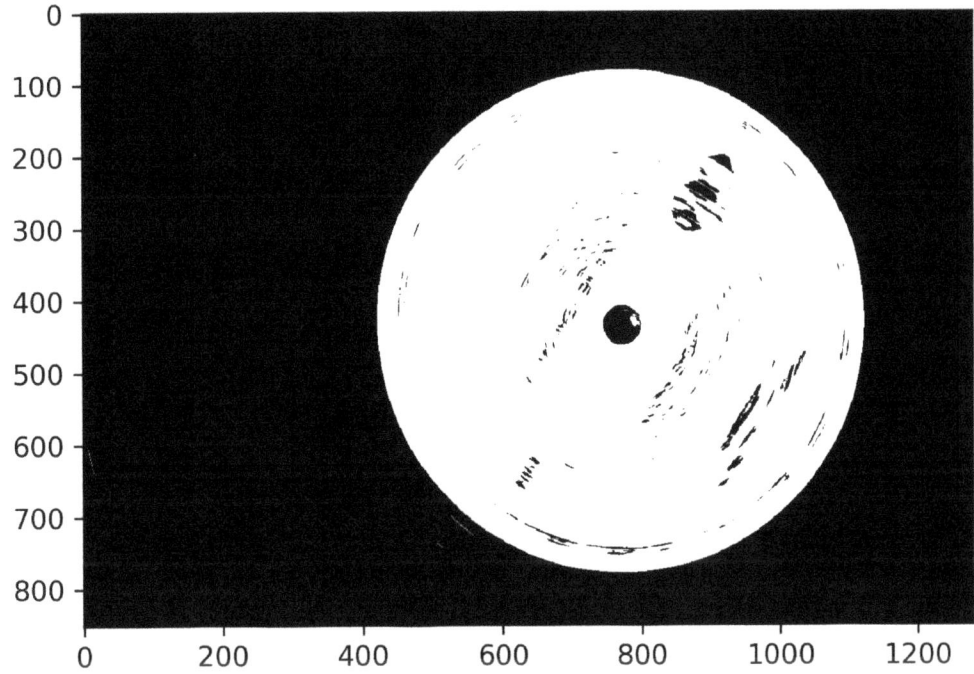

Figure 2.54: Disk image after thresholding

7. Load the image of the zebras and convert it into grayscale.

> **NOTE**
> Be sure to change the path to the image (highlighted) based on where the image is saved in your system.

The code for this is as follows:

```
foreground = cv2.imread("../data/zebra.jpg")
foreground = cv2.cvtColor(foreground, cv2.COLOR_BGR2GRAY)
```

8. Display the image, as follows:

```
plt.imshow(foreground, cmap='gray')
plt.show()
```

The output is as follows. The X and Y axes refer to the width and height of the image, respectively:

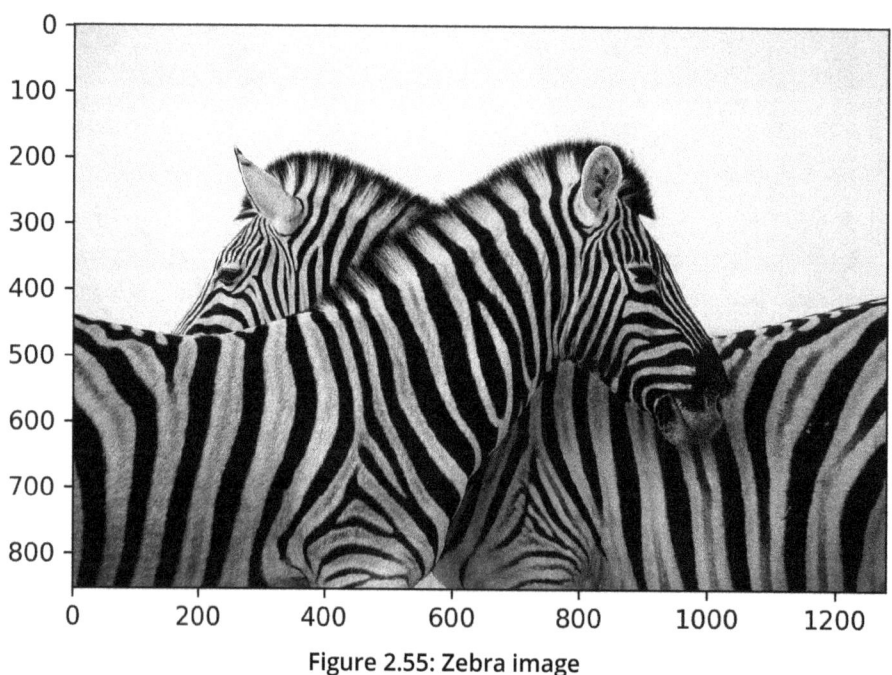

Figure 2.55: Zebra image

9. Display the shape of the image. Check the shape of the **foreground** image, as follows:

```
foreground.shape
```

The shape is **(853, 1280)**.

10. Check the shape of the **dst** image, as follows:

    ```
    dst.shape
    ```

    The shape is **(851, 1280)**.

    We can see that the images have different dimensions and need to be resized.

11. Resize both images using the **cv2.resize** function. We know that the output image should have the shape 1,280×800:

    ```
    # resize both to 800 px height and 1280 px width
    dst = cv2.resize(dst, (1280,800), \
                    interpolation = cv2.INTER_LINEAR)
    foreground = cv2.resize(foreground, (1280,800), \
                    interpolation = cv2.INTER_LINEAR)
    ```

    Verify whether the dimensions of the images have changed or not.

12. Check the shape of the **foreground** image, as follows:

    ```
    foreground.shape
    ```

    The shape of the image is **(800, 1280)**.

13. Check the shape of the **dst** image, as follows:

    ```
    dst.shape
    ```

    The shape of the image is **(800, 1280)**.

14. Finally, use NumPy's **where** function to carry out masking and display the result:

    ```
    result = np.where(dst, foreground, 0)
    plt.imshow(result, cmap='gray')
    plt.show()
    ```

The output is as follows. The *X* and *Y* axes refer to the width and height of the image, respectively:

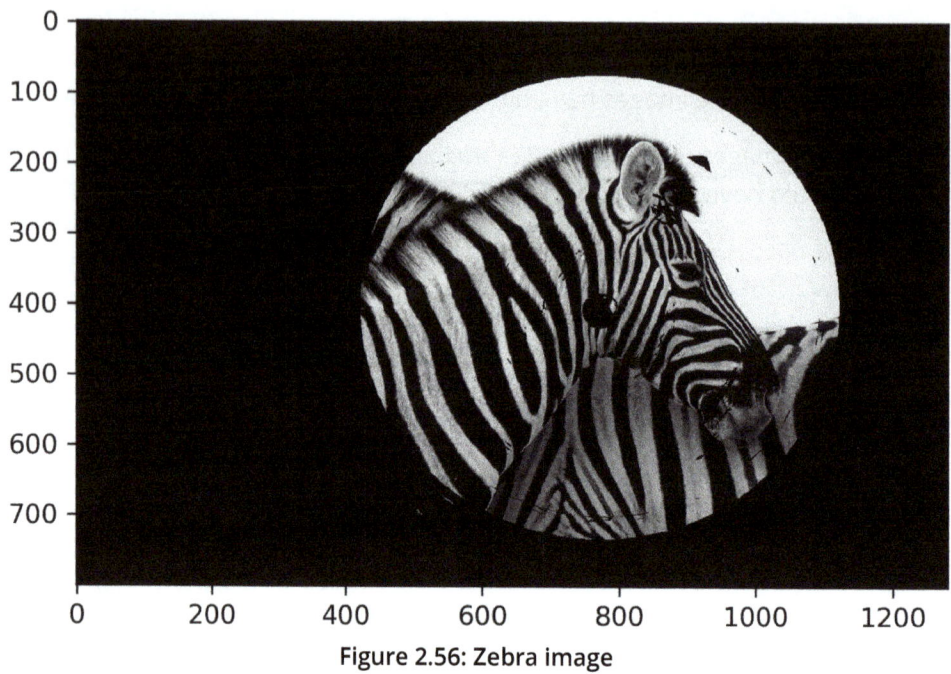

Figure 2.56: Zebra image

> **NOTE**
>
> To access the source code for this specific section, please refer to https://packt.live/3eQUoTU.

CHAPTER 3: WORKING WITH HISTOGRAMS

ACTIVITY 3.01: ENHANCING IMAGES USING HISTOGRAM EQUALIZATION AND CLAHE

Solution:

1. Open a new file in a Jupyter notebook in the folder you want to work on. Remember that your relevant finger veins image file must also be in the same folder so that when reading the image, you won't need to give a path along with the filename.

2. Import the OpenCV library and the **pyplot** module from the Matplotlib library:

   ```
   import cv2
   import matplotlib.pyplot as plt
   ```

3. Read the image as grayscale:

   ```
   img= cv2.imread('fingervein.bmp', 0)
   ```

4. Display the image. Let's use Matplotlib to display the images in this example. You can also use OpenCV's **cv2.imshow()** if you want:

   ```
   imgplot = plt.imshow(img , cmap="gray")
   plt.title('Original image')
   plt.show()
   ```

   The plot looks as follows:

   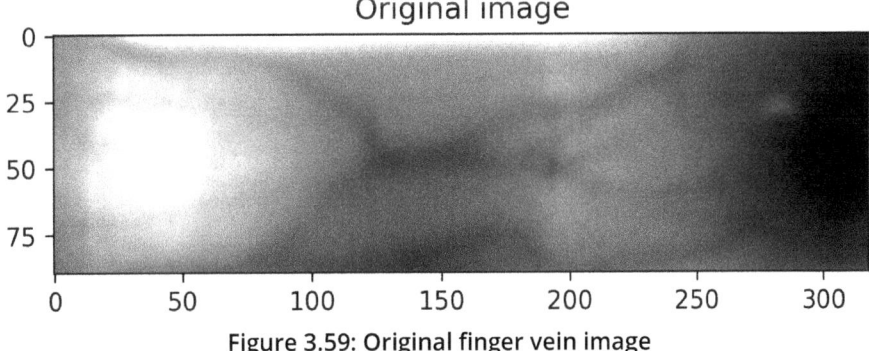

   Figure 3.59: Original finger vein image

   The values on the X and Y axes show the width and height of the displayed image.

5. Plot the histogram:

```
plt.hist(img.ravel(), bins= 256)
plt.title('Histogram of Original Image')
plt.show()
```

The histogram looks as follows:

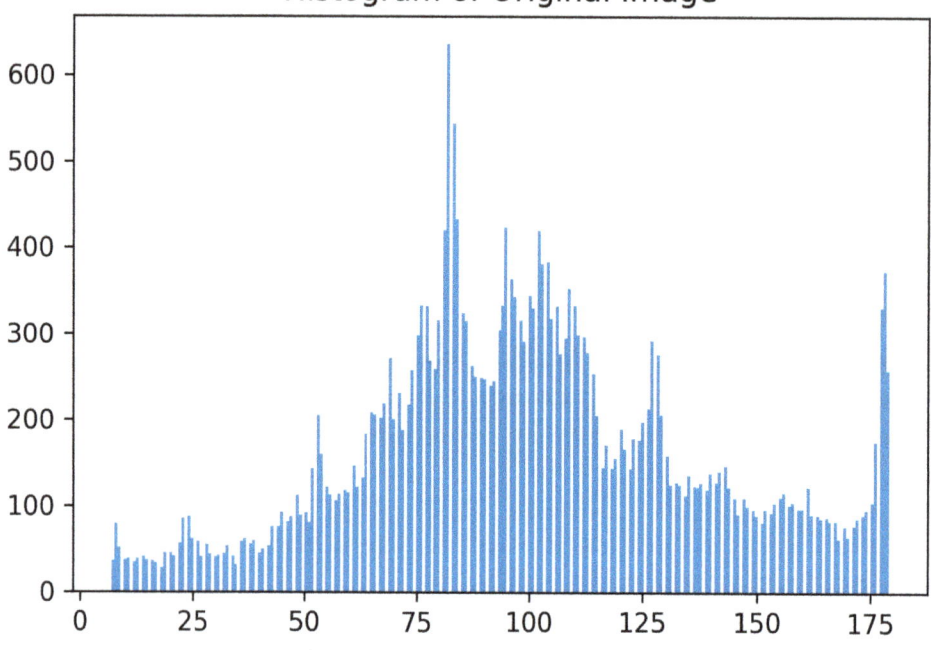

Figure 3.60: Computed histogram

6. Let's now perform histogram equalization. Apply histogram equalization to this image:

```
histequ = cv2.equalizeHist(img)
```

7. Display the histogram equalized image:

```
plt.imshow(histequ , cmap="gray")
plt.title('Histogram equalized image')
plt.show()
```

The histogram equalized image looks as follows:

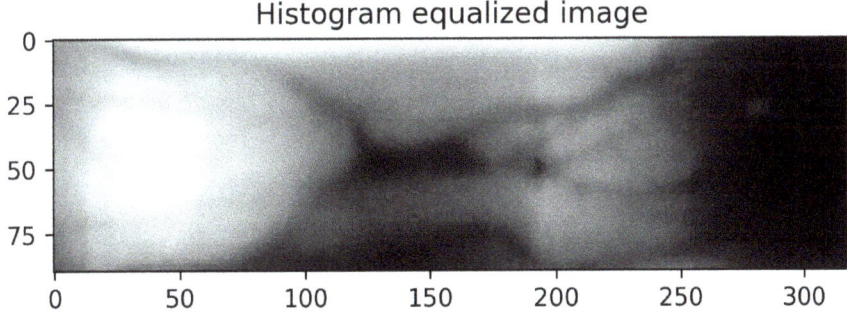

Figure 3.61: Finger vein image after applying histogram equalization

8. Compute its histogram and plot it:

```
plt.hist(histequ.ravel(), bins= 256)
plt.title('Histogram of histogram equalized image')
plt.show()
```

The histogram of the histogram equalized image looks as follows:

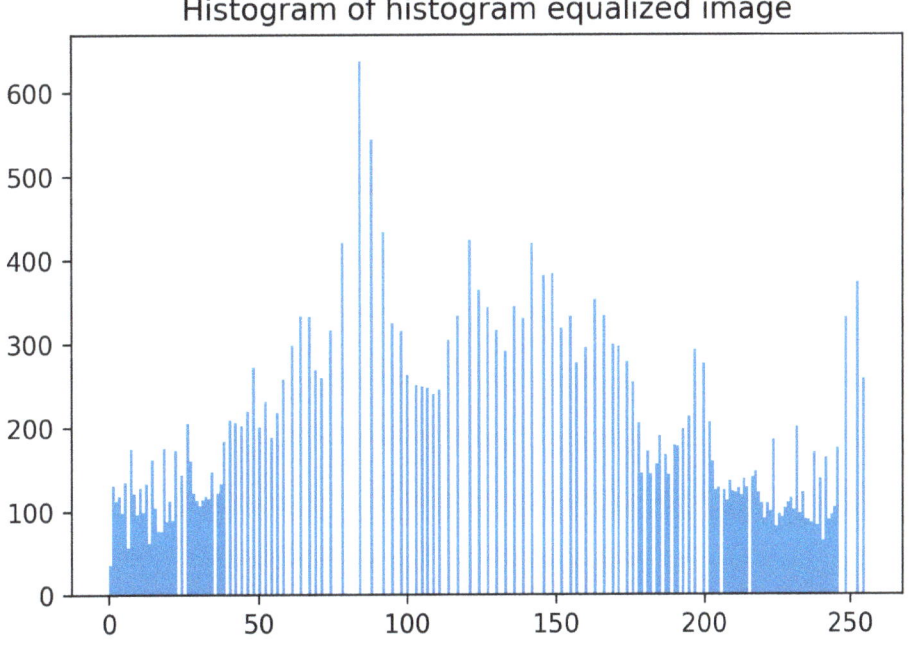

Figure 3.62: Computed histogram after applying histogram equalization

9. Make a CLAHE object:

```
clahe = cv2.createCLAHE(clipLimit=4.0, \
                        tileGridSize=(16,16))
```

10. Apply it to the original image that you read from the image file:

```
clahe_img = clahe.apply(img)
```

11. Display the image:

```
plt.imshow(clahe_img , cmap="gray")
plt.title('CLAHE image')
plt.show()
```

The image looks as follows:

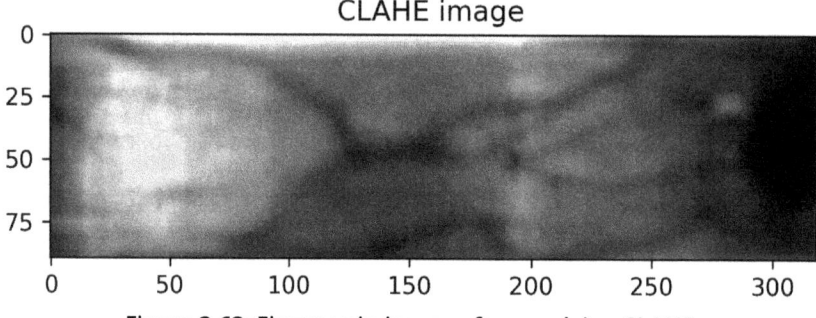

Figure 3.63: Finger vein image after applying CLAHE

12. Display the histogram of the image as follows:

```
plt.hist(clahe_img.ravel(), bins= 256)
plt.title('Histogram of CLAHE Image')
plt.show()
```

The histogram looks as follows:

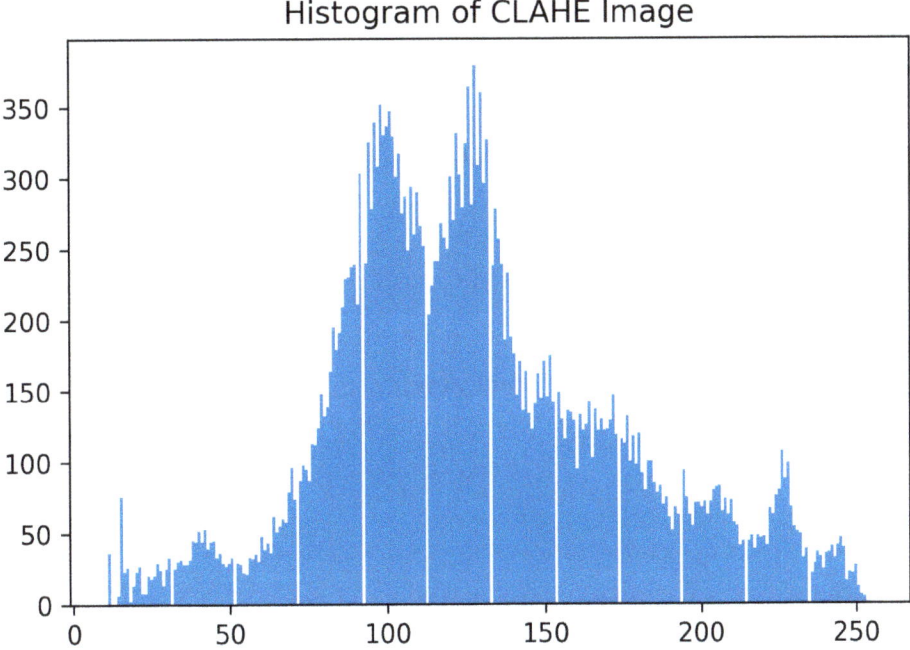

Figure 3.64: Computed histogram after applying CLAHE

> **NOTE**
>
> To access the source code for this specific section, please refer to https://packt.live/3iiify7.

484 | Appendix

ACTIVITY 3.02: IMAGE ENHANCEMENT IN A USER-DEFINED ROI

Solution

1. Open a new file in your Jupyter Notebook and name it anything you like.

2. Import OpenCV for image processing operations and Matplotlib for plotting. You will also need the **numpy** library to create the binary mask, so you can also import that at the start of the code:

```
import cv2
import numpy as np
import matplotlib.pyplot as plt
```

3. Read the image provided to you and view it:

```
im = cv2.imread('ocean.jpg')
origrgb = cv2.cvtColor(im,cv2.COLOR_BGR2RGB);
imgplot = plt.imshow(origrgb)
plt.title('Original image')
plt.show()
```

The output is as follows:

Figure 3.65: Displaying the original image

4. Prompt the user to draw a rectangular bounding box on the image to mark the ROI:

```
x,y,w,h = cv2.selectROI(im, fromCenter=0)
cv2.destroyAllWindows();
```

The ROI selector looks as follows:

Figure 3.66: ROI selector

Suppose the box is drawn like this:

Figure 3.67: The user will draw a bounding box to select the ROI

5. Use **numpy** to create a totally blank image with all pixels being black (**0**). This new image will have the same dimensions in terms of width and height as the original image:

```
mymask= np.zeros(im.shape[:2],  dtype = "uint8")
```

6. Within the mask's ROI, make all the pixels white (**255**). This is now the mask image:

```
mymask[int(y):int(y+h) , int(x):int(x+w)]= 255
```

7. Display the mask:

   ```
   mask_rgb = cv2.cvtColor(mymask,cv2.COLOR_BGR2RGB);
   imgplot = plt.imshow(mask_rgb)
   plt.title('Created mask')
   plt.show()
   ```

   The mask will look as follows:

   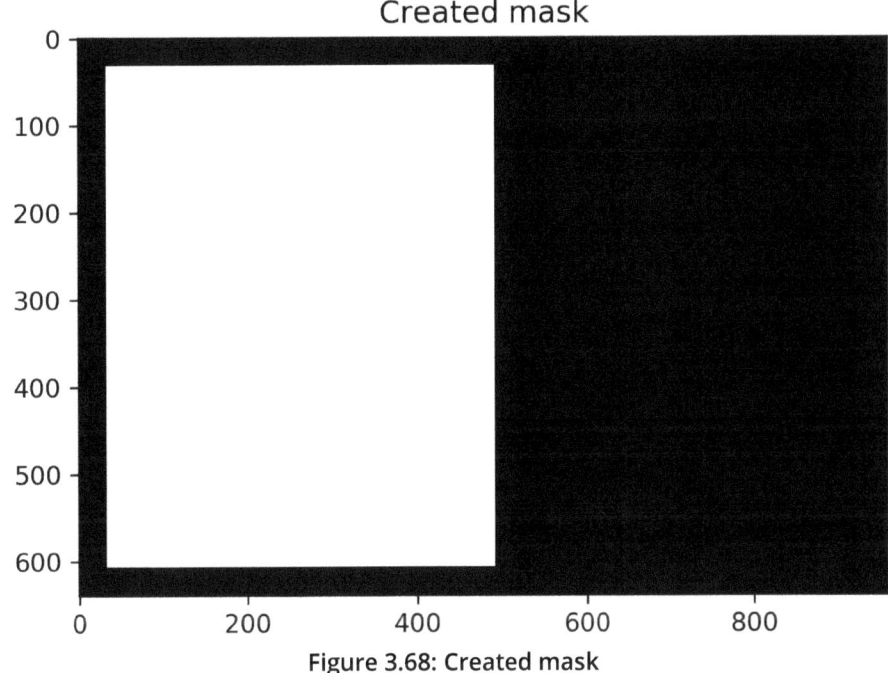

   Figure 3.68: Created mask

8. Save the mask to a .png file:

   ```
   cv2.imwrite('mask_image.png' , mymask)
   ```

   You will get the output as **True**. This indicates that the image of the mask is saved successfully.

9. Convert the image to the HSV space:

   ```
   imgHSV = cv2.cvtColor(im,cv2.COLOR_BGR2HSV);
   ```

10. Compute a histogram of the masked region in the **V** plane (channel **2**):

    ```
    hist = cv2.calcHist([imgHSV],[2],mymask,[256],[0,255])
    ```

11. Plot this histogram:

```
plt.plot(hist)
plt.title('V plane')
```

The histogram looks as follows:

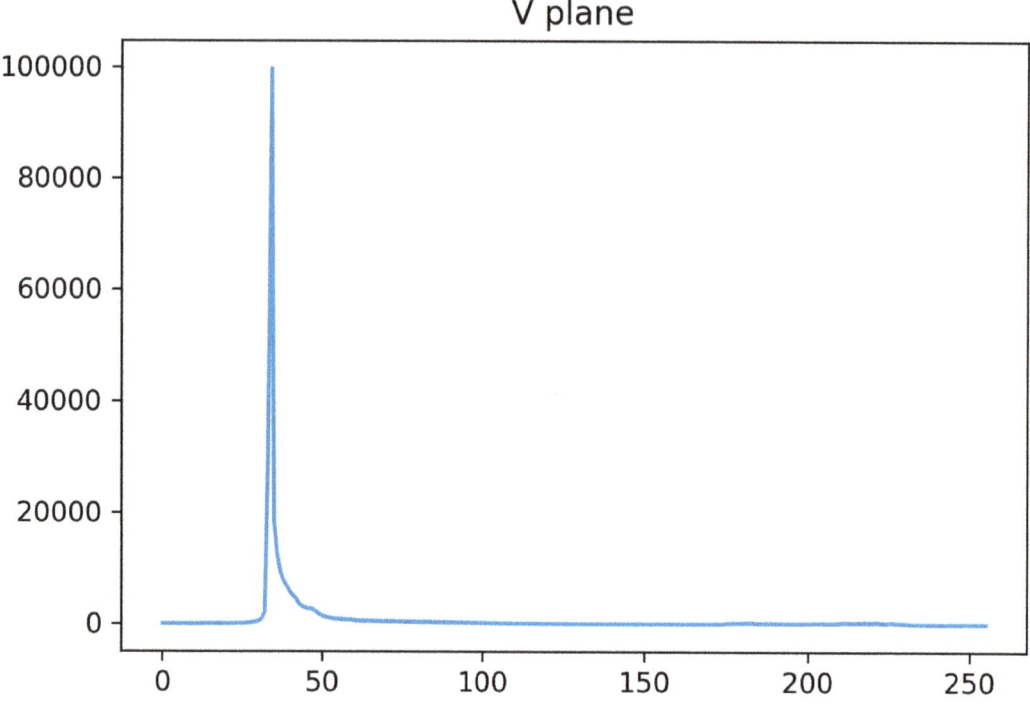

Figure 3.69: Histogram of the V plane of the selected ROI

12. Make a variable, **v**, to access the **V** channel of this HSV image, such that any change you make to **v** will automatically be made to the **V** channel of the image. For this, use the following code:

```
v= imgHSV[:,:,2]
```

> **NOTE**
>
> If you use the `.copy()` method, then any change to the **copy** variable will not be automatically applied to the original data, which is why we did not use the `.copy()` method here.

13. Create a CLAHE object:

    ```
    clahe = cv2.createCLAHE(clipLimit=7.0, \
                            tileGridSize=(8,8))
    ```

14. Access the area within the ROI of the **v** plane and apply CLAHE to it:

    ```
    v[int(y):int(y+h) , \
      int(x):int(x+w)]= clahe.apply(v[int(y):int(y+h) , \
                                     int(x):int(x+w)])
    ```

15. Compute the histogram of the area within the ROI of the modified V plane:

    ```
    hist = cv2.calcHist([imgHSV],[2],mymask,[256],[0,255])
    ```

16. Plot the histogram:

    ```
    plt.plot(hist)
    plt.title('enhanced V plane')
    ```

    The histogram looks as follows:

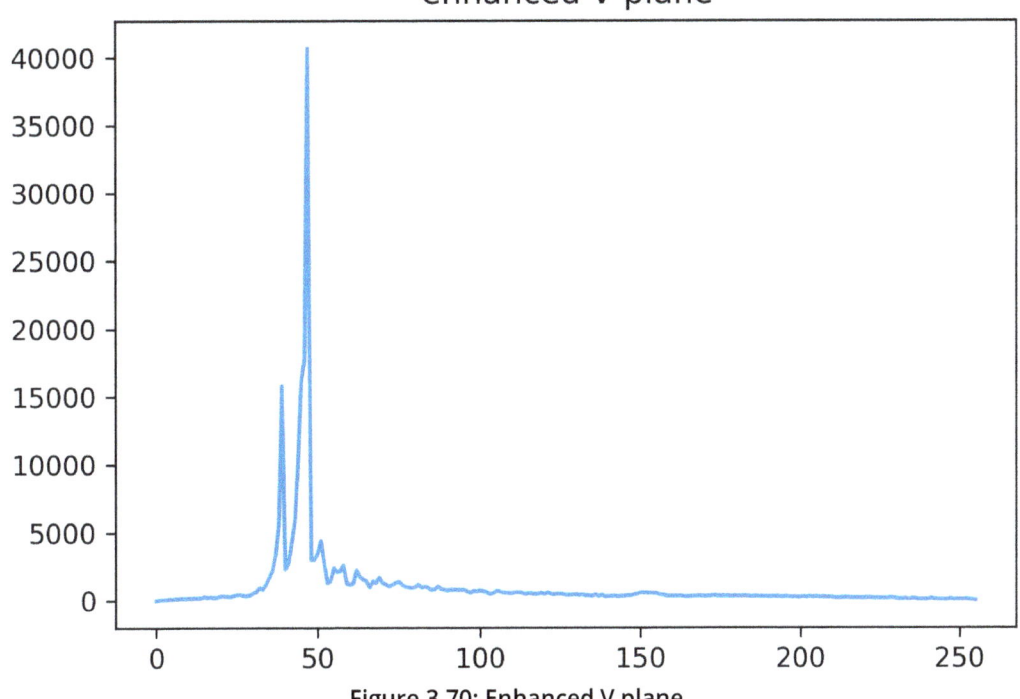

Figure 3.70: Enhanced V plane

17. Bring the image back to the RGB space and use Matplotlib to display it:

```
enhanced = cv2.cvtColor(imgHSV,cv2.COLOR_HSV2RGB);
imgplot = plt.imshow(enhanced)
plt.title('Enhanced ROI in the image')
plt.show()
```

The image will look like this:

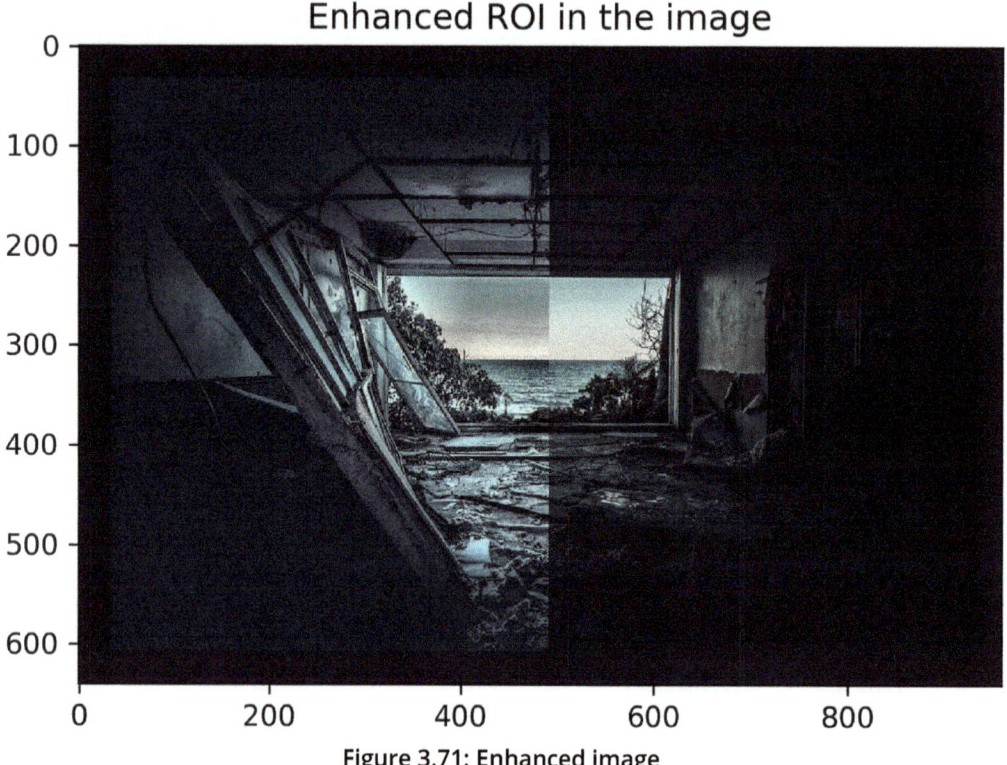

Figure 3.71: Enhanced image

> **NOTE**
>
> To access the source code for this specific section, please refer to https://packt.live/31yC3r1.

CHAPTER 4: WORKING WITH CONTOURS

ACTIVITY 4.01: IDENTIFYING A CHARACTER ON A MIRRORED DOCUMENT

Solution:

1. Open Jupyter Notebook. Go to **File** | **New File**. A new file will open. Save it as **Activity4.01.ipynb**.

2. Import the OpenCV library:

```
import cv2
```

3. Read the image of the handwritten note as follows:

```
image = cv2.imread('phrase_handwritten.png')
```

4. Save a copy of this image. You will mark the bounding box on this copy at the end of the activity:

```
imagecopy= image.copy()
```

5. Convert the image to grayscale as follows:

```
gray_image = cv2.cvtColor(image, cv2.COLOR_BGR2GRAY)
```

6. Now, convert the image into binary and display it as follows:

```
ret,binary_im = cv2.threshold(gray_image,0,255,\
                              cv2.THRESH_OTSU)
cv2.imshow('binary image', binary_im)
cv2.waitKey(0)
cv2.destroyAllWindows()
```

The preceding code produces the following output:

Figure 4.62: Binary image

7. Now, find all the contours on this image. Which retrieval method do you think you should use to detect these contours? Note that the letter 'B' has two holes inside it. If you use any method other than **cv2.RETR_EXTERNAL**, then 'B' would be detected as three contours instead of one (one contour would encompass the outer boundary of B, and the other two would be for its two holes). For this task, we want the shape of 'B' to be detected as a single contour, which is why we will use the **cv2.RETR_EXTERNAL** method of contour detection:

```
contours_list,_ = cv2.findContours(binary_im, \
                    cv2.RETR_EXTERNAL, \
                    cv2.CHAIN_APPROX_SIMPLE)
```

We only require the second output from this function, so we have used the _ symbol to receive the other output (the hierarchy). This allows us to save some memory by not allocating any variable to that output.

8. Now, draw all the detected contours on the image using rectangular bounding boxes. Let's use yellow (its color code is 0, 255, 255) and a thickness of **2**.

   Since we want to do it for all contours, we will do it in a **for** loop. Also, we will draw each bounding box inside that **for** loop as we go, so that after the **for** loop is finished executing, we will have an image with all the bounding boxes drawn on it around the detected contours:

   ```
   for cnt in contours_list:
       x,y,w,h = cv2.boundingRect(cnt)
       cv2.rectangle(image,(x,y),(x+w,y+h),(0, 255, 255),2)
   cv2.imshow('Contours marked on RGB image', image)
   cv2.waitKey(0)
   cv2.destroyAllWindows()
   ```

   The output is as follows:

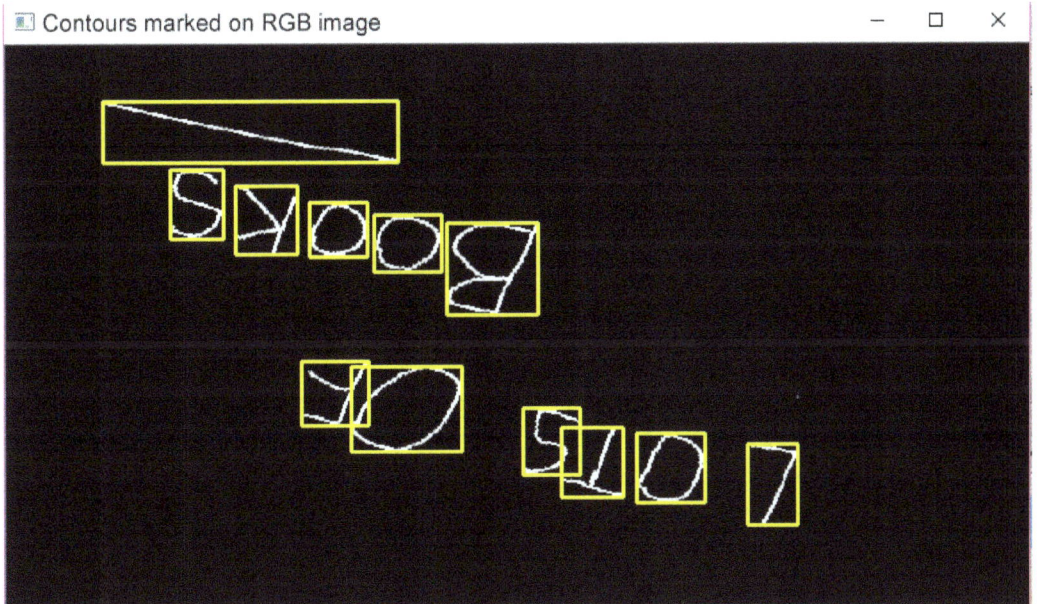

   Figure 4.63: All detected external contours

9. Now that we have detected all external blobs on the handwritten note, the next step is to compare each detected blob with our reference shape to find the closest match. The first step in that direction is to find the contour on the reference. For that, read the reference image in grayscale:

   ```
   ref_gray = cv2.imread('typed_B.png', cv2.IMREAD_GRAYSCALE)
   ```

10. Convert the image into binary and display the result. It's your choice whether you want to use a fixed numerical threshold for this binary segmentation, or Otsu's method, or any other appropriate method. We usually use Otsu's method because it saves us the trouble of finding a suitable threshold by trial and error:

```
ret, ref_binary = cv2.threshold(ref_gray,0, 255, \
                                cv2.THRESH_OTSU)
cv2.imshow('Reference image', ref_binary)
cv2.waitKey(0)
cv2.destroyAllWindows()
```

The output is as follows:

Figure 4.64: Reference image - this is the blob whose closest match we will find on the handwritten note

11. Find all the contours on this reference. Since this is your reference contour image, it should only have one contour – the outer shape of the letter 'B'. Note that here, we are again using the **cv2.RETR_EXTERNAL** retrieval mode because this will give you a single outline on the boundary shape. All other retrieval modes would have given you two additional contours, which lie on the inner holes of 'B' – these we do not need:

```
ref_contour_list,_ = cv2.findContours(ref_binary, \
                         cv2.RETR_EXTERNAL, \
                         cv2.CHAIN_APPROX_SIMPLE)
```

12. Next, as a good programming practice, make sure that there is one, and only one detected contour in the **ref_contour_list** list. If there is not, then that means that there is something fishy going on and the program execution should immediately be stopped so that you can debug it. To terminate a program's execution through code, Python has the **sys.exit()** command that you can use after importing the **sys** library:

```
if len(ref_contour_list)==1:
    ref_contour= ref_contour_list[0]
else: # stop the program
    import sys
    print('Reference image contains more than 1 contour. '\
        'Please check!')
    sys.exit()
```

13. Now, compare each contour on the handwritten note with the detected contour on the reference image using a **for** loop.

    Before this **for** loop, initiate an empty list by the name of **dist_list**. As you get the distance of each contour from the reference contour, append that distance to this list. Also, initialize a counter variable before the **for** loop. Its value will increment after each iteration of this loop. In the following steps, this variable will help you to retrieve the index of your required contour:

```
ctr= 0
dist_list= [ ]
for cnt in contours_list:
    retval = cv2.matchShapes(cnt, ref_contour, \
            cv2.CONTOURS_MATCH_I1,0)
    dist_list.append(retval)
    ctr= ctr+1
```

14. Find the minimum distance in **dist_list**:

```
min_dist= min(dist_list)
```

15. Find the index of this minimum distance in **dist_list**. This is the index of that contour in **contours_list** that is most similar to the reference contour:

```
ind_min_dist= dist_list.index(min_dist)
```

16. Get the contour of interest from the list of contours. This is the contour on the **index= ind_min_dist** list:

```
required_cnt= contours_list[ind_min_dist]
```

17. Get a bounding box that will enclose this contour:

   ```
   x,y,w,h = cv2.boundingRect(required_cnt)
   ```

18. Draw a bounding box around this contour on a fresh copy of the image using any color (say, blue) and any line thickness (say, **2**):

   ```
   cv2.rectangle(imagecopy,(x,y),(x+w,y+h),\
                 (255, 0, 0),2)
   cv2.imshow('Detected B', imagecopy)
   cv2.waitKey(0)
   cv2.destroyAllWindows()
   ```

   Now, display the result as follows:

Figure 4.65: Detected 'B' on a handwritten note

> **NOTE**
>
> To access the source code for this specific section, please refer to https://packt.live/3dOq4rY.

CHAPTER 5: FACE PROCESSING IN IMAGE AND VIDEO

ACTIVITY 5.01: EYE DETECTION USING MULTIPLE CASCADES

Solution:

Perform the following steps to complete this activity:

1. Open your Jupyter notebook and create a new file called `ctivity5.01.ipynb`. Write your code in this file.

2. Import the required libraries:

   ```
   import cv2
   import numpy as np
   ```

3. Use the `detectionUsingCascades` function that we wrote in *Exercise 5.02, Eye Detection Using Cascades*, as the starting point and modify it to incorporate the second cascade classifier as well.

4. Revise the arguments of the function since we will be providing the paths of both cascade classifiers now:

   ```
   def detectionUsingCascades(imageFile, cascadeFile1, \
                              cascadeFile2):
   ```

5. Load the image and create a copy of it:

   ```
   image = cv2.imread(imageFile)
   # Create a copy of the image
   imageCopy = image.copy()
   ```

6. Convert the image from BGR into grayscale:

   ```
   # Step 2 - Convert the image from BGR to Grayscale
   gray = cv2.cvtColor(image, cv2.COLOR_BGR2GRAY)
   ```

7. Add the code to load both the cascades as *Step 3* of the `detectionUsingCascades` custom function:

   ```
   # Step 3 - Load the cascade
   haarCascade1 = cv2.CascadeClassifier(cascadeFile1)
   haarCascade2 = cv2.CascadeClassifier(cascadeFile2)
   ```

   Mention the paths for **cascadeFile1** and **cascadeFile2**, as per their location in your system.

8. Keep *Step 4* of the **detectionUsingCascades** custom function as it was in *Exercise 5.02, Eye Detection Using Cascades*:

   ```
   # Step 4 - Perform multi-scale detection
   detectedObjects = haarCascade1.detectMultiScale(gray, 1.2, 9)
   ```

9. Use the detected bounding boxes from the first cascade classifier (the frontal face classifier) to crop out the face region as *Step 5* of the **detectionUsingCascades** custom function:

   ```
   # Step 5 - Draw bounding boxes
   for bbox in detectedObjects:
       # Each bbox is a rectangle representing
       # the bounding box around the detected object
       x, y, w, h = bbox
       cv2.rectangle(image, (x, y), (x+w, y+h), \
                     (0, 0, 255), 3)
       # Cropped object
       crop = gray[y:y+h, x:x+w]
       imgCrop = imageCopy[y:y+h, x:x+w]
       cv2.imwrite("crop.png", imgCrop)
   ```

10. Use the second classifier (the eye detection classifier) to detect eyes in the cropped face between *Step 5* and *Step 6* of the **detectionUsingCascades** custom function and use the **cv2.rectangle** function to draw a bounding box over the detected eye region:

    ```
            # Perform multi-scale detection
            cropDetectedObjects = haarCascade2.detectMultiScale(crop,
    1.2, 2)
            for bbox2 in cropDetectedObjects:
                X, Y, W, H = bbox2
                cv2.rectangle(image, (x+X, y+Y), \
                              (x+X+W, y+Y+H), (255,0,0), 3)
    ```

    Note how we have used **(x+X, y+Y)** and **(x+X+W, y+Y+H)** as the coordinates of the bounding box for eyes to take into account the fact that (X,Y) correspond to the coordinates for the cropped region, whereas we want to draw the bounding box on the main image.

11. Keep the rest of the **detectionUsingCascades** custom function the same as what we used in *Exercise 5.02, Eye Detection Using Cascades*:

    ```
    # Step 6 - Display the output
    
    cv2.imshow("Object Detection", image)
    cv2.waitKey(0)
    cv2.imwrite("eyes-combined-result.png",image)
    cv2.destroyAllWindows()
    
    # Step 7 - Return the bounding boxes
    return detectedObjects
    ```

12. Supply the paths of both cascade classifiers while calling the **detectionUsingCascades** function:

    > **NOTE**
    >
    > Before proceeding, ensure that you can change the path to the image and the files (highlighted) based on where they are saved in the system.

    ```
    eyeDetection = detectionUsingCascades("../data/eyes.jpeg", \
                  "../data/haarcascade_frontalface_default.xml", \
                  "../data/haarcascade_eye.xml")
    ```

This preceding code produces the following output:

Figure 5.40: Result obtained by using multiple cascades

> **NOTE**
>
> To access the source code for this specific section, please refer to https://packt.live/3f1nlag.

ACTIVITY 5.02: SMILE DETECTION USING HAAR CASCADES

Solution:

Perform the following steps to complete this activity:

1. Open your Jupyter notebook and create a new file called **Activity5.02.ipynb**. We will be writing our code in this file.

2. Import the necessary modules:

```
import cv2
import numpy as np
```

3. Use the same function that we wrote in our **Activity5.01.ipynb** code – **detectionUsingCascades**. Let's start with *Steps 1* to *4* in the code:

```
def detectionUsingCascades(imageFile, cascadeFile1, cascadeFile2):

    # Step 1 - Load the image
    image = cv2.imread(imageFile)
    # Create a copy of the image
    imageCopy = image.copy()
    # Step 2 - Convert the image from BGR to Grayscale
    gray = cv2.cvtColor(image, cv2.COLOR_BGR2GRAY)
    # Step 3 - Load the cascade
    haarCascade1 = cv2.CascadeClassifier(cascadeFile1)
    haarCascade2 = cv2.CascadeClassifier(cascadeFile2)
    # Step 4 - Perform multi-scale detection
    detectedObjects = haarCascade1\
                      .detectMultiScale(gray, 1.2, 9)
```

4. Next, use the same *Step 5* of the `detectionUsingCascades` custom function for drawing bounding boxes:

```
# Step 5 - Draw bounding boxes
for bbox in detectedObjects:
    # Each bbox is a rectangle representing
    # the bounding box around the detected object
    x, y, w, h = bbox
    cv2.rectangle(image, (x, y), (x+w, y+h), (0, 0, 255), 3)
    # Cropped object
    crop = gray[y:y+h, x:x+w]
    imgCrop = imageCopy[y:y+h, x:x+w]
    cv2.imwrite("crop.png",imgCrop)
    # Perform multi-scale detection
    cropDetectedObjects = haarCascade2.detectMultiScale(crop, 1.2, 9)
    for bbox2 in cropDetectedObjects:
        X, Y, W, H = bbox2
        cv2.rectangle(image, (x+X,y+Y), \
                      (x+X+W, y+Y+H), \
                      (255,0,0), 3)
```

5. Finally, we will just display the output and return the detected objects:

```
# Step 6 - Display the output
cv2.imshow("Object Detection", image)
cv2.waitKey(0)
cv2.imwrite("smile-combined-result.png",image)
cv2.destroyAllWindows()

# Step 7 - Return the bounding boxes
return detectedObjects
```

6. The only modification required is the parameter values. Use a `scaleFactor` of `1.2` and `minNeighbors` set to `9` for both `detectMultiScale` functions.

502 | Appendix

7. Once you are done with parameter modification, you can call the function by providing the paths of the frontal face cascade and smile detection cascade XML files:

> **NOTE**
>
> Before proceeding, ensure that you can change the path to the image and the files (highlighted) based on where they are saved in the system.

```
smileDetection = detectionUsingCascades("../data/eyes.jpeg", \
                 "../data/haarcascade_frontalface_default.xml", \
                 "../data/haarcascade_smile.xml")
```

This preceding code produces the following output:

Figure 5.41: Result obtained by using the frontal face and smile cascade classifiers

> **NOTE**
>
> To access the source code for this specific section, please refer to https://packt.live/38gi6q8.

ACTIVITY 5.03: SKIN SEGMENTATION USING GRABCUT

Solution:

Perform the following steps to complete this activity:

1. Open your Jupyter notebook and create a new file called **Activity5.03.ipynb**.

2. Import both the NumPy and OpenCV modules:

```
import cv2
import numpy as np
```

3. Copy the **Sketcher** class from the **Exercise5.04.ipynb** file:

Activity5.03.ipynb

```
# OpenCV Utility Class for Mouse Handling
class Sketcher:
    def __init__(self, windowname, dests, colors_func):
        self.prev_pt = None
        self.windowname = windowname
        self.dests = dests
        self.colors_func = colors_func
        self.dirty = False
        self.show()
        cv2.setMouseCallback(self.windowname, self.on_mouse)
```

The complete code for this step can be found at https://packt.live/2Bv4wU3.

4. Read the input image and convert it into grayscale.

> **NOTE**
>
> Before proceeding, ensure that you can change the path to the image (highlighted) based on where the image is saved in your system.

The code is as follows:

```
# Read image
img = cv2.imread("../data/grabcut.jpg")
# Convert image to grayscale
gray = cv2.cvtColor(img, cv2.COLOR_BGR2GRAY)
```

5. Next, load the frontal face cascade using the **cv2.CascadeClassifier** function:

   > **NOTE**
   > Before proceeding, ensure that you can change the path to the file (highlighted) based on where the file is saved in your system.

   The code is as follows:

   ```
   # Load haar cascade for frontal face
   haarCascadePath = "../data/haarcascade_frontalface_default.xml"
   haarCascade = cv2.CascadeClassifier(haarCascadePath)
   ```

6. Now, detect faces using the **detectMultiScale** function:

   ```
   # Detect face using Haar Cascade Classifier
   detectedFaces = haarCascade.detectMultiScale(gray, 1.2, 9)
   ```

7. Directly use the detected face region to crop out the face from the image, considering that we have only a face in our image:

   ```
   # Crop face
   x,y,w,h = detectedFaces[0]
   img = img[y:y+h, x:x+w]
   imgCopy = img.copy()
   ```

8. Next, resize the cropped face to increase its size by **2** times. To do that, calculate the new dimensions of the face:

   ```
   # Resize the cropped face
   # Increase the size of the face by 200%
   scale_percent = 200
   # New width
   width = int(img.shape[1] * scale_percent/100)
   # New height
   height = int(img.shape[0] * scale_percent/100)
   # New dimensions
   dim = (width, height)
   ```

9. Next, use the **cv2.resize** function to resize the cropped face:

    ```
    # Resize the face using cv2.resize function
    img = cv2.resize(img, dim, interpolation = cv2.INTER_AREA)
    cv2.imshow("Face Detected", img)
    cv2.waitKey(0)
    ```

    The rest of the steps will stay the same as the ones we used in **Exercise5.03.ipynb**. We have to use the same code to carry out the GrabCut operation on the resized face and then add a finishing touch to the mask to only keep the skin region. Once you've done this, press the *Esc* key to exit and save the results.

10. Create the mask filled with zeros:

    ```
    # Create a mask
    mask = np.zeros(img.shape[:2], np.uint8)
    ```

11. Create two placeholders for the background and foreground models:

    ```
    # Temporary arrays
    bgdModel = np.zeros((1,65),np.float64)
    fgdModel = np.zeros((1,65),np.float64)
    ```

12. Select the ROI and draw a rectangle around the image:

    ```
    # Select ROI
    rect = cv2.selectROI(img)
    # Draw rectangle
    x,y,w,h = rect
    cv2.rectangle(imgCopy, (x, y), (x+w, y+h), \
                  (0, 0, 255), 3);
    ```

13. Next, perform the first step of GrabCut:

    ```
    # Perform grabcut
    cv2.grabCut(img,mask,rect,bgdModel,\
                fgdModel,5,cv2.GC_INIT_WITH_RECT)
    ```

14. Using the preceding GrabCut code, obtain a new mask:

    ```
    mask2 = np.where((mask==2)|(mask==0),0,1).\
                    astype('uint8')
    cv2.imshow("Mask",mask*80)
    cv2.imshow("Mask2",mask2*255)
    cv2.imwrite("mask.png",mask*80)
    cv2.imwrite("mask2.png",mask2*255)
    cv2.waitKey(0)
    cv2.destroyAllWindows()
    ```

15. Next, create a new image and mask using the **Sketcher** class:

    ```
    img = img*mask2[:,:,np.newaxis]
    img_mask = img.copy()
    mask2 = mask2*255
    mask_copy = mask2.copy()
    # Create sketch using OpenCV Utility Class: Sketcher
    sketch = Sketcher('image', [img_mask, mask2], \
                lambda : ((255,0,0), 255))
    ```

16. Use the **cv2.waitKey()** function to keep a record of the key that was pressed:

    ```
    while True:
        ch = cv2.waitKey()
        # Quit
        if ch == 27:
            print("exiting...")
            cv2.imwrite("img_mask_grabcut.png",img_mask)
            cv2.imwrite("mask_grabcut.png",mask2)
            break
        # Reset
        elif ch == ord('r'):
            print("resetting...")
            img_mask = img.copy()
            mask2 = mask_copy.copy()
            sketch = Sketcher('image', [img_mask, mask2], \
                    lambda : ((255,0,0), 255))
            sketch.show()
    ```

If the *Esc* key is pressed, we will break out of the while loop after saving both the mask and the image and if the *R* key is pressed, we will reset the mask and the image. In the `while` loop, we will have to take care based on which key is pressed, starting off with the *Esc* key.

17. Switch to background revision if the **b** key is pressed:

```
# Change to background
elif ch == ord('b'):
    print("drawing background...")
    sketch = Sketcher('image', [img_mask, mask2], \
            lambda : ((0,0,255), 0))
    sketch.show()
```

18. Similarly, switch to foreground revision if the **f** key is pressed:

```
# Change to foreground
elif ch == ord('f'):
    print("drawing foreground...")
    sketch = Sketcher('image', [img_mask, mask2], \
            lambda : ((255,0,0), 255))
    sketch.show()
```

19. Perform GrabCut if any other key is pressed:

```
else:
    print("performing grabcut...")
    mask2 = mask2//255
    cv2.grabCut(img,mask2,None,bgdModel,\
            fgdModel,5,cv2.GC_INIT_WITH_MASK)
    mask2 = np.where((mask2==2)\
            |(mask2==0),0,1).astype('uint8')
    img_mask = img*mask2[:,:,np.newaxis]
    mask2 = mask2*255
    print("switching bank to foreground...")
    sketch = Sketcher('image', [img_mask, mask2], \
            lambda : ((255,0,0), 255))
    sketch.show()
```

20. Finally, close all the display windows once you are out of the `while` loop:

    ```
    cv2.destroyAllWindows()
    ```

    The segmented skin region will look as follows:

Figure 5.42: Segmented skin region

Now that we have obtained the skin region, we can perform operations such as skin smoothing, which has extensive use in cosmetic-based applications.

> **NOTE**
>
> To access the source code for this specific section, please refer to https://packt.live/2Bv4wU3.

ACTIVITY 5.04: EMOJI FILTER

Solution:

Perform the following steps to complete this activity:

1. Open your Jupyter notebook and create a new file called **Activity5.04.ipynb**.

2. First, import the required modules:

```
import cv2
import numpy as np
```

3. Next, we will modify the **detectionUsingCascades** function that we used in **Exercise5.02.ipynb**.

4. Rename the function to **emojiFilter** and add a new argument called **emojiFile**, which will store the path of the emoji image:

```
def emojiFilter(image, cascadeFile, emojiFile):
```

5. Modify *Step 1* of the function to read the emoji image. We will provide **–1** as the second argument to the **cv2.imread** function since we want to read the alpha channel of the image as well:

```
# Step 1 - Read the emoji
emoji = cv2.imread(emojiFile,-1)
```

6. We will keep the following steps as they are:

```
# Step 2 - Convert the image from BGR to Grayscale
gray = cv2.cvtColor(image, cv2.COLOR_BGR2GRAY)
# Step 3 - Load the cascade
haarCascade = cv2.CascadeClassifier(cascadeFile)
# Step 4 - Perform multi-scale detection
detectedObjects = haarCascade.detectMultiScale(\
                    gray, 1.2, 3)
```

7. Next, we will modify *Step 4* to check if any faces were detected or not. In case no faces were detected, we will return **None**:

```
# If no objects detected, return None
if len(detectedObjects) == 0:
    return None
```

8. Now, for *Step 5*, we will add the code to resize the emoji to match the size of the face:

    ```
    # Step 5 - Draw bounding boxes
    for bbox in detectedObjects:

        # Each bbox is a rectangle representing
        # the bounding box around the
        # detected object
        x, y, w, h = bbox

        # Resize emoji to match size of face
        emoji_resized = cv2.resize(emoji, (w,h), \
        interpolation = cv2.INTER_AREA)
    ```

9. Now comes the most important part of the code. We have to overwrite the face pixels with the pixels from the emoji, but only where the transparency of the emoji image is non-zero. We do this using NumPy's **where** function:

    ```
    (image[y:y+h, x:x+w])\
    [np.where(emoji_resized[:,:,3]!=0)] = \
    (emoji_resized[:,:,:3])\
    [np.where(emoji_resized[:,:,3]!=0)]
    ```

10. We will also revise *Step 6* in the function in order to return the revised image instead of the detected faces:

    ```
    # Step 6 - Return the revised image
    return image
    ```

11. Now comes the part for the webcam. First, create a **VideoCapture** object and pass **0** as the argument since we want to take the video input from the webcam:

    ```
    # Start webcam
    cap = cv2.VideoCapture(0)
    ```

12. Check whether the webcam was opened successfully or not:

    ```
    # Check if camera opened successfully
    if (cap.isOpened() == False):
        print("Error opening webcam")
    ```

13. Now, create an infinite **while** loop that will be responsible for taking in input from the webcam, processing the frame, and displaying it:

    ```
    while True:
    ```

14. First, capture a frame from the webcam:

    ```
    # Capture frame
    ret, frame = cap.read()
    ```

15. Next, check whether we were able to successfully capture the frame or not and initialize the **emojiFilterResult** variable to **None**:

    ```
    # Check if frame read successfully
    emojiFilterResult = None
    if ret == True:
    ```

    This variable will contain the results of the **emojiFilter** function.

16. If the frame was read successfully, we will call our **emojiFilter** function:

    > **NOTE**
    >
    > Before proceeding, ensure that you can change the path to the image and file (highlighted) based on where they are saved in your system.

    ```
    # Emoji filter
    emojiFilterResult = emojiFilter(frame, \
    "../data/haarcascade_frontalface_default.xml", \
    "../data/emoji.png")
    ```

17. If the frame was not read successfully, we will break from the **while** loop:

    ```
    else:
        break
    ```

512 | Appendix

18. Also, check whether the **emojiFilter** function returned **None**, to indicate that no face was detected:

```
if emojiFilterResult is None:
    continue
else:
    cv2.imshow("Emoji Filter", emojiFilterResult)
    k = cv2.waitKey(25)
    if k == 27:
        break
```

19. Finally, outside of the **while** loop, we will close the webcam and close all the open display windows:

```
cap.release()
cv2.destroyAllWindows()
```

The activity produces the following output:

Figure 5.43: Sample output of the emoji filter on a single frame

> **NOTE**
>
> To access the source code for this specific section, please refer to https://packt.live/3ihGLiN.

CHAPTER 6: OBJECT TRACKING

ACTIVITY 6.01: IMPLEMENTING AUTOFOCUS USING OBJECT TRACKING

Solution:

1. Create a new Jupyter Notebook called **Activity6.01.ipynb**. We will be writing our code in this.

2. First, import the modules we will need:

```
# Import modules
import cv2
import numpy as np
import dlib
```

3. Next, create a **VideoCapture** object and check whether it opened successfully or not:

> **NOTE**
>
> The video can be downloaded from https://packt.live/2NHXwpp. Before proceeding, ensure that you can change the path to the video (highlighted) based on where the video is saved in your system.

```
# Create a VideoCapture Object
video = cv2.VideoCapture("../data/soccer2.mp4")
# Check if video opened successfully
if video.isOpened() == False:
    print("Could not open video!")
```

4. Next, capture the first frame of the video and check whether we are able to read the frame successfully or not:

```
# Read first frame
ret, frame = video.read()
# Check if frame read successfully
if ret == False:
    print("Cannot read video")
```

5. Now that we have the first frame ready, we will display it after selecting the ROI:

```
# Show the first frame
cv2.imshow("First Frame", frame)
cv2.waitKey(0)
# Specify the initial bounding box
bbox = cv2.selectROI(frame)
cv2.destroyAllWindows()
```

6. Now comes the object tracker part. Using experimentation, we found that it was best to go for the TLD object tracker. So, let's create and initialize it:

```
# Create tracker
# TLD Tracker
OpenCV_Tracker = cv2.TrackerTLD_create()
# Initialize OpenCV tracker
ret = OpenCV_Tracker.init(frame, bbox)
```

7. Create a display window. This is where we will show our object tracking results:

```
# Create a new window where we will
# display the results
cv2.namedWindow("Tracker")
# Display the first frame
cv2.imshow("Tracker", frame)
```

You will see the following image, which is the output of the preceding code snippet:

Figure 6.27: ROI selection

8. Now comes the `while` loop. Let's read the frame and check whether it was read successfully or not:

```
while True:
    # Read next frame
    ret, frame = video.read()
    # Check if frame was read
    if ret == False:
        break
```

9. Now, let's update our tracker and see whether we were able to find the object:

```
    # Update tracker
    found, bbox = OpenCV_Tracker.update(frame)
```

10. If we were able to find the object, display a bounding box around it:

    ```
    # If object found, draw bbox
    if found:
        # Top left corner
        topLeft = (int(bbox[0]), int(bbox[1]))
        # Bottom right corner
        bottomRight = (int(bbox[0]+bbox[2]), int(bbox[1]+bbox[3]))
        # Display bounding box
        cv2.rectangle(frame, topLeft, \
                        bottomRight, (0,0,255), 2)
    ```

11. If we were not able to find the object, display a message on the window saying that the object was not found:

    ```
    else:
        # Display status
        cv2.putText(frame, "Object not found", (20,70), \
        cv2.FONT_HERSHEY_SIMPLEX, 0.75, (0,0,255), 2)
    ```

12. Finally, display the frame on the window:

    ```
    # Display frame
    cv2.imshow("Tracker", frame)
    k = cv2.waitKey(5)
    if k == 27:
        break
    ```

13. Once we are done with tracking, we will close the display window:

    ```
    cv2.destroyAllWindows()
    ```

Chapter 6: Object Tracking | 517

A sample output for a frame in the video is shown in the following image:

Figure 6.28: Output of the object tracker

> **NOTE**
>
> To access the source code for this specific section, please refer to https://packt.live/31zGf9O.

CHAPTER 7: OBJECT DETECTION AND FACE RECOGNITION

ACTIVITY 7.01: OBJECT DETECTION IN A VIDEO USING MOBILENET SSD

Solution:

1. Firstly, open an untitled Jupyter Notebook and name the file **Activity7.01**.

2. We will start by importing all the required libraries for this activity:

```
import cv2
import numpy as np
import matplotlib.pyplot as plt
```

3. The pre-trained model can detect a list of object classes. Let's define those classes:

```
categories = {0:'background', 1:'aeroplane', 2:'bicycle', \
              3:'bird', 4:'boat', 5:'bottle', \
              6:'bus', 7:'car', 8:'cat', 9:'chair', \
              10:'cow', 11: 'diningtable', 12: 'dog', \
              13: 'horse', 14:'motorbike', \
              15:'person', 16:'pottedplant', \
              17:'sheep', 18:'sofa', 19:'train',\
              20:'tvmonitor'}
```

4. Next, let's specify the path of the pre-trained model of MobileNet SSD. We will use this model to detect objects in a new image. Please make sure that you have downloaded the **prototxt** and **caffemodel** files and saved them in the code folder.

> **NOTE**
>
> The **prototxt** and **caffemodel** files can be downloaded from https://packt.live/2Zvqpdl.

The code is as follows:

```
net = cv2.dnn.readNetFromCaffe(\
      'MobileNetSSD_deploy.prototxt.txt', \
      'MobileNetSSD_deploy.caffemodel')
```

5. For each object detected in the given input image, you need to create a rectangular box. These rectangular boxes help to locate the instances of objects. This step will help to select random colors for the boxes:

   ```
   colors = np.random.uniform(255, 0, \
                size=(len(categories), 3))
   ```

6. Use OpenCV to read each frame of the input video.

   > **NOTE**
   >
   > The `walking.avi` input video can be downloaded from https://packt.live/3gecS12.

   The code is as follows:

   ```
   # read input video
   #.cap = cv2.VideoCapture('walking.avi')

   """
   use cap = cv2.VideoCapture(0), for webcam
   """

   cap = cv2.VideoCapture(0)
   while True:
       ret, image = cap.read()
   ```

7. Read the input image and construct a blob for the image. Note that MobileNet requires fixed dimensions for input images, so we will first resize the images to 300 x 300 pixels and then we will normalize them:

   ```
   resized_image = cv2.resize(image, (300,300))
   blob = cv2.dnn.blobFromImage(resized_image, \
              0.007843, (300, 300), \
              (127.5, 127.5, 127.5), False)
   ```

8. The next two lines feed the constructed blob to the model and store the output of the model in the **detections** variable:

    ```
    net.setInput(blob)
    detections = net.forward()
    # to calculate scale factor
    (h, w) = resized_image.shape[:2]
    ```

9. Now, once we have the result from the model, we can check the probability of each detected object. If a probability is greater than **0.2**, then we will find the coordinates of those objects; otherwise, we will discard it. Note that the SSD model returns nearly 100 detected instances. The **threshold** value helps us to discard the irrelevant results:

    ```
    for i in range(detections.shape[2]):
        confidence = detections[0, 0, i, 2]
        if confidence > 0.2:
            class_id = int(detections[0, 0, i, 1])
            startX = int(detections[0, 0, i, 3] * w)
            startY = int(detections[0, 0, i, 4] * h)
            endX = int(detections[0, 0, i, 5] * w)
            endY = int(detections[0, 0, i, 6] * h)
    ```

10. Recall that, in an earlier step, the image was resized to **(300, 300)**. In this step, we will map the detected coordinates to the original image:

    ```
    # estimate factors by which resized images needs to scaled.
            heightFactor = image.shape[0]/300.0
            widthFactor = image.shape[1]/300.0

            startX = int(widthFactor * startX)
            startY = int(heightFactor * startY)
            endX   = int(widthFactor * endX)
            endY   = int(heightFactor * endY)
    ```

11. Visualize the detected objects. Use OpenCV to create a rectangular box around each object:

    ```
    cv2.rectangle(image, (startX, startY), \
    (endX, endY), (0, 255, 0))
    ```

12. In addition to the rectangular boxes, we will use OpenCV's built-in function to write a label and category for each object in the image:

    ```
    if class_id in categories:
        label = categories[class_id] + ": " + \
                str(confidence)
        labelSize, baseLine = cv2.getTextSize(label, \
        cv2.FONT_HERSHEY_SIMPLEX, 0.5, 1)
        startY = max(startY, labelSize[1])
        # draw the rectangular with labels
        cv2.rectangle(image, \
        (startX, startY - labelSize[1]), \
        (startX + labelSize[0], startY + baseLine), \
        (0, 255, 0), cv2.FILLED)
        # write a label for each object
        cv2.putText(image, label, (startX, startY), \
        cv2.FONT_HERSHEY_SIMPLEX, 0.5, (0, 0, 255))
    ```

13. Now, it's time to visualize the result. We will use OpenCV to display the result. The window will automatically be destroyed once it iterates over all the frames. If the webcam is in use, then the window can be destroyed by pressing *q*:

    ```
    cv2.imshow("object detection result", image)
    if cv2.waitKey(1) & 0xFF == ord('q'):
        break
    cap.release()
    cv2.destroyAllWindows()
    ```

The preceding code produces the following output:

Figure 7.32: Sample output of the prediction results for one of the frames of the input video

Figure 7.32 shows the prediction results for one of the frames of the input video.

Figure 7.33 is a sample of how the result would look if you were to use a webcam. For representation purposes, an image of a train was placed in front of the webcam. You are advised to bring any object in front of the webcam from the list that we created in *Step 4* of *Exercise 7.04, Object Detection Using MobileNet SSD*:

In this activity, you have learned how to implement object detection models for real-time scenarios and recorded videos. Now, you can compare the performance of the SSD algorithm and the Haar-based method by using the same input video. You can extend this solution to business use cases such as crowd estimation (to count footfall in malls or near advertisement boards), traffic estimation on streets, and more.

> **NOTE**
>
> To access the source code for this specific section, please refer to https://packt.live/3ggKOtX.

ACTIVITY 7.02: REAL-TIME FACIAL RECOGNITION USING LBPH

Solution:

1. Firstly, open an untitled Jupyter Notebook and name the file **Activity7.02**.

2. Start by importing all the required libraries for this activity:

```
import numpy as np
import cv2
import os
```

3. Specify the path of the OpenCV XML files that will be used to detect faces in each image. Create a folder named **opencv_xml_files** and save **lbpcascade_frontalface.xml**.

> **NOTE**
>
> The **opencv_xml_files** folder is available at https://packt.live/2WkeAX5.

The code is as follows:

```
lbpcascade_frontalface = 'opencv_xml_files/lbpcascade_frontalface.xml'
```

4. Use this function to detect the face in an input. This function will convert a three-dimensional image into grayscale (a two-dimensional image) and return the face in the image with coordinates:

```python
def detect_face(input_img):
    """
    detect faces from an input image
    return: detected face and its postions that is x,y,w,h
    """
    image = cv2.cvtColor(input_img, cv2.COLOR_BGR2GRAY)
    face_cascade = cv2.CascadeClassifier(\
                lbpcascade_frontalface)
    faces = face_cascade.detectMultiScale(image, \
            scaleFactor=1.2, minNeighbors=5)
    if (len(faces) == 0):
        return -1, -1
    (x, y, w, h) = faces[0]
    return image[y:y+w, x:x+h], faces[0]
```

5. The **draw_rectangle** function will create a bounding box around the detected objects and the **draw_text** function will put labels on those objects:

```python
def draw_rectangle(img, rect):
    """
    draws rectangular bounding box
    around detected face
    """
    (x, y, w, h) = rect
    cv2.rectangle(img, (x, y), (x+w, y+h), \
                (0, 255, 0), 2)
def draw_text(img, text, x, y):
    """
    put label above the box
    """
    cv2.putText(img, text, (x, y), \
    cv2.FONT_HERSHEY_PLAIN, 1.5, (0, 255, 0), 2)
```

6. Open your webcam and read each frame. Each frame is passed through the face detection function and finally, the detected faces are saved in a folder that will be used for training. Note that you have to create a folder path – **detected_faces/1**. Each detected face will be stored in that folder:

```
cap = cv2.VideoCapture(0)
count = 0
while True:
    _, frame = cap.read()
    face, rect = detect_face(frame)
    if face is not -1:
        count +=1
        draw_rectangle(frame, rect)
        # change the folder path as per requirement
        cv2.imwrite("detected_faces/1/"+str(count)+ \
        ".jpg", face)
    cv2.imshow("frame", frame)
    # destroy window on pressing key 'q'
    if cv2.waitKey(1) & 0xFF == ord('q'):
        break
cap.release()
cv2.destroyAllWindows()
```

7. This step and *Step 8* cover the data preparation steps. In this step, initialize two empty lists that will store detected faces and their respective categories, which will eventually be used for training:

```
detected_faces = []
face_labels = []
```

8. In this step, use the data path used in *Step 3* to locate the training images folder and iterate over all the images sequentially. Note that the LBPH method does not expect all training images to be the same size:

```
def prepare_training_data(training_data_folder_path):
    """
    read images from folder and prepare training dataset
    return list of detected face and labels
    """
    training_image_dirs = os.listdir(\
        training_data_folder_path)
```

```
            for dir_name in traning_image_dirs:
                label = int(dir_name)
                training_image_path = training_data_folder_path + \
                    "/" + dir_name
                training_images_names = os.listdir(\
                    training_image_path)

                for image_name in training_images_names:
                    image_path = training_image_path + \
                                "/" + image_name

                    # read an input image
                    image = cv2.imread(image_path, 0)
                    detected_faces.append(image)
                    face_labels.append(label)
        return detected_faces, face_labels
```

9. In this step, call the function from the previous step to get the detected faces and labels:

```
detected_faces, face_labels = \
    prepare_training_data('detected_faces')
```

10. This step helps to verify the number of faces and labels. The numbers should be equal:

```
print("Total faces: ", len(detected_faces))
print("Total labels: ", len(face_labels))
```

The preceding code produces the following output:

```
Total faces: 124
Total labels: 124
```

11. As discussed before, OpenCV is equipped with face recognizer modules. Use the LBPH face recognizer module from OpenCV:

```
lbphfaces_recognizer = \
    cv2.face.LBPHFaceRecognizer_create(\
    radius=1, neighbors=8)
```

12. In the previous two steps, we have prepared the training data and initialized a face recognizer. Now, in this step, train the face recognizer. Note that labels are first converted into a NumPy array before being passed into the recognizer because OpenCV expects labels to be a NumPy array:

```
lbphfaces_recognizer.train(detected_faces, \
    np.array(face_labels))
```

13. In this step, again, use the webcam and read each frame. Then, each frame is passed through the trained model and the model outputs the label with a confidence value. Check the confidence values and set the threshold value appropriately. The threshold value depends on the quality of the data as well. Note that the higher the confidence value, the lower the similarity between images is:

```
cap = cv2.VideoCapture(0)
while True:
    _, frame = cap.read()
    face, rect = detect_face(frame)
    if face is not -1:
        label= lbphfaces_recognizer.predict(face)
        if label[1]<75:
            label_text = str(label[0])
        else:
            label_text = 'unknown'
        draw_rectangle(frame, rect)
        if label[1]<75:
            draw_text(frame, label_text, \
                      rect[0], rect[1]-5)
        else:
            draw_text(frame, label_text, \
                      rect[0], rect[1]-5)
    cv2.imshow("frame", frame)
    if cv2.waitKey(1) & 0xFF == ord('q'):
        break
cap.release()
cv2.destroyAllWindows()
```

Once all the steps are done successfully, you should be able to get an image of a person with name. *Figure 7.34* is a representation of the output:

Figure 7.34: Example output for this activity

> **NOTE**
>
> To access the source code for this specific section, please refer to https://packt.live/3gh96nz.

CHAPTER 8: OPENVINO WITH OPENCV

ACTIVITY 8.01: FACE DETECTION USING OPENVINO AND OPENCV

Solution:

1. We will create a new Jupyter notebook and name it **Activity8.01.ipynb**. We will be writing our code in this notebook.

2. First, we will load the required modules:

```
# Load the required modules
import os
import cv2
import numpy as np
```

3. Next, let's specify the path to the downloader script:

```
# Specify the path to the downloader script
DOWNLOADER_SCRIPT = "/opt/intel/openvino/deployment_tools/"\
                    "open_model_zoo/tools/"\
                    "downloader/downloader.py"
```

4. We will also specify the model name, output directory, and precision.

   The code for this is as follows:

```
# Specify the model name which we want to download
MODEL_NAME = "face-detection-retail-0004"
# Specify where you want to store the files
OUTPUT_DIR = "/home/hp/workfolder/
The-Computer-Vision-Workshop/Chapter08/data/"
# Specify the precision
PRECISION = "FP16"
```

> **NOTE**
>
> Make sure you change the output directory path as per your system.

5. We will also specify the command that we want to run to download the model:

   ```
   # Specify the command we need to run
   DOWNLOAD_COMMAND = "{} --name {} \
   --precisions {} -o {}"\
   .format(DOWNLOADER_SCRIPT, MODEL_NAME, PRECISION, OUTPUT_DIR)
   ```

6. As we did in *Exercise 8.01, Downloading the Pedestrian and Vehicle Detection Model*, we will create an environment variable for this command and then run it:

   ```
   # Create environment variable
   os.environ["DOWNLOAD_COMMAND"] = DOWNLOAD_COMMAND
   %%bash
   $DOWNLOAD_COMMAND
   ```

   The output is as follows:

   ```
   ################|| Downloading models ||################

   ========== Downloading /home/hp/workfolder/The-Computer-Vision-Workshop/Chapter08/data/intel/face-detectio
   n-retail-0004/FP16/face-detection-retail-0004.xml
   ... 100%, 99 KB, 203 KB/s, 0 seconds passed

   ========== Downloading /home/hp/workfolder/The-Computer-Vision-Workshop/Chapter08/data/intel/face-detectio
   n-retail-0004/FP16/face-detection-retail-0004.bin
   ... 89%, 1024 KB, 1781 KB/s, 0 seconds passed
   ... 100%, 1148 KB, 1795 KB/s, 0 seconds passed

   ################|| Post-processing ||################
   ```

   Figure 8.20: Downloading the model files

7. Once the model has been downloaded, we will remove the environment variable:

   ```
   # Remove the environment variable
   os.environ.pop("DOWNLOAD_COMMAND")
   ```

   The output is as follows:

   ```
   '/opt/intel/openvino/deployment_tools/open_model_zoo/tools/
   downloader/downloader.py --name face-detection-retail-0004
   --precisions FP16 -o /home/hp/workfolder/The-Computer-Vision-
   Workshop/Chapter08/data/'
   ```

8. We can follow the same process that we followed in *Exercise 8.01, Downloading the Pedestrian and Vehicle Detection Model*, to find out whether the model files were downloaded. Let's start by listing the files present in the output directory:

   ```
   # List files present in the output directory
   os.listdir(OUTPUT_DIR)
   ```

   The output is as follows:

   ```
   ['mobilenet_v2.caffemodel',
    'haarcascade_frontalface_default.xml',
    'intel',
    'mobilenet_v2.prototxt',
    'crowd.jpg',
    'pedestrians.jpg']
   ```

9. Next, let's list the files present in the **intel** directory:

   ```
   # List files in intel folder
   os.listdir("{}/intel".format(OUTPUT_DIR))
   ```

   The output is as follows:

   ```
   ['pedestrian-and-vehicle-detector-adas-0001', 'face-detection-retail-0004']
   ```

10. Let's visit the **model** directory and list the files:

    ```
    # Go to the model directory
    os.listdir("{}/intel/{}".format(OUTPUT_DIR,MODEL_NAME))
    ```

    The output is **['FP16']**.

11. Finally, let's go to the **FP16** directory and list the files present there:

    ```
    # Go to the precisions directory
    os.listdir("{}/intel/{}/{}"\
            .format(OUTPUT_DIR,MODEL_NAME,PRECISION))
    ```

    The output is as follows:

    ```
    ['face-detection-retail-0004.xml', 'face-detection-retail-0004.bin']
    ```

12. Next comes model inference. First, we will specify the paths of the XML and `.bin` files.

   > **NOTE**
   >
   > Before proceeding, ensure that you change the paths to the XML and `.bin` files based on where they are saved on your system.

   The code is as follows:

   ```
   # Specify the path to IR
   xmlFile = "../data/intel/{}/{}/{}.xml"\
           .format(MODEL_NAME, PRECISION,MODEL_NAME)
   binFile = "../data/intel/{}/{}/{}.bin"\
           .format(MODEL_NAME, PRECISION,MODEL_NAME)
   ```

13. Next, we will read the image and create a copy of it.

   The code is as follows:

   ```
   # Image path
   imgPath = "../data/crowd.jpg"
   # Read image
   img = cv2.imread(imgPath)
   imgCopy = img.copy()
   ```

   > **NOTE**
   >
   > The path should be specified based on the location of the folder on your system.

14. Now, let's load the model using the XML and `.bin` files:

   ```
   # Load the network
   net = cv2.dnn.readNet(xmlFile,binFile)
   ```

15. We will now preprocess the image and resize it to **384x672** pixels, as specified by the model documentation (https://docs.openvinotoolkit.org/latest/_models_intel_face_detection_retail_0004_description_face_detection_retail_0004.html):

```
# Pre-process the image
# We want to resize the image to 384x672
blob = cv2.dnn.blobFromImage(img, size=(672,384))
```

16. We can verify whether the size of the processed image matches the input shape or not:

```
# Check the blob shape
blob.shape
```

The shape is **(1, 3, 384, 672)**.

17. Now, we will set the image as the network input and carry out a forward step:

```
# Set the image as network input
net.setInput(blob)
# Carry a forward propagation
out = net.forward()
```

18. Check the shape of the output, as follows:

```
out.shape
```

The shape is **(1, 1, 200, 7)**.

19. Now, we can print the number of objects (faces) that were detected by the model:

```
# Compare this with [1,1,N,7]
print("Number of objects found = {}"\
      .format(out.shape[2]))
```

The output is as follows:

```
Number of objects found = 200
```

20. Now, we need to draw the bounding boxes around the detected objects. To do this, we will reshape the output array:

```
# Reshape the output
detection = out.reshape(-1,7)
```

21. Check the shape, as follows:

```
detection.shape
```

The shape is **(200, 7)**.

22. Now, we just have to iterate over each detected object; find its confidence, label, and bounding box coordinates; and draw a bounding box if the confidence is at least **15%**. This is exactly the same as what we did in *Exercise 8.03, Vehicle and Pedestrian Detection*:

`Activity8.01.ipynb`

```
for detectedObject in detection:
    # Find label
    label = int(detectedObject[1])
    # Choose color of bounding box
    if label == 1:
        # Green color
        color = (0,255,0)
    else:
        # Red color
        color = (0,0,255)
    # Find confidence
    confidence = float(detectedObject[2])
```

The complete code for this step can be found at https://packt.live/31FUBWa.

23. We can display the output image using the following code:

```
# Display image
cv2.imshow("Output Image",img)
cv2.imwrite("image.png",img)
cv2.waitKey(0)
cv2.destroyAllWindows()
```

The output is as follows:

Figure 8.21: Displaying the output image

24. Now comes the Haar Cascade part. First, we will specify the Haar Cascade path:

```
# Haar Cascade Path
haarCascadePath = \
    "../data/haarcascade_frontalface_default.xml"
```

> **NOTE**
>
> The **haarcascade_frontalface_default.xml** file can be downloaded from https://packt.live/2NOxz7m. Also, the path should be specified based on the location of the folder on your system.

25. Let's convert the image into grayscale mode. Note that we will have to use the copy of the image from now on because the original image has already been modified:

```
# Convert image to grayscale
grayInputImage = cv2.cvtColor(imgCopy,\
                 cv2.COLOR_BGR2GRAY)
```

26. Next, let's load the Cascade classifier and perform multi-scale detection:

```
# Load Haar Cascade
haarCascade = cv2.CascadeClassifier(haarCascadePath)
detectedFaces = haarCascade.detectMultiScale(\
                grayInputImage, 1.2, 1)
```

27. Since multi-scale detection already gives the bounding boxes around the objects, we can draw them directly:

```
for face in detectedFaces:
    # Each face is a rectangle representing
    # the bounding box around the detected face
    x, y, w, h = face
    cv2.rectangle(imgCopy, (x, y), \
                  (x+w, y+h), (0, 0, 255))
```

28. Finally, we will display the results of performing Haar Cascade-based face detection:

```
cv2.imshow("Faces Detected", imgCopy)
cv2.imwrite("haarcascade_result.png",imgCopy)
cv2.waitKey(0)
cv2.destroyAllWindows()
```

The output is as follows:

Figure 8.22: Result for Haar Cascade-based face detection

NOTE

To access the source code for this specific section, please refer to https://packt.live/31FUBWa.

INDEX

A

access: 1, 10, 35, 38, 45, 50, 60, 79, 83, 91, 96, 101, 108, 118-119, 124, 127, 140, 146, 158, 164, 170, 174, 181-182, 194, 197, 199, 205, 213-215, 224, 231, 238, 240, 243, 259, 268, 285, 297, 316, 337, 346, 369, 383, 394, 404, 413, 417, 422, 426, 442, 448, 458
adaboost: 249, 413
alexnet: 354
algorithm: 148, 278-280, 305, 308, 311, 320-322, 330, 339, 342, 346, 354-355, 358, 360-361, 374-376, 383, 386-387, 395, 397, 400, 405-407, 410-411, 413, 416, 423
alvaro: 363-364, 368
analysis: 97, 113, 142, 224, 357, 371
annotate: 437
argument: 4, 28-29, 67, 73, 99, 111, 186, 251, 258, 266, 274-275, 280, 301, 303-304, 359, 373, 385
attribute: 17

B

backslash: 6-7
bashrc: 433
bilinear: 72, 148, 272, 385
binary: 53, 55, 96-98, 100-102, 105-106, 109, 111, 113-114, 119, 127-128, 131, 137-138, 176, 182-185, 191-194, 196-197, 204, 208, 210-213, 216, 219-220, 227, 230-231, 234-235, 242, 280, 314, 353, 355, 383-385, 405, 407, 409, 428, 450
brackets: 131
building: 2, 8, 10-12, 38, 246, 263, 425
built-in: 368, 382, 393

C

caffemodel: 329, 401, 418, 427, 441, 451
calchist: 131, 134-135, 141
camshift: 307, 321-322, 324-325, 350
carsten: 278
cascade: 249-251, 253, 256-257, 259-266, 269-272, 274-275, 298, 301, 305, 307-308, 353, 355, 365, 369, 378, 390, 395, 414, 423, 461-463
categorize: 413
centroid: 201
classes: 276-277, 359, 369, 371-372, 383, 394, 398, 401, 403, 407, 434
closed: 182, 202, 204, 415
clubbing: 270, 274-275, 291, 299
column: 38, 57-58, 206, 210, 214
command: 4, 36, 40, 47, 86, 111-112, 119-121, 129, 131, 134-136, 140-142, 145, 151-152, 161, 166, 168, 170, 184, 186-187, 189-190, 196, 200-201, 205, 208, 224, 226-227, 233, 235-236, 338-339, 341-342, 432, 435, 438-440, 451
commit: 298
component: 357, 428-429, 448, 452, 463
compute: 119, 121, 131, 134-135, 200, 372, 400, 402, 405, 434
condition: 301
connection: 431
consist: 308, 437
contain: 189, 217, 252, 370-371, 395

contours: 2, 178, 181-187, 193-202, 204-206, 209-211, 213-215, 220-222, 224-228, 231, 235-240, 242-243
convbnrelu: 450
coordinate: 12-14, 19, 38, 58, 185, 200, 271, 308, 445
correlated: 135, 357
create: 4-10, 24, 30, 35, 40-41, 46-50, 57-58, 74-75, 79, 81, 86, 89, 92, 95, 99, 103, 111, 137-139, 141, 154, 163, 169, 176, 184, 242-243, 245-246, 248, 257, 263-265, 268, 271, 274-275, 279, 281-282, 287-288, 291, 293, 298, 301, 303, 307-308, 311-312, 326, 330-334, 341, 343-347, 359, 362, 364, 366, 371, 373, 376, 378, 380, 383, 385, 388, 390-392, 400, 402-403, 407, 414, 418-419, 421, 438-440, 460

D

database: 369
define: 12, 120, 365, 367, 379-380, 390, 392, 401, 405, 447

depending: 17, 19, 66, 151, 267, 326, 328, 338, 432-433, 449
detect: 2, 18, 108, 181-182, 184-188, 191, 194, 210, 213, 215, 220, 224, 228, 238-239, 242-243, 245-246, 249-250, 256, 259-264, 270-273, 275, 298, 308, 310, 317, 335-336, 355, 358, 364-365, 367, 369, 378-379, 381, 390-392, 401, 406, 414-415, 417-418, 421, 435
detector: 305, 353, 395, 442
developer: 426, 428
directory: 189, 363-364, 378, 389, 431-432, 438-439, 441, 451
dlibrect: 341, 344
domain: 2, 20, 67, 346, 425, 427

E

eigenface: 355-356, 360, 367, 369-371, 383, 423
extension: 197, 232, 239, 428
extract: 14, 38, 48, 54, 83, 278, 358, 371, 407, 409, 456

F

factor: 67-68, 260, 264, 400
feature: 120, 224, 249, 346, 354, 357-358, 370-372, 383, 385, 395, 398-399, 405-407, 409, 412-413
fetched: 184
fgdmodel: 282-283, 288-289, 295
flatten: 135, 224
frames: 3, 24, 200, 249, 303, 305, 312, 314, 330, 334, 341, 344, 348-350, 414-416, 420, 435
framework: 398, 400, 442, 451
function: 4, 6, 27-29, 31, 34-35, 37, 48, 67, 73-74, 78-79, 84-85, 91-93, 98, 100, 111-112, 125, 151, 184-186, 200, 204, 224, 226, 251-252, 255-258, 260, 263, 265-268, 271-272, 274-275, 279-280, 282, 286, 288, 292-293, 298-299, 301, 303-304, 320-321, 326, 331-332, 337, 341, 347-348, 359-361, 365-368, 371, 373, 375-376, 378-379, 381-383,

385, 387, 390-393,
400, 403, 405-407,
412, 414, 435,
438, 444-445, 448,
452-455, 460

G

grabcut: 245-246,
276-281, 283,
285-290, 293, 295,
297-299, 301, 305
grayscale: 17, 19-22,
25-26, 28, 37,
77, 96, 98-100,
104-105, 111-112,
117, 119-120,
122-127, 131, 134,
141-143, 149,
152-153, 160-161,
178, 182-183,
189-190, 194, 204,
208, 212, 218-219,
226, 229, 233-234,
251, 257, 263,
265-266, 271, 274,
280, 298, 360-361,
365, 375-376, 378,
381, 383, 387, 390,
407, 409, 412, 415,
418, 444, 461

H

hershey: 334, 367,
380, 392, 403,
409, 416
histogram: 114,
117-119, 121-122,
124-125, 127-128,
130-132, 134-136,
141-148, 150-156,
159-168, 170,
172, 174-175,
177-178, 182, 321,
385, 407-409
hstack: 145

I

imagine: 35, 67, 84,
249, 263, 427,
434-435, 437
imgplot: 120, 161,
163-164, 166, 171
imread: 27-28, 30,
40, 76, 81, 86, 93,
99, 104, 122, 126,
130, 134, 138, 142,
152, 161, 166, 171,
189, 207-208, 211,
217, 229, 233, 257,
265, 281, 287, 298,
301, 363, 365, 368,
377, 379, 381, 389,
391-392, 402, 407,
409, 446, 454
imshow: 29, 31-33,
36-37, 40-41, 43-44,
76-78, 81, 83,
86-88, 90, 93-95,
99, 101, 104-107,
120, 122, 125-126,
139, 143-145, 152,
154, 161, 163-164,
166, 168-169,
171-173, 189,
191-193, 196, 198,
200, 207, 211-212,
215, 217-220,
222-223, 229-231,
233-235, 238-239,
258, 267, 283, 285,
290, 292, 304,
314, 331, 334-335,
343-345, 363, 368,
377, 382, 389, 393,
403, 409, 416, 458
imutils: 406
imwrite: 29, 34, 45,
140, 168, 259,
279, 282-283,
285, 289-290,
293, 331, 458
initialize: 237, 282,
288, 318, 326,
333, 344, 347,
365, 379, 405
iterate: 201, 258,
271-272, 275, 347,
402, 407, 419, 456

K

kolmogorov: 278

L

libraries: 25, 40, 50,
86, 92, 111, 119,
122, 126, 217, 250,
257, 264, 271, 274,
281, 287, 346, 362,
376, 388, 400,
406, 414, 427-428,
434-435, 460
linear: 72-73, 78, 112,
148, 357, 371, 383,
405-406, 408, 411
location: 12, 14, 19,

38, 56, 111, 210, 225, 321-322, 349, 395, 399, 431, 439-440, 451

M

mapping: 357
matplotlib: 1, 25, 35-38, 40, 47, 50, 76, 81, 83, 86, 92-93, 99, 103-104, 106, 111, 119-122, 125-126, 131, 134, 138, 143, 153, 161-163, 166, 168-169, 171-173, 264, 268-269, 362, 368, 376-377, 382, 388-389, 393, 400, 403, 406
matrix: 25-26, 55-62, 64, 67, 73-74, 79, 82, 114, 206, 209-210, 358
median: 325, 330, 335-337, 342, 346, 350
merging: 28-29, 38, 48, 168-169
mobilenet: 353, 355, 399-402, 404-405, 418, 420, 423, 441-442, 451
models: 67, 261, 263, 269-271, 276, 325, 350, 395, 413, 423, 428-429, 436-438, 442, 445-446, 450-452, 459-460, 463
modifying: 27, 54, 142
module: 3-5, 26, 30, 35, 58, 60, 119, 251, 264, 269, 338-340, 359-360, 366, 373, 375, 380, 385, 387, 391, 400, 404-405, 418, 421, 423, 434-436, 438, 452-453

N

network: 395, 398-400, 402, 427-428, 434-435, 452-455
nparray: 4-10

O

object: 2, 18, 39, 67, 151-152, 154, 163, 169, 173, 181-182, 202, 205, 216, 235, 243, 249, 251, 261-263, 265-267, 270, 274, 292, 301, 303, 305, 307-311, 314, 317-322, 325-332, 334-350, 353-355, 395, 399-406, 411, 413-414, 418-420, 423, 426-427, 436, 444-445, 451, 456-457
occlusion: 261, 326, 336, 342
opencv: 1, 3, 12, 19, 25-30, 35, 37-38, 40, 47, 50, 53, 55, 57, 67, 72-73, 75, 79-80, 84-85, 88-92, 94, 96, 98, 111, 114, 119-121, 125, 130-131, 138, 142, 151-152, 161, 166, 168, 171, 182, 186, 189, 191-192, 194, 196, 200-202, 204, 208, 211, 217, 224, 226, 228, 230, 235, 243, 246, 251-252, 255, 263, 274, 276, 280, 282, 292-293, 298, 301, 305, 307-308, 311, 317, 325, 329-330, 337, 340-342, 346-348, 350, 354, 359-361, 364-366, 368, 373-375, 378, 380, 382-383, 385-387, 390-391, 393, 395, 400, 403-405, 409, 414, 418, 421, 423, 425, 427-428, 430, 434-437, 444-446, 448, 452-454, 459-460, 463

operations: 2, 38, 53, 55, 84, 96-97, 101-102, 113-114, 118, 138, 151, 246, 269-270, 279, 311, 399, 449-450
optimizer: 425, 428, 430, 434-435, 448-452

original: 39, 42, 44, 49, 54, 64, 66, 71, 89, 95, 120, 139, 143-147, 150-154, 161, 163, 166, 171, 173-174, 193, 207, 211-212, 217, 223, 229, 231, 237, 282, 357-358, 368, 382, 393, 419

P

package: 427, 431
parameter: 187, 226, 253-254, 260, 266, 268-269, 272, 286, 298, 320, 366, 373, 380, 385, 394-395, 405, 407
pattern: 147, 159-160, 353, 383, 405, 407, 409
perform: 2, 60, 91, 98, 108, 112-113, 118, 122, 126, 138, 152, 160-161, 164, 172, 176, 189, 198, 217, 226, 228, 233, 242, 246, 258, 264, 266, 270-272, 274-275, 278, 281, 283, 287-289, 301, 308, 311, 342, 371, 395, 400, 414, 419, 430, 435, 446, 452-453, 462
plotting: 37, 119-122, 125, 138, 142, 152, 161, 166, 171, 182, 186, 204, 220, 255
position: 35, 44, 225, 238-239, 317-318, 341, 345, 348, 397, 402
practical: 38, 182, 215, 243, 346, 353
predict: 276, 317-318, 359, 367-369, 373, 381-383, 385, 392-394, 399, 402, 409, 411
program: 29, 137, 190-191, 241-242, 258, 272, 314, 326, 349, 426
provide: 4, 29, 99, 120-121, 131-132, 274, 279, 284, 286, 303-304, 341, 349-350, 354, 373, 419, 437, 444-445
pyplot: 35, 40, 76, 81, 86, 92, 99, 103, 119, 121-122, 126, 143, 153, 161, 166, 171, 264, 268, 362, 376, 388, 400, 406
pytorch: 427-428, 434, 450

R

reading: 27-28, 30, 36, 122, 125, 142, 152, 301, 312, 347, 460
rectangle: 64, 176, 200-201, 221, 223, 239, 251-252, 254-255, 258, 266, 280, 282, 289, 295, 334, 341-342, 344-345, 367, 380-381, 392, 403, 415, 457
rendered: 148
repository: 263, 274, 429
reshape: 409, 444-445, 448, 455-456
retrieve: 184, 202, 238, 369
rotate: 67, 75, 77, 79

S

scanning: 83, 240
scattered: 320
segment: 127, 131, 134, 148, 183, 276, 279, 281
shapes: 5, 112, 181, 188-189, 191, 195, 197, 224-228, 284, 397
skimage: 405-407
sklearn: 406
software: 263, 426, 428, 430, 433
source: 10-11, 25, 35, 45, 60, 79, 83, 91, 96, 101, 108, 124, 127, 140, 146, 158, 164, 170, 174, 184, 194, 197, 199, 215, 224, 231, 238, 240, 259, 268, 285, 297, 305, 316, 337, 346, 369, 383, 394, 404, 413, 417, 433,

442, 448, 458
splitting: 28, 38
structure: 206, 248, 363, 383
subplot: 268
suitable: 160, 212, 230, 234, 346, 353
swapping: 453
switch: 294, 349

T

technique: 54, 97, 113, 118, 159, 178, 228, 238, 242, 246, 249, 276-278, 286, 298, 301, 305, 311, 317, 330, 339-340, 342, 354, 357, 371-372, 395, 413, 452
tracked: 311, 319-320, 330, 335, 350
transform: 64, 73, 168-169, 172-173, 357

U

uniform: 150-151, 402, 407, 409, 423
utilities: 428, 435

V

validation: 437
values: 1, 4, 19-22, 25, 38, 54, 67, 69-72, 84, 88, 96-98, 120-121, 131, 135-136, 142, 162, 186, 200, 209-210, 224-225, 248-249, 253-254, 260, 266, 268, 271-272, 280, 290, 298, 303, 310, 313, 326, 332, 366, 384-385, 397, 404-405, 407-408, 439, 444, 463
variable: 136, 184, 186, 189-190, 198, 206, 213, 217, 226-227, 237, 274-275, 357, 367, 381, 392, 408, 415, 419, 435, 438-440
variance: 357-359, 371-372
vector: 121, 224-225, 398, 405, 414, 417
versions: 16, 226, 405, 427

W

waitkey: 29, 31-33, 122, 126, 139, 143-145, 152, 154, 190-193, 196, 198, 200, 207, 211-212, 215, 217-220, 222-223, 229-231, 233-235, 238-239, 258, 267, 283, 285, 290, 293, 304, 314, 331, 335, 343, 345, 403, 409, 416, 458
whereas: 19, 35, 37, 64, 72, 125, 197, 202, 238, 280, 310, 313, 341, 434, 439, 449
wrapped: 350

Z

zooming: 11, 355